World Sustainability Series

Series Editor

Walter Leal Filho, European School of Sustainability Science
and Research, Research and Transfer Centre "Sustainable Development
and Climate Change Management", Hamburg University of Applied Sciences,
Hamburg, Germany

Due to its scope and nature, sustainable development is a matter which is very interdisciplinary, and draws from knowledge and inputs from the social sciences and environmental sciences on the one hand, but also from physical sciences and arts on the other. As such, there is a perceived need to foster integrative approaches, whereby the combination of inputs from various fields may contribute to a better understanding of what sustainability is, and means to people. But despite the need for and the relevance of integrative approaches towards sustainable development, there is a paucity of literature which address matters related to sustainability in an integrated way.

Notes on the quality assurance and peer review of this publication

Prior to publication, the works published in this book are initially assessed and reviewed by an in-house editor. If suitable for publication, manuscripts are sent for further review, which includes a combined effort by the editorial board and appointed subject experts, who provide independent peer-review. The feedback obtained in this way was communicated to authors, and with manuscripts checked upon return before finally accepted. The peer-reviewed nature of the books in the "World Sustainability Series" means that contributions to them have, over many years, been officially accepted for tenure and promotion purposes.

Rubee Singh · Amit Joshi · Walter Leal Filho · Shahbaz Khan
Editors

Zero Carbon Industry, Eco-Innovation and Environmental Sustainability

Springer

Editors
Rubee Singh
Institute of Business Management
GLA University
Chaumuhan, Mathura, Uttar Pradesh, India

Amit Joshi
Department of Management Studies
Kumaun University
Nainital, Uttarakhand, India

Walter Leal Filho
Faculty of Life Sciences
Hamburg University of Applied Sciences
Hamburg, Germany

Shahbaz Khan
Department of Industrial Engineering
University of Tabuk
Tabuk, Saudi Arabia

ISSN 2199-7373 ISSN 2199-7381 (electronic)
World Sustainability Series
ISBN 978-3-031-80219-5 ISBN 978-3-031-80220-1 (eBook)
https://doi.org/10.1007/978-3-031-80220-1

This Springer imprint is published by the registered company Springer Nature Switzerland AG
The registered company address is: Gewerbestrasse 11, 6330 Cham, Switzerland

Preface

Zero carbon industry represents one of the most crucial steps in our collective effort to combat climate change and ensure sustainable economic growth. This book, *Zero Carbon Industry, Eco-Innovation and Environmental Sustainability*, seeks to provide a comprehensive understanding of the principles, challenges, and strategies involved in achieving a zero carbon future. It addresses the urgency of responding to the UN Sustainable Development Goals (UN-SDGs) as part of the global mandate. It aims to present a unique perspective by exploring and integrating innovative multidisciplinary ideas, techniques, and tools critical for implementing zero carbon emissions, innovative ideas, technologies, and tools for environmental sustainability. By combining the academic expertise of emerging and established scholars from diverse academic fields and countries, the book provides an integrated understanding of zero carbon emissions, eco-innovation, and environmental sustainability from a broader perspective.

To achieve sustainability, it is necessary to fundamentally transform the zero carbon industry and the eco-innovation ecosystem. The industrial sector, being one of the largest contributors to greenhouse gas emissions, stands at the forefront of this transformation. This book explores the multifaceted aspects of zero carbon industry, emphasizing that this shift is not just an environmental necessity but also a driver of economic and social innovation. In the book, the transformation of sustainable business models for the manufacturing industry is discussed, focusing on the integration of zero carbon principles. It also, describes how the resilience of supply chains can be enhanced through the adoption of Industry 4.0 technologies, providing a comprehensive assessment of greenhouse gas emissions. The book also presents innovative solutions and zero carbon production practices that pave the way for a sustainable future. Sustainable supply chain management in the zero carbon era is explored, highlighting the strategies and practices necessary to align supply chains with zero carbon goals.

Human factors play an indispensable role in the adoption of zero carbon practices. This book underscores the significance of fostering human elements within the industry, focusing on sustainable practices that can be adopted by the workforce. The book also focuses on the sustainability driven eco-innovation, examining the critical role of human capital in the eco-innovation ecosystem. Sustainable HR practices are assessed for their impact on achieving environmental sustainability. The book also addresses the adverse impacts of innovation and development from a global perspective, emphasizing the importance of sustainable strategies in mitigating these effects. This book explores the various technologies that are essential in this transformation, providing a roadmap for their implementation. The decarbonization of supply chains and its impact on the environment and society is analyzed, providing insights into how businesses can integrate sustainability for success in a zero carbon world. Additionally, the development of performance assessment frameworks for zero carbon supply chains is discussed, offering tools for measuring and improving sustainability performance.

The book explores the issues and challenges inherent in the zero carbon industry and innovation. Understanding these factors is crucial for businesses and policymakers as they work towards achieving zero carbon goals. Innovation and environmental sustainability are inextricably linked, and this relationship is analyzed in detail. It investigates the barriers, drivers, and critical success factors of the zero carbon industry, offering a detailed examination of key performance indicators and best practices for zero carbon leadership. The drivers for innovation in environmental sustainability are illustrated through experiences from the South Asian apparel industry, shedding light on the unique challenges faced by this sector. The dark side of innovation and environmental sustainability in developing nations is explored, highlighting the unintended consequences and challenges these regions face. The book concludes with a discussion on the challenges and opportunities in enhancing the urban water sector for environmental sustainability and innovation, emphasizing the need for innovative solutions in this critical area.

This book will be an invaluable resource for readers to learn how the zero carbon industry is transforming innovation ecosystems and promoting environmental sustainability. It will teach readers how to navigate the zero carbon industry and eco-innovation in the era of sustainability, offering sustainable innovation recommendations for adopting zero carbon practices. Readers will gain insights into the role of policy frameworks, financing mechanisms, and international collaboration in facilitating the adoption of zero carbon industry and eco-innovation towards global sustainability. It will serve as a guide for scholars, practitioners, policymakers, and anyone interested in fostering a sustainable and resilient industrial ecosystem. Furthermore, the book is beneficial for industrial professionals working in the areas of net zero emissions, innovations, sustainable development, environmental sustainability, eco-friendly business management, and entrepreneurship. By exploring the intricate and essential aspects of zero carbon industry, eco-innovation, and environmental sustainability, this book aims to provide a comprehensive resource for all stakeholders committed to a sustainable future.

Book has divided in 3 parts in which there is total 16 chapters from different countries such as India, Sri Lanka, United Kingdom, Dubai, and Malaysia. This book is the result of a collective effort by scholars and experts from various fields, bringing together diverse perspectives and in-depth research. We extend our heartfelt gratitude to all contributors and reviewers whose efforts have made this publication possible.

Special thanks go to Springer Nature for their support and dedication to advancing scholarly work in environmental sustainability. In conclusion, *Zero Carbon Industry, Eco-Innovation and Environmental Sustainability* aims to inspire and equip readers with the knowledge and tools necessary to drive sustainable world and environmental stewardship. It is our hope that this book serves as a valuable resource, guiding the global community towards a zero carbon future and fostering a sustainable and resilient industrial ecosystem.

Chaumuhan, Mathura, India — Rubee Singh
Nainital, India — Amit Joshi
Hamburg, Germany — Walter Leal Filho
Tabuk, Saudi Arabia — Shahbaz Khan

Acknowledgements

We would like to express our gratitude to all those who have contributed to the creation of this book. Firstly, we thank the researchers and experts who have shared their knowledge and insights with us, providing valuable information on the current state of *Zero Carbon Industry, Eco-Innovation and Environmental Sustainability*. We also extend our appreciation to the organizations and institutions that have supported our work, including government agencies, non-governmental organizations, and academic institutions.

Finally, we would like to acknowledge the invaluable contributions of our colleagues and editors, who have worked tirelessly to ensure that this book is informative, engaging, and accessible to a wide range of readers. We hope that this book will inspire further research and action towards a Zero Carbon Policies and Eco-Innovation System to promote Environmental Sustainability as per United Nations agenda. We would like to express our sincere gratitude to the experts and stakeholders who contributed their valuable insights and perspectives to this book. Their contributions have been instrumental in shaping our understanding of the challenges and opportunities involved in achieving Zero Carbon Industry with use of eco-innovation and environmental sustainability at globally. We hope that this book will serve as a useful resource for policymakers, practitioners, and researchers working towards a zero carbon future for the continent.

We are immensely thankful to *Book Series-Editors, Prof. Walter Leal Filho, Editorial Director/Publishing Editor, Dr. Christian Witschel and* the entire team at Springer Nature for bringing out this publication in a very professional and timely manner.

Rubee Singh
Amit Joshi
Walter Leal Filho
Shahbaz Khan

About This Book

This book presents the latest research on the Zero Carbon Industry and its practical implications for eco-innovation and environmental sustainability in the current business environment. As the world aims to achieve zero carbon industry by 2050, the economic output will shift progressively towards goods and services that do not emit greenhouse gases, fundamentally transforming entire value chains and industries. This transformation necessitates changes in our economies and societies, with businesses taking eco-innovation and environmental sustainability seriously.

The book explores the complex and interconnected domains of zero carbon industry, eco-innovation, and environmental sustainability by combining the academic expertise of emerging and established scholars. It emphasizes the need for Industry 4.0 technologies, such as Artificial Intelligence, the Internet of Things, Blockchain, and Cloud Technologies, to drive sustainability. The book first provides insights into transforming sustainable business models, enhancing supply chain resilience, and innovative zero carbon production practices. It then examines the role of human capital in the eco-innovation ecosystem, assessing the impact of sustainable HR practices, and addressing the adverse impacts of innovation and development. Finally, it explores the challenges and opportunities in the zero carbon industry, focusing on leadership, key performance indicators, and the integration of sustainability strategies.

The book addresses the urgency of responding to the UN Sustainable Development Goals (UN-SDGs) and presents a unique perspective on integrating multidisciplinary ideas, techniques, and tools critical for environmental sustainability. It also provides insights into successful initiatives, best practices, and lessons learned in zero carbon emissions, entrepreneurship, and innovation ecosystems. This comprehensive resource targets undergraduate and postgraduate students, practitioners, industry experts, and research scholars working in the area of Climate Change, sustainability, environmental sciences, management, and related fields. It is also beneficial for industrial professionals working in net zero emissions, innovations, sustainable development, and eco-friendly business management, offering valuable guidance and inspiration for a sustainable future.

Contents

About the Editors

Dr. Rubee Singh (Ph.D.) is an Assistant Professor of Human Resource at Institute of Business Management, GLA University India. She is a Post-Doctoral Researcher (D.Lit.) at Department of Management Studies, Kumaun University, Nainital, India. She holds Ph.D. in Management from Noida International University India, an MBA in HR from Dr. A. P. J Abdul Kalam Technical University India. She has more than 10 years of experience in research and teaching. Apart from teaching and research, she has handled many administrative positions in premier institutes affiliated with A.K.T.U. She has conducted a number of FDPs, Conferences, Workshops, and HR Conclaves in the area of Sustainability, CSR, AI, HR, Human Factors, Women Empowerment, and Human rights. She is a United Nations Volunteer, and a contributing author for Voices of Youth (UNICEF). She is an honorary life member of Women Economic Forum and National Secretary of IATM (Youth Wing) International Anti-Terrorism Movement India. Dr Singh is a Managing Editor of IJARSH India. She is serving on the Editorial Board of more than 30 international journals. She serves as the Advisory Board Member of the Heliyon Journal, Elsevier, Research in Globalization, and reviewer of many reputed journals like European Journal of Innovation Management (EJIM), British food journal, Sustainability, Social Responsibility Journal, etc.

She is life member of International Association of Engineers, Thailand, IMRTC USA, ISRD U.K etc. She is a bestselling author with significant experience as an excellent Editor, Eminent researcher, great academician, Global Ambassador (Children's) & Social Activist. She has presented many papers at national and international conferences, published 20 academic books including Springer, Elsevier, De Gruyter Nova Science Publishers, and more than 75+ research papers in SCI, ABDC-A, Scopus & refereed/ UGC index journals. Her areas of research are sustainability, Digital HRM, Sustainable HRM, Circular Economy, Digital CE, Net Zero Emissions, Climate Change, Industry 4.0, Artificial Intelligence, Blockchain Technology, and Corporate Social Responsibility. Being a thought leader, her books are honored by being added to the library, Ohio Christian University, University of California, Corporacion Universitaria Unisabanta, UMass Chan Medical School, Tunix Community College, USA, Aston University UK, Dow University of Health Sciences, Pakistan, and University of Port Harcourt, Nigeria. Her book titled "Training & Development" was added to the MBA syllabus of "The Institute of Aeronautical Engineering" (Jawaharlal Nehru Technological University, Hyderabad). Dr Singh has been honored with many prestigious awards, which include Most Influential Woman Leaders in 2022, Best Researchers Award of the Year-2022–23 by GLA University, India, Child Ambassador Award of The Year 2022, Iconic Faculty Award 2019, International Young Researchers Award-2021, Young Scientist Award 2020, and Nari Ratan Award 2023, India Book of Records-2020 & Asia Book of Records Holder for 2021. The list of awards and appreciation is long, but Dr. Singh has always prioritized her work over recognition, making her unique and commendable in her field.

Prof. Amit Joshi (Ph.D.), holding a Ph.D. and an MBA, is a distinguished academician and visionary leader with a wealth of experience in both industry and academia. Currently serving as Professor and Head of the Department of Management Studies at Kumaun University, Nainital, his teaching and research expertise encompasses Marketing, Strategic Management, and Knowledge Management. Before transitioning to academia, Prof. Joshi garnered valuable experience in export finance, working capital management, merchant banking, market research, and concept selling. His teaching journey extends beyond India, having shared his knowledge and insights at universities in Saudi Arabia, Malaysia, and Bahrain. Renowned as a captivating speaker, he has been invited to deliver lectures at esteemed institutions across several countries, reflecting his global influence and impact. Prof. Joshi's interdisciplinary research interests have led him to explore various fields, with a particular focus on tourism, evident through his numerous publications. As Head of Department, he provides invaluable support to a large student and faculty community, fostering academic excellence and growth. His administrative prowess is evident through his roles as Convener of the Board of Studies and Research Degree Committee at Kumaun University, along with his previous directorial positions. Prof. Joshi's commitment to student welfare is commendable, as he serves as a mentor and career counsellor, guiding countless students towards success.

Furthermore, Prof. Joshi actively contributes to research activities, seminars, and initiatives on both national and global platforms. His expertise in academic administration and quality assurance has been recognized internationally, showcasing his dedication to excellence in education. Prof. Joshi has published several research papers in UGC Index, refereed journals, Scopus, and book chapters in Springer Nature, Nova Sciences Publishers, and Emerald Publishers. He is an Editor for Edited Book of Springer nature and Emerald Publishing. An accomplished author and reviewer, Prof. Joshi's scholarly contributions extend to several international and national journals, further solidifying his reputation as a thought leader in academia. His leadership qualities are evident through his strategic collaborations with universities, institutions, and businesses, aimed at driving positive change and innovation. Prof. Amit Joshi's remarkable journey, characterized by academic excellence, leadership, and a passion for knowledge dissemination, continues to inspire and empower individuals within and beyond the academic realm.

Prof. Walter Leal Filho (D.Lit., Ph.D.) holds the Chairs of Climate Change Management at the Hamburg University of Applied Sciences (Germany), and Environment and Technology at Manchester Metropolitan University (UK). He directs the Research and Transfer Centre *Sustainability Development and Climate Change Management*. His main research interests are in the fields of sustainable development and climate change, also including aspects of climate change and health. Professor Leal has been author, co-author, and editor or co-editor in many publications on matters related to environment, sustainable development, and climate change issues. He has in excess of 700 publications to his credit, which includes 193 books, 102 book chapters, and over 430 papers. His field experience in over 70 countries, involves missions undertaken on behalf of various international organizations (e.g. European Union, World Bank, OECD, UNESCO, UNEP) and attendance to specialized events in North America, Latin America/Caribbean, Europe, Africa, Asia and Pacific Region, and the Middle East. Among the many awards he has received, it can be mentioned the 1999 "Leading Editor of the Year Award" and the 2000 "Editor of the Year Award", by MCB University Press, the "Eco-Citizen Award" (2001) in Rio de Janeiro and the Aurelio Giuseppe Prize awarded by the Italian Council of Ministers for his environmental work. Moreover, he was awarded the "Golden Page Award" (April 2003) in London for his editorial work and the North Sea Star Award (2006) by the EU-Interreg Secretariat for his project work in the North Sea Region. In the field of sustainable development, Professor Walter Leal is the founder of the European School of Sustainability Science and Research and of the Inter-University Sustainable Development Research Programme (IUSDRP). He holds a record of publications on sustainability issues, some of which are available here: https://www.haw-hamburg.de/en/ftz-nk/programmes/iusdrp/ (please scroll down and click on "sharing"). He is also editor of the World Sustainability Series with Springer. In addition, he is the founding editor of the International Journal of Sustainability in Higher Education (IJSHE) launched in the year 2000. IJSHE is the only journal which fully focuses on sustainability in a higher education context. Professor Walter Leal is also Editor-in-Chief of the Encyclopedia of the UN Sustainable Development Goals currently being prepared. With 17 volumes, one devoted to each SDG, more than 1.700 chapters and in excess of 2.300 authors, this is the largest editorial project on matters related to sustainable development, ever undertaken.

Dr. Shahbaz Khan (Ph.D.) is an Assistant Professor at the Department of Industrial Engineering, University of Tabuk, Kingdom of Saudi Arabia. He has published more than 80 research articles in many reputed journals including Scopus, SCI, and ABDC-A categories such as Business Strategy and Environment, Annals of Operation Research, Operations Management Research, etc. He has an excellent research track record with 3000 citations on Google Scholar and H-index-30. He has obtained his Ph.D. and M.Tech. degree with specialization in Industrial and Production Engineering from Jamia Millia Islamia India. He is a gold medalist in M.Tech. He is a member of the International Association of Engineers, Thailand. His research interests are Supply Chain Management, Digital Supply chain management, Circular

Economy & Sustainability, Industry 4.0, Blockchain, Artificial Intelligence, benchmarking, and decision modelling for sustainable business. He has published 5 books including springer, Elsevier, Springer and De Gruyter. He is serving as the Advisory Board Member of the Heliyon Journal, Elsevier and Associate Editor in Green Technologies and Sustainability. He is the reviewer of many reputed journals such as the Journal of Business Research, Trends in Food Science & Technology, International Journal of Fuzzy Systems, Engineering Applications of Artificial Intelligence, Soft Computing, sustainability, etc.

Zero Carbon Industry and Environmental Sustainability

Transforming Sustainable Business Models for Manufacturing Industry to Build Zero Carbon Industry

Rubee Singh, Amit Joshi, and Seema Rani

Abstract Sustainability has garnered considerable interest in the field of manufacturing research in recent decades. It has emerged as a major catalyst for the advancement of innovative technology and management systems. Businesses must continuously work to achieve standards encompassing environmental, economic, and social objectives, which are together known as the triple bottom line. The manufacturing industry has made significant advancements by implementing sustainable business strategies in this field. This chapter aims explore the widely adopted sustainable practices to identify the sustainable business models in the manufacturing industry. Through a thorough examination of existing literature, this chapter presents a comprehensive framework that emphasizes four interrelated approaches in manufacturing sector namely additive manufacturing, lean manufacturing, waste management, and sustainable product design and development. Literature suggests that there has been empirical research conducted to examine the impact of sustainability in manufacturing on social and environmental aspects along with economic output. In general, there is a continuous shift towards implementing sustainable business practices throughout the manufacturing industry. The chapter offers valuable information to scholars and professionals regarding the implementation of sustainable business models across manufacturing organizations and how it benefits their stakeholders. Additionally, it offers opportunities for exploring uncharted approaches to sustainable business models through research.

Keywords Manufacturing industry · Zero carbon emissions · Sustainability · Sustainable business model

R. Singh (✉) · A. Joshi
Department of Management Studies, Kumaun University, Nainital, Uttarakhand, India
e-mail: dr.rubeerajput@gmail.com

R. Singh
Institute of Business Management, GLA University, Mathura, Uttar Pradesh, India

S. Rani
Department of Commerce, Kurukshetra University, Kurukshetra, India

R. Singh et al. (eds.), *Zero Carbon Industry, Eco-Innovation and Environmental Sustainability*, World Sustainability Series,
https://doi.org/10.1007/978-3-031-80220-1_1

1 Introduction

The recent worldwide financial and economic issues have prompted research into the effects of contemporary business strategies on the world economy and society in the long run. The occurrence of these crises has led to the urging of international agencies and scholars to reassess the potential contributions of firms towards sustainable development. Sustainable development was originally described almost three decades ago as "development that fulfills the requirements of the current generation without jeopardizing the capacity of the next ones to fulfill their own requirements" (Glavič and Lukman 2007). At the organization level, the idea of sustainable growth has given rise to various domains, where sustainability is associate with social business, innovation, management, entrepreneurship, and corporate sustainability (Singh et al. 2021). In case of Sustainability management involves addressing not only economic concerns but it also deals with the environmental and social issues using a holistic approach to reshape organizations for actively contributing towards development of the society and economy simultaneously, while adhering to the ecological boundaries (Bonfanti et al. 2023).

The impact of globalization and climate change on economic activity has been substantial. The different stakeholders along with the government, are exerting pressure to establish a sustainable business model (SBM) that optimizes resource consumption and minimizes adverse environmental effects (Comin et al. 2020). Several organizations have been prioritizing sustainable and cleaner production by implementing net-zero economy principles (Singh et al. 2024). Currently, individuals in positions of leadership, management, and entrepreneurship are being called upon to actively participate in promoting sustainable development not only at the organizational but the societal levels as well (Ritala et al. 2018). Researchers and professionals are now investigating whether altered or entirely new business models can contribute to overall growth of economy by significantly reducing the adverse environmental and societal impacts, or by generating positive effects (Godina et al. 2020; Ciliberto et al. 2021; Delaney et al. 2022; Kurien and Qureshi 2018). This viewpoint encompasses both established organizations and their adaptations of business models, as well as innovative models introduced by entrepreneurs.

Manufacturing industry, being a significant participant in several aspects of human existence, has a crucial function to fulfill in promoting sustainability as it has a major impact on overall environmental (Bonfanti et al. 2023). Simultaneously, as production processes get more precise at increasingly leading to small scale operations, the usage of energy associated with these systems is growing at a fast pace, along with overall estimations of continued growth in energy use (Zhang et al. 2020). In addition to its immediate environmental effects, the discrete product manufacturing industry also has a significant and intricate impact on the resource usage of its products during their entire lifespan, making it a crucial factor in sustainability (Ahmad et al. 2018). Despite the steady in the manufacturing sector and improved work environments over the past few decades, there are still persistently poor working conditions that lead to around 3 lakh of annual work-related deaths coupled with the monetary lose of equivalent to

4% of the GDP of the Europe (WHO 2016). On a global scale, businesses account for 7.2% of child labor, which amounts to around 12 million individuals. However, the significance of manufacturing in the growth and advancement of economies cannot be undermined. The anticipated value added is US$8.6 trillion by the end of 2024 in the Manufacturing industry with a compound annual growth rate of 1.09% between 2024 and 2029 (Statista 2024). The anticipated output in the Manufacturing market is estimated to reach a total of US$34.8 trillion by the year 2024. Further, the estimated workforce size in the sector is expected to reach 209.80 million employees while the compound annual growth rate for the period 2024–2029 is 1.70%. In summary, manufacturing activities are essential for the advancement of modern economies.

The manufacturing sector significantly contributes to global greenhouse gas (GHGs) emissions with a share of around 30% towards greenhouse gases (World Economic Forum 2024). To attain net-zero emissions by 2050, the manufacturing industry must focus to develop specific business plans, considering the significant variations in the climate and the sustainable development goals (Bonfanti et al. 2023). Consequently, this chapter seeks to explore and critically analyze the practices to promote the implementation of innovative business models, which in turn contributes towards sustainability and facilitates the attainment of a net-zero economy. A comprehensive analysis of literature pointed out different business models however this chapter throws light on the most discussed sustainable practices which includes additive manufacturing, lean manufacturing, waste management, and sustainable product design and development. This chapter can offer valuable insights for professionals in the industry and academics, offering substantial insights into specific sustainable business models (SBMs) in manufacturing industry to lead the way towards sustainability.

The chapter is divided into several sections for clear understanding. Initially, the literature review section is given that develops our basic understanding on sustainable business models and the linkage between sustainability and manufacturing. This section also comprehensively discusses the widely discussed business models in the manufacturing sector throughout the reviewed literature. Then, there is a research methodology section followed by implications of the study. These implications highlights that how sustainable practices are essential for formulating net-zero strategies customized for manufacturing sector and attaining net-zero goals within the specified time period. Finally, conclusion of the study is presented along with its limitations and avenues for research in this area.

2 Literature Review

2.1 Sustainable Business Models (SBMs)

Earlier companies or organizations followed the common business models utilizing the mechanism to generate and enhance the usefulness of the goods or services they provide prior to final sales (Zott et al. 2011). Each business model has unique propositions for defining the process of identifying the value chain, targeting markets, positioning the organization, merging channels to support distribution, and specifying revenue and cost composition (Chesbrough 2010). Business models have been recognized as a competitive tool for gaining strategic advantage if we look into the development of organizational theory and these models sets one organization apart from the similar ones (McGrath 2010). Nevertheless, the primary focus on generating revenue in conventional businesses has been associated with societal and environmental challenges because the companies were mainly focused on the financial aspects and hence ignored the non-financial domain relating to environment and society as whole.

A business model is developed on the basis of a comprehensive framework of ideas, with a purpose of guiding an organization as a part of social system of the entire economy, with a focus on society and the environment (Upward and Jones 2016). Value creation has specifically advanced beyond a narrow focus on the end-consumer and now encompasses a business model that benefits all stakeholders together (Tolkamp et al. 2018). The expenses as well as advantages for parties involved are encompassed by this concept, extending beyond simply the customers and the company. It includes everyone from suppliers to customers, partners to employees, the environment, shareholders to investors, and society to environment (Donaldson and Preston 1995). Such changes in business practices entail the active user participation in the formation of value for products (Saebi and Foss 2015). From a sustainability standpoint, extending a company model is primarily dependent on this crucial issue (Tolkamp et al. 2018). The old business model were focused on forming, storing, and delivering value to generate revenue in new markets. Contrastingly, the modern SBM combine economic and social values with environmental value by implementing a series of ongoing practices. Nevertheless, there is no agreement on the optimal strategy for companies to adopt for addressing the challenge sustainability in their business operations (Boons and Lüdeke-Freund 2013).

With each historical period, new ways arise, that give rise the innovative ideas. The literature discussed SBMs in the context of the circular economy. Development of new frameworks play a crucial role in innovating existing business models and emphasize the need of shared and collective efforts as key elements (Kozlowski et al. 2018). Previous studies on SBMs reveals a multitude of concepts regarding potential approaches to addressing the economic issues along with social and environmental concerns by facilitating the spread of innovations, technology, and organizational structures this enables organizations to enhance the overall functioning by adopting sustainable practices in their operations (Boons and Lüdeke-Freund 2013).

2.2 Sustainability in Manufacturing

Previously, certain academics depicted sustainable manufacturing (SM) as approaches or technologies for enhancing not only the economic performance but also improve the environmental outcomes. Madu (2001) defined SM as a method for maximizing the value of products by efficiently utilizing the earth's limited resources, minimizing harm to the environment, and aiming for environmentally friendly production techniques. As per Allwood (2005), sustainable manufacturing as the process of utilizing technology that can change resources without exerting adverse impact on environment, using limited natural renewable or poisonous materials, or generating waste. In the words of the US Department of Commerce (2007) SM means production of products that are manufactured using environment friendly technologies that preserve natural resources and low energy usage, while also being affordable and ensuring the safety of workers, consumers, and society. Ritala et al. (2018) defined sustainable management in manufacturing context as the integration of economic and social aspects with the environmental domain, into a firm's operational and business activities as shown in Fig. 1 SM aims to pursue financial and environmental improvements while promoting and supporting social well-being.

In the past years, the evolution of sustainability and SM concepts resulted into the discovery of various environmentally friendly methods in manufacturing industries. These practices range from using technology to control and treat pollution to implementing integrated production systems that foster collaboration among employees and between various industries (OECD 2010). The creation of sustainable manufacturing practices (SMP) can be understood by considering three levels including product, process, and system, rather than relying solely on individual approaches. The shift from a linear economy system with open-loop system to a circular system of closed-loop system has transformed the classic 3R idea of reduce, reuse, recycle into a wide 6R approach having three more components in the form of recover, redesign, remanufacture at the product level (Jayal et al. 2010). While many attempts have been made to optimize technological advancements and process planning in order to decrease resource usage and waste production simultaneously enhancing useful life of the product, by shifting the focus from the organization level to a wider industrial level (Jawahir and Dillon 2007).

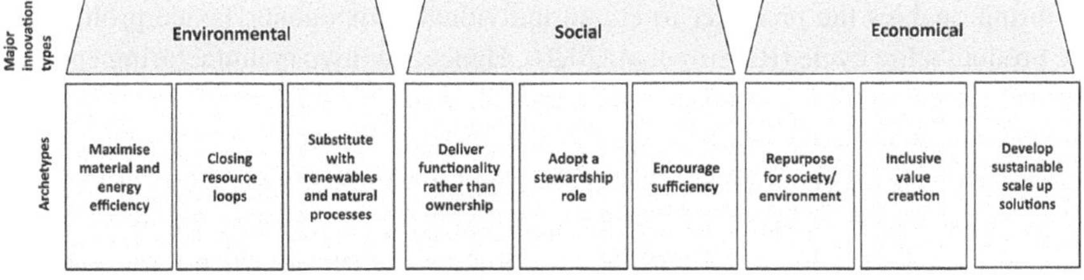

Fig. 1 Aspects of sustainable management. *Source* Ritala et al. (2018)

2.3 Additive Manufacturing (AM)

The European Commission Report (2016) states that additive manufacturing is a prominent technology in the digital era, which has a substantial influence on the sector. This revolves around the prospective advantages it offers, specifically, the simplification of distribution networks and the enhancement of effectiveness in the manufacturing value chains by reducing the costs associated with the process of developing, designing, and testing. Additionally, it significantly improves the product customization in a cost-effective manner. This product customization, known as individualization when undertaken at a large scale, is considered a significant issue in this technology (Rejeski et al. 2018). It takes advantage of the optimization of digitization processes that are evolving in the framework of Industry 4.0 (Dilberoglu et al. 2017). For instance, the utilization of additive manufacturing technologies, specifically 3D printers, in conjunction with the Internet of Things and control systems using computer numerical, can play a crucial role in automation while still retaining the ability to customize products on an individual level (Qin et al. 2017). These system scans offer significant advantages by integrating AM technology with other concepts of technology innovation. This integration leads to improved efficiency in managing production, including stock control, managing logistics, and monitoring of production settings including component and material use (Rosienkiewicz et al. 2018).

According to Gutowski et al. (2017), there are two crucial factors contributing towards environmentally benign nature of AM and make it a sustainable manufacturing process. Firstly, additive manufacturing minimizes waste by utilizing only the required material throughout the layer-by-layer addition process, in contrast to subtractive manufacturing. Consequently, the amount of waste generated is significantly reduced. Moreover, its capacity of repurposing plastic waste through transforming the latter into printing filaments and manufacturing novel items. One important aspect is the increased ease of use of additive manufacturing technology, which has been made possible by Industry 4.0 (Baumers et al. 2017). This allows producers to build items directly on-site, resulting in lower costs for transportation and logistics. The longer the useful life of a product, the lesser would be its environmental effect due to low manufacturing rate. If a product consisting of multiple components is manufactured using injection molding and if any of the components is destroyed, typically, a replacement product needs to be acquired. In contrast, additive manufacturing enables the producer to create individual components, hence prolonging the product's life cycle (Ribeiro et al. 2020). Hence, additive manufacturing enables the incorporation of new components or the substitution with superior ones, thereby enhancing and prolonging the lifespan of the initial product.

Moreover, there is potential for recycling and reusing tools in additive manufacturing to prolong their lifespan. This is supported by the fact that several researchers have already begun investigating the recycling of specific metal powders (Oros-Daraban et al. 2019). Cost is an important factor in assessing the feasibility of the product or technology when making decisions. The AM have the substantial economic implications for the industry as whole (Baumers et al. 2016). By adopting additive manufacturing, waste is effectively minimized as only the essential material is utilized to create the desired shape, and one can fully recycle any faulty parts. It enhances the adoption of closed loop system and thereby promotes the sustainability of manufacturing operations (Fitzsimons et al. 2020). By making design improvements, the material utilized can be optimized, resulting in parts that are both stronger and lighter (Niaki et al. 2019). These can be manufactured through a single operation, which reduces burden of the final processes for assembling and thus, enhancing the final product's dependability. The potential for local production at a lower cost necessitates a significant overhaul of the existing supply chain, resulting in substantial reductions in transportation routes (Caviggioli and Ughetto 2019; Kunovjanek and Reiner 2020).

2.4 Lean Manufacturing

Lean manufacturing (LM) is known as a system that decreases waste by minimizing the level of variations in the internal processes and those related with customers and suppliers following socio-technical improvements (Shah and Ward 2007). The approach is multi-dimensional and includes a various practices related to the management that work together and support each other. The categorization of these practices includes: Human Resource Management (HRM), Total Preventive Maintenance (TPM), Total Quality Management (TQM), and Just-in-Time (JIT). The implementation of LM theory and techniques has been thoroughly investigated by (Belekoukias et al. 2014). Furthermore, when examining the connection between the two, the understanding of LM is usually limited to JIT, which is a component of LM but not identical to it. For the incorporation of LM principles and the preservation of the environment several authors proposed that organizations might strategically utilize LM to enhance their environmental practices (Dues et al. 2013). They positive impact of LM on the environment is described as the "beneficial public spillover of lean", considering it a logical progression or initial step (Franchetti et al. 2009). Significance of this linkage is referred as a mutually beneficial relationship between environmental and operational management (Carvalho and Cruz-Machado 2009).

Lean production is a strategic approach with the objective of meeting customer requirements, through creating value and value streams, eliminating waste, pursuing excellence, ensuring reliability in all stages of production, and improving all processes continuously (Womack and Jones 1996). The examination of lean manufacturing demonstrates that it embodies a sophisticated manufacturing approach that ensures enhanced efficiency (Ohno 1988; Resta et al. 2017). In recent times, the individuals or groups responsible for the process of producing and delivering a product or service have been anticipating a stronger connection between how well a company performs and competes in the market and how the environment and society is affected by its operations (Gupta 2016; Martínez Leon and Calvo-Amodio 2017). It is already being utilized more frequently in intricate socio-technical systems that are marked by significant uncertainty, diversity, and dynamic interactions. As a result, LM is already well-suited to address the complex challenges associated with sustainability difficulties (Azadegan et al. 2013).

Nevertheless, whereas previous investigations have established that lean practices and beneficial environmental outcomes are inter-linked (Moreira et al. 2010; Vinodh et al. 2011), some researchers have identified certain places where the integration of both of these approaches is not feasible and where conflicts and trade-offs may arise (Rothenberg et al. 2001). To provide consistent and conclusive findings, there is a need to analyze that how LM is related to the environment in a comprehensive manner, instead of individually focusing on a single lean practice or a limited number of environmental parameters because regarding the environmental aspect, the existing literature does not offer conclusive findings (Martnez-Jurado and Moyano-Fuentes 2014). LM has been shown to have a favorable impact on the attitude of employees. This is attributed to the introduction of more diverse tasks, enhanced autonomy with responsibility, and a boost in intrinsic motivation (Treville and Antonakis 2006). Additionally, there is undue pressure on individuals (Hines et al., 2004). However, there is a general agreement regarding the beneficial impact of LM on employees' health and safety. This is primarily attributed to the use of standards for designing the workstations, which ultimately enhances the working conditions of employees (Taubitz 2001).

2.5 *Waste Management*

Waste management plays an important role in sustainability. Waste prevention is centered on the idea of eradication of garbage prior to its creation. Waste prevention refers to the activities, processes, and procedures that effectively minimize, prevent and erase waste directly at its origin (Williams 2005). It is typically confined within the boundaries of a production facility. Waste minimization is closely associated with waste prevention. Waste minimization and waste prevention are more effective methods of protecting the environment than simply reducing trash at the process completion stage (Zamorano et al. 2011). Re-use refers to the practice of utilizing a product in its original state, even after it has beyond its planned lifespan, by processes such as cleaning, mending, reclaiming, and overhauling. Further, re-manufacturing can reduce the adverse impact of a product on the environment while maintaining its original shape (Khan et al. 2022). This strategy might be considered similar to the "waste prevention" concept as it ensures that products can be used again instead of transforming into waste. According to Pongrácz and Pohjola (2004), any product or the by-product or residuals of production process should not be considered as waste if they are intended to be directly reused in its current form in another process.

Further, up-cycling involves enhancing the value of obsolete product material rather than diminishing it. The objective is to transform waste resources into new goods or materials that have higher quality and worth, while causing minimal negative influence on the environment. Up-cycling decreases the utilization of new raw materials, leading to a decrease in pollution and the formation of waste (Braungart and McDonough 2002). Nevertheless, up-cycling is positioned lower than reclamation and re-use in the hierarchy of waste management due to the fact that it involves the dismantling of a product into its fundamental components, which are subsequently re-processed for utilization in new items (Lodder et al. 2014). Composting is also a kind of recycling in natural system. It breaks down the biodegradable portion of the waste into substances like plant fertilizer and soil conditioners (Kurien and Qureshi 2018). Utilizing biodegradable substances as the primary material, implementing effective waste segregation, and establishing a robust composting system are the key methods for achieving environmentally friendly waste management.

Composting is a process that provides nourishment to nature and, as a result, decreases overall entropy. On the other hand, down-recycling for industrial purposes does not decrease entropy, but typically leads to an increase in entropy. Alternative methods of waste management include harnessing energy from trash incineration and utilizing landfill gasses as a fuel source (Delaney et al. 2022). Landfill is the least favored choice in the waste management hierarchy. Over time, landfills undergo stabilization and transformation into inert material as a result of biological and chemical processes occurring within the waste (Hogland and Stenis 2000). Nevertheless, the duration of this stabilization process is lengthy, and during this period of deterioration and stabilization, greenhouse gasses such as methane and carbon dioxide are consistently released from landfills. Managing leachate, which refers to the liquid fluid that contains undesired chemicals that drains from landfills, is a significant difficulty in landfill operations (Williams 2005).

A waste strategy is implemented by the UK Government in 2000, which establishes a structure to guarantee for minimizing, reusing, recovering, and subsequently the disposing of trash (Chick and Micklethwaite 2004). This approach ensured sustainable consumption and disposal throughout the manufacturing process. Designers must adhere to this paradigm in order to enhance their environmental sustainability. Graedel and Guth (1990) pinpointed the elements that for assisting designers in identifying areas for development over the lifespan of a product. It is also argued that in addition to waste arising out of production process, designers should take into account packaging waste and work towards reducing it in order to achieve ecologically sustainable production (Monteiro et al. 2019). In their study, Deutz et al. (2010) examined various methods for incorporating such systems of waste management into the process of product designing as a necessary component for environmental sustainability. This approach can assist designers in effectively minimizing waste during the process of new product development.

2.6 Sustainable Product Design and Development

The stage of product designing has a great impact on the following phases in terms of long-term effects (Rebitzer et al. 2004). Hallstedt (2008) stressed the importance of sustainability incorporation into the product design phase as early as feasible to advance society towards sustainability. Various phrases, such as design for sustainability, design for the environment, and eco-design, are used to define this technique (Khor and Udin 2013). Multiple extant investigations emphasize that product designing and development with a focus on sustainability aimed at either eliminating or lowering harmful materials, minimizing waste, enhanced preservation and efficiency, increased recovery of resources through recycle process, reusability, and remanufacturing, and enhancing sustainability (Khan et al. 2022). Ma and Kremer (2016) propose that in order to adequately address GHG emissions in the design phase, designers should take into account all forms of emissions and strive to minimize them. There is ISO 14000 which has a collection of regulations aimed at promoting sustainability. It provides designers with tools to meet emission limits and effectively control and manage emissions of GHGs. The persistent attention given to GHG emissions may be attributed to government initiatives, such as the implementation of carbon caps for factories (Gao et al. 2018). Manufacturers and design firms must acknowledge and effectively handle product development by alerting designers about greenhouse gas emission standards that they must comply with during the process of designing of a new product.

Life Cycle Assessment (LCA) methods are utilized to verify the viability of environmentally friendly design choices with regards to energy consumption, material utilization, and factors related to the entire process (Brezet et al. 1999). The LCA is an approach for evaluating the designs consisting of three steps: stock parameters, assessment about environmental consequences, and a methodology for improving useful life of the product (Stuart et al. 1999). Traditionally, life cycle assessments have mostly focused on analyzing individual products. However, there has been a recent shift towards incorporating economic, technological, and environmental limitations into the design of processes. This broader approach to LCA's takes into account the various factors that can impact the overall sustainability and efficiency of a process. Sakao (2007) suggests that this methodology is relatively objective because it requires quantitative data. The term "cradle-to-grave analysis" is also employed to refer to Life Cycle Assessment (LCA) by Glavič and Lukman (2007). Nevertheless, it has been shown that LCA methods necessitate the expertise of an individual in order to achieve success while use, along with requiring comprehensive insights to accurately estimate the environment related aspects that need evaluation (Raoufi et al. 2019). This can impede their performance and efficacy in evaluating environmental effects across the entire life cycle. In addition, it has been emphasized that LCA's necessitate more comprehensive data and would arguably be most effectively incorporated at each step of the designing process, rather than in the initial ones (Devanathan et al. 2010). The integration of Computer-Aided Design with LCA is being implemented in the industry at a growing rate. This integration aims to help designers evaluate environmental factors during the design phase. However, due to its novelty, there are still limitations to its effectiveness, and further research is needed.

Industry utilizes sustainability checklists to evaluate products from perspective of impact on the environment. This approach identifies ways of enhancement across the entire lifecycle of the products (Sakao 2007). This approach might be employed to assess areas of divergence and identify the underlying causes of environmental problems, which have the potential to impact the manufacturing process (De Ron 1998). The aspects that are taken into account while using a sustainability checklist include the selection and usage of materials, the characteristics of waste, energy consumption, recyclability, and disposal methods (De Ron1998; Sakao2007; Zhang et al. 2020). The application of the checklist approach is potentially has higher subjectivity involved when compared to other methodologies (Devanathan et al. 2010). Its criteria may be influenced by commercial objectives rather than an objective evaluation of environmental sustainability (Sakao 2007). Schöggl et al. (2017) contended that checklists may suffer from excessive generality, which might be detrimental due to the potential lack of clarity regarding the designer's objectives or assessments. Sustainability checklists may include inquiries for designers about specific factors chosen for evaluation, such as "Does the product exhibit resistance to natural degradation?" (Zhang et al. 2020) suggests that while this strategy is useful, it can only be implemented after the design phase, making it difficult to reverse earlier selections once they have been made.

3 Research Methodology

To develop a concise and adequate comprehension of the current stage of adoption of sustainability practices in the manufacturing sector, a review of the existing research has been undertaken. The concept of sustainable business models is discussed to bring clarity about the basic idea behind these models. Sustainability in special reference to the manufacturing sector is also discussed to understand the importance of this sector in achieving net zero goals. Finally, the widely discussed sustainable business models are comprehensively discussed to provide implications for research and practice.

4 Implications

Decarbonizing industry involves the redesigning of products and industrial processes to make them more environmentally friendly. This requires a strategic change in all the production functions. Additive manufacturing is an innovative component of industrial revolution which is closely connected to the digital realm. This process represents a novel method of manufacturing and in the long term, this approach will enable us to eliminate several tools and adjust the process of production as per the market demands with greater flexibility. The field of additive manufacturing has had significant advancements, allowing for the replication of intricate and fully functional industrial equipment. Hence, the introduction of additive manufacturing is giving rise to innovative and transformative business models in various industries. The implications AM are evaluated by considering its impact on economy, society, and environment. The current AM technologies remain sluggish, inefficient, and have excessively high per-part costs, making them unable to rival with existing methods. Nevertheless, AM holds a highly ambitious outlook for the future, and several endeavors are being undertaken in various domains to ensure its economic, social, and environmental viability. The utilization of additive technology and the current state of many emerging business models being in their nascent stages necessitates further research.

The scientific contribution of this work pertains to the understanding that comprehensively and systematically elucidates how the implementation of linear manufacturing principles and instruments facilitates the attainment of sustainability objectives. Managers can utilize these findings to comprehend the opportunities presented by the combination of lean and sustainable principles. The integration of principles of lean manufacturing also has significant environmental and social impacts apart from economic benefits. The data unequivocally demonstrate that lean has a distinctly good effect on the economic aspect. HRM techniques are seen as a strategic tool for sustaining long-term beneficial outcomes by fostering a culture of efficiency within the firm. In contrast, it has been observed that the impact on the remaining two elements can be detrimental in certain situations. This is a result of the absence of coordination between the sustainability strategy and implementation process of lean system, or even worse, the complete exclusion of the sustainability aspect. Although lean and social activities may initially appear to be distinct ideas with separate goals, a closer examination of the elements and the outcomes they yield for the production system reveals that they are not as unlike as they may seem. This commitment aligns with the principles of social sustainability and the reduction of waste, resulting in significant environmental benefits.

This study examines the existing and cutting-edge methods of handling industrial waste and suggests a fresh perspective for achieving sustainable management of the industrial waste. The adoption of a waste management system that incorporate sustainability plays a crucial role in enhancing the overall sustainability of the organization. Minimizing waste leads to more efficient utilization of resources and prolongs the lifespan of products, resulting in decreased expenses and mitigated hazards. Therefore, the objectives of safeguarding the environment, preserving resources, and providing economic advantages can be achieved through effective waste management strategies. Conquering all forms of waste can contribute to the improvement of manufacturing sustainability. Relying solely on waste identification technologies may not provide for a comprehensive garbage removal procedure. In order to completely eliminate all sorts of waste, this chapter suggests a waste removal architecture that has three sequential steps. These phases can be further customized to align with different sizes and types of industrial processes. In order to improve the process, formulas for prioritizing garbage have been developed. In future research, it is advisable to undertake pre- and post-measurements to determine the effectiveness of waste removal in connection to manufacturing sustainability.

The chapter emphasizes the areas where the product designer should take into account the criteria of environmental sustainability by considering relevant aspects at each stage, ensuring a feasible implementation. In addition, the framework incorporates a section on business considerations. This highlights the reality about the limited influence of designers on certain factors related to the environmental sustainability, which need to be addressed from a perspective of organization rather than on a case-by-case basis during each design process. The discussion would be applicable to several designing processes, allowing the designers to concentrate on different elements throughout the process. The importance of sustainable practices may vary based on the unique product, but the solutions offered in this chapter can be adjusted to meet the situational requirements. The suggestions are expected to serve as a reference for designers to incorporate sustainable elements into their standard process. We provides a clear insight for considering environmental sustainability factors at each stage of the design process. By using the parameters discussed in literature as a checklist or reminder system, designers can effectively apply their existing knowledge to maximize the potential for environmental sustainability in each stage of the new product development. They can also deepen their understanding of each factor and implement this information effectively in their product development processes.

5 Conclusion

In manufacturing sector, sustainability has garnered significant attention specifically during the last two decades. In the present era, there is a requirement that the manufacturing industry to prioritize environmental preservation while still ensuring economic progress and social improvement. The manufacturers are now acknowledging the necessity of implementing proactive measures to enhance their sustainability performance by integrating sustainability principles into their organization's plan of action. Sustainable manufacturing is a viable technique for enhancing sustainability performance. This study conducted a comprehensive analysis of existing literature that specifically examines the sustainable business adopted in manufacturing sector. This has facilitated the comprehension of crucial elements of environmental sustainability, for which manufacturers bear accountability, and to understand that how they should be incorporated into the production process. Manufacturers must comprehend the various potential choices available to achieve net zero emissions, and this chapter serves that purpose by offering insights into the most prevalent sustainable models encompassing the utilization of both established methodologies and technology, as well as novel ones. The qualitative aspect of the research limits the adoption of insights drawn. To investigate the actual effectiveness of different sustainable practices, it is recommended to conduct empirical research on a wide range of organizations with varying levels of sustainability implementation, ranging from "beginners" to highly advanced systems.

References

Ahmad S, Wong KY, Tseng ML, Wong WP (2018) Sustainable product design and development: a review of tools, applications and research prospects. Resour Conserv Recycl 132:49–61

Allwood J (2005) What is sustainable manufacturing? In: Sustainability: what are the issues? (Sustainable manufacturing seminar series, 16 February 2005). IfM, University of Cambridge

Azadegan A, Patel PC, Zangoueinezhad A, Linderman K (2013) The effect of environmental complexity and environmental dynamism on lean practices. J Oper Manag 31:193–212. https://doi.org/10.1016/j.jom.2013.03.002

Baumers M, Dickens P, Tuck C, Hague R (2016) The cost of additive manufacturing: machine productivity, economies of scale and technology-push. Technol Forecast Soc Chang 102:193–201

Baumers M, Duflou JR, Flanagan W, Gutowski TG, Kellens K (2017) Charting the environmental dimensions of additive manufacturing and 3D printing: environmental dimensions of AM. J Ind Ecol 21(S1):S9–S14

Belekoukias I, Garza-Reyes JA, Kumar V (2014) The impact of lean methods and tools on the operational performance of manufacturing organisations. Int J Prod Res 52(18):5347–5366

Bonfanti A, Mion G, Brunetti F, Vargas-Sánchez A (2023) The contribution of manufacturing companies to the achievement of sustainable development goals: an empirical analysis of the operationalization of sustainable business models. Bus Strateg Environ 32(4):2490–2508

Boons F, Lüdeke-Freund F (2013) Business models for sustainable innovation: state-of-the-art and steps towards a research agenda. J Clean Prod 45(8):9–19. https://doi.org/10.1016/j.jclepro.2012.07.007

Braungart M, McDonough W (2002) Design for the triple top line. Corp Environ Strateg 9(3):251–258

Brezet H, Stevels A, Rombouts J (1999) LCA for ecodesign: the Dutch experience. In: Proceedings first international symposium on environmentally conscious design and inverse manufacturing, pp 36–40. IEEE

Carvalho H, Cruz-Machado V (2009) Integrating lean, agile, resilience and green paradigms in supply chain management. In: Xu J, Qu ZT, Yang ST (eds) Proceedings of the third international conference on management science and engineering management, pp 1031–1038

Caviggioli F, Ughetto E (2019) A bibliometric analysis of the research dealing with the impact of additive manufacturing on industry, business and society. Int J Prod Econ 208:254–268

Chesbrough H (2010) Business model innovation: opportunities and barriers. Long Range Plan 43(2–3):354–363. https://doi.org/10.1016/j.lrp.2009.07.010

Chick A, Micklethwaite P (2004) Specifying recycled: understanding UK architects' and designers' practices and experience. Des Stud 25:251–273

Ciliberto C, Szopik-Depczyńska K, Tarczyńska-Łuniewska M, Ruggieri A, Ioppolo G (2021) Enabling the circular economy transition: a sustainable lean manufacturing recipe for Industry 4.0. Bus Strat Environ 30(7):3255–3272

Comin LC, Aguiar CC, Sehnem S, Yusliza MY, Cazella CF, Julkovski DJ (2020) Sustainable business models: a literature review. Benchmarking Int J 27(7):2028–2047

De Ron AJ (1998) Sustainable production: the ultimate result of a continuous improvement. Int J Prod Econ 56:99–110

Delaney E, Liu W, Zhu Z, Xu Y, Dai JS (2022) The investigation of environmental sustainability within product design: a critical review. Des Sci 8:e15

Deutz P, Neighbour G, McGuire M (2010) Integrating sustainable waste management into product design: sustainability as a functional requirement. Sustain Dev 18:229–239

Devanathan S, Ramanujan D, Bernstein WZ, Zhao F, Ramani K (2010) Integration of sustainability into early design through the function impact matrix. J Mech Des 132:1–8

Dilberoglu UM, Gharehpapagh B, Yaman U, Dolen M (2017) The role of additive manufacturing in the era of Industry 4.0. Procedia Manuf 11:545–554

Donaldson T, Preston LE (1995) The stakeholder theory of the corporation: concepts, evidence, and implications. Acad Manag Rev 20(1):65–91. https://doi.org/10.2307/258887

Dues CM, Tan KH, Lim M (2013) Green as the new lean: how to use lean practices as a catalyst to greening your supply chain. J Clean Prod 40:93–100

European Commission (2016) Identifying current and future application areas, existing industrial value chains and missing competences in the EU, in the area of additive manufacturing (3D-Printing)—final report. European Commission

Fitzsimons L, McNamara G, Obeidi M, Brabazon D (2020) The circular economy: additive manufacturing and impacts for materials processing. In: Hashmi S, Choudhury IA (eds) Encyclopedia of renewable and sustainable materials. Elsevier, pp 81–92

Franchetti M, Bedal K, Ulloa J, Grodek S (2009) Lean and green: Industrial engineering methods are natural stepping stones to green engineering. Ind Eng 41(9):24–29

Gao J, Xiao Z, Wei H, Zhou G (2018) Active or passive? Sustainable manufacturing in the direct-channel green supply chain: a perspective of two types of green product designs. Transp Res Part D Transp Environ 65:332–354

Glavič P, Lukman R (2007) Review of sustainability terms and their definitions. J Clean Prod 15:1875–1885

Godina R, Ribeiro I, Matos FT, Ferreira B, Carvalho H, Peças P (2020) Impact assessment of additive manufacturing on sustainable business models in Industry 4.0 context. Sustainability 12(17):7066

Graedel TE, Guth LA (1990) The impact of environmental issues on materials and processes. AT&T Tech J 69:129–140

Gupta SM (2016) Lean manufacturing, green manufacturing and sustainability. J Jpn Ind Manag Assoc 67:102–105

Gutowski T, Jiang S, Cooper D, Corman G, Hausmann M, Manson J-A, Schudeleit T, Wegener K, Sabelle M, Ramos-Grez J (2017) Note on the rate and energy efficiency limits for additive manufacturing: rate and energy efficiency limits for AM. J Ind Ecol 21(S1):S69–S79

Hallstedt S (2008) A foundation for sustainable product development (Doctoral dissertation). Blekinge Institute of Technology, Sweden

Hines P, Holweg M, Rich N (2004) Learning to evolve: a review of contemporary lean thinking. Int J Oper Prod Manag 24(10):994–1011

Hogland W, Stenis J (2000) Assessment and system analysis of industrial waste management. Waste Manage 20(7):537–543

Jawahir IS, Dillon Jr. OW (2007) Sustainable manufacturing processes: new challenges for developing predictive models and optimization techniques. In: 1st International conference on sustainable manufacturing. Montreal, Canada

Jayal AD, Badurdeen F, Dillon OW Jr, Jawahir IS (2010) Sustainable manufacturing: modeling and optimization challenges at the product, process and system levels. CIRP J Manuf Sci Technol 2(3):144–152

Khan S, Ali SS, Singh R (2022) Determinants of remanufacturing adoption for circular economy: a causal relationship evaluation framework. Appl Syst Innov 5(4):62. https://doi.org/10.3390/asi5040062

Khor KS, Udin ZM (2013) Reverse logistics in Malaysia: investigating the effect of green product design and resource commitment. Resour Conserv Recycl 81:71–80

Kozlowski A, Searcy C, Bardecki M (2018) The reDesign canvas: fashion design as a tool for sustainability. J Clean Prod 183(14):194–207. https://doi.org/10.1016/j.jclepro.2018.02.014

Kunovjanek M, Reiner G (2020) How will the diffusion of additive manufacturing impact the raw material supply chain process? Int J Prod Res 58:1540–1554

Kurien GP, Qureshi MN (2018) House of sustainable waste management: an implementation framework. Int J Sustain Manuf 4(1):79–96

Lodder M, Huffenreuter RL, Braungart M, Den Held D (2014) Regenerative sustainable development: towards a triple top line approach and increasing positive externalities. In 5th International sustainability transitions conference, pp 1–13

Ma J, Kremer GEO (2016) A systematic literature review of modular product design (MPD) from the perspective of sustainability. Int J Adv Manuf Technol 86:1509–1539

Madu CN (2001) Handbook of environmentally conscious manufacturing. Kluwer Academic Publishers

Martínez Leon HC, Calvo-Amodio J (2017) Towards lean for sustainability: understanding the interrelationships between lean and sustainability from a systems thinking perspective. J Clean Prod 142:4384–4402. https://doi.org/10.1016/j.jclepro.2016.11.132

Martínez-Jurado PJ, Moyano-Fuentes J (2014) Lean management, supply chain management and sustainability: a literature review. J Clean Prod 85:134–150

McGrath RG (2010) Business models: a discovery driven approach. Long Range Plan 43(2–3):247–261. https://doi.org/10.1016/j.lrp.2009.07.005

Monteiro J, Silva F, Ramos S, Campilho R, Fonseca A (2019) Eco-design and sustainability in packaging: a survey. Procedia Manuf 38:1741–1749

Moreira F, Alves AC, Sousa RM (2010) Towards eco-efficient lean production systems. In: Gonçalves P, Gasquet PG, Marzani A (eds) Balanced automation systems for future manufacturing networks, vol 322, pp 211–219. Springer, Berlin

Niaki MK, Torabi SA, Nonino F (2019) Why manufacturers adopt additive manufacturing technologies: the role of sustainability. J Clean Prod 222:381–392

OECD. (2010). Eco-innovation in industry: Enabling green growth. OECD Publishing.

Ohno T (1988) Toyota production system: beyond large-scale production, 1st edn. Productivity Press. https://books.google.it/books/about/Toyota_Production_System.html?id=7_67SshOy8C&redir_esc=y

Oros-Daraban AE, Negrea CS, Artimon FGP, Angelescu D, Popan G, Gheorghe SI, Gheorghe M (2019) A deep look at metal additive manufacturing recycling and use tools for sustainability performance. Sustainability 11:5494

Pongrácz E, Pohjola VJ (2004) Re-defining waste, the concept of ownership and the role of waste management. Resour Conserv Recycl 40(2):141–153

Qin J, Liu Y, Grosvenor R (2017) A framework of energy consumption modelling for additive manufacturing using Internet of Things. Procedia CIRP 63:307–312

Raoufi K, Park K, Khan MTH, Haapala KR, Psenka CE, Jackson KL, Kremer GEO, Kim K-Y (2019) A cyber learning platform for enhancing undergraduate engineering education in sustainable product design. J Clean Prod 211:730–741

Rebitzer G, Hunkeler D, Norris G, Rydberg T, Finkbeiner M (2004) Review of life cycle assessment part 1: framework, goal, scope definition, inventory analysis and applications. Environ Int 30:701–720

Rejeski D, Zhao F, Huang Y (2018) Research needs and recommendations on environmental implications of additive manufacturing. Addit Manuf 19:21–28

Resta B, Dotti S, Gaiardelli P, Boffelli A (2017) How lean manufacturing affects the creation of sustainable value: an integrated model. Int J Autom Technol 11(4):542–551. https://doi.org/10.20965/ijat.2017.p0542

Ribeiro I, Matos F, Jacinto C, Salman H, Cardeal G, Carvalho H, Godina R, Peças P (2020) Framework for life cycle sustainability assessment of additive manufacturing. Sustainability 12:929

Ritala P, Huotari P, Bocken N, Albareda L, Puumalainen K (2018) Sustainable business model adoption among S&P 500 firms: a longitudinal content analysis study. J Clean Prod 170:216–226. https://doi.org/10.1016/j.jclepro.2017.09.159

Rosienkiewicz M, Gabka J, Helman J, Kowalski A, Susz S (2018) Additive manufacturing technologies cost calculation as a crucial factor in Industry 4.0. In: Hamrol A, Ciszak O, Legutko S, Jurczyk M (eds) Advances in manufacturing, pp 171–183. Springer International Publishing

Rothenberg S, Pil FK, Maxwell J (2001) Lean, green, and the quest for superior environmental performance. Prod Oper Manag 10(3):228–243

Saebi T, Foss N (2015) Business models for open innovation: matching heterogeneous open innovation strategies with business model dimensions. Eur Manag J 33(3):201–213. https://doi.org/10.1016/j.emj.2014.11.002
Sakao T (2007) AQFD-centred design methodology for environmentally conscious product design. Int J Prod Res 45:4143–4162
Schöggl J-P, Baumgartner RJ, Hofer D (2017) Improving sustainability performance in early phases of product design: a checklist for sustainable product development tested in the automotive industry. J Clean Prod 140:1602–1617
Shah R, Ward PT (2007) Defining and developing measures of lean production. J Oper Manag 25(4):785–805
Singh R, DSilva J, Kumar RS (2021) Modelling the CSR initiatives on firm performance: a context of emerging economies. Empir Econ Lett 20(3):221–228
Singh R, Khan S, Kumar A, Luthra S, Chokshi H (2024) Net zero economy, corporate social responsibility and sustainable value creation. CSR, sustainability, ethics & governance. Springer, Cham. https://doi.org/10.1007/978-3-031-55779-8_1
Statista (2024) Manufacturing–worldwide. Statista. https://www.statista.com/outlook/io/manufacturing/worldwide#value-added
Stuart JA, Ammons JC, Turbini LJ (1999) A product and process selection model with multidisciplinary environmental considerations. Oper Res 47:221–234
Taubitz MA (2010) Lean, green & safe: integrating safety into the lean, green and sustainability movement. Prof Saf 55(5):32–39
Tolkamp J, Huijben JCCM, Mourik RM, Verbong GPJ, Bouwknegt R (2018) User-centred sustainable business model design: the case of energy efficiency services in the Netherlands. J Clean Prod 182(13):755–764. https://doi.org/10.1016/j.jclepro.2018.02.032
Treville S, Antonakis J (2006) Could lean production job design be intrinsically motivating? Contextual, configurational, and levels of analysis issues. J Oper Manag 24(2):99–123
Upward A, Jones P (2016) An ontology for strongly sustainable business models: defining an enterprise framework compatible with natural and social science. Organ Environ 29(1):97–123. https://doi.org/10.1177/1086026615592933
US Department of Commerce (2007) Promoting competitiveness: partnerships and progress of the office of manufacturing and services. International Trade Administration
Vinodh S, Arvind KR, Somanaathan M (2011) Tools and techniques for enabling sustainability through lean initiatives. Clean Technol Environ Policy 13(3):469–479
WHO (2016) Occupational health. World Health Organization—Regional Office for Europe. http://www.euro.who.int/en/health-topics/environment-and-health/occupational-health
Williams PT (2005) Waste treatment and disposal. Wiley
Womack JP, Jones DT (1996) Lean thinking, banish waste and create wealth in your corporation, 1st edn. Productivity Press. https://www.worldcat.org/title/lean-thinking-banish-waste-and-create-wealth-in-your-corporation/oclc/51031471
World Economic Forum (2024) How the industrial sector is turning net zero goals into practice. World Economic Forum. https://www.weforum.org/agenda/2024/06/industrial-sector-turning-net-zero-goals-into-practice/
Zamorano M, Grindlay A, Molero E, Rodríguez MI (2011) Diagnosis and proposals for waste management in industrial areas in the service sector: case study in the metropolitan area of Granada (Spain). J Clean Prod 19(17–18):1946–1955
Zhang X, Zhang L, Fung KY, Bakshi BR, Ng KM (2020) Sustainable product design: a life-cycle approach. Chem Eng Sci 217:115508
Zott C, Amit R, Massa L (2011) The business model: recent developments and future research. J Manag 37(4):1019–1042. https://doi.org/10.1177/0149206311406265

Enhancing Supply Chain Resilience Through Industry 4.0 for Zero Carbon Industry

Sonal Khurana, Bisma Mannan, Vanita Bhardwaj, and Jayati Ahuja

Abstract The last few decades have seen a rise in organizational rivalry due to globalization and the rapid advancement of information technology. In order to draw in new clients, businesses now a days are always wants to expand their market geographically. In light of this, businesses prioritize meeting the needs of their clients, which lengthens and complicates the supply chain (SC). Consequently, managing SC and limiting risks and disruptions becomes more challenging. Resilient supply chain essentially rests on the ability to respond, recover from operational disruption, to maintain operational procedures operational procedures, and recover to its pre-disturbance state after an event. Additionally, by building their SC with the capacity to respond swiftly to any disturbances in workflow, businesses may boost stability and solidify their position in the market. Analysts have been studying strategies for enhancing supply chain resilience, include boosting and fostering cooperation between SC partners or raising the organization's profile by keeping an eye on SC events and trends. Research has indicated that Industry 4.0 can enhance production efficiency. Industry 4.0 involves usage of cutting edge technologies for strengthening the resiliency of supply chain and work towards a zero-carbon future. Industry 4.0 makes it easier to monitor real-time monitoring, optimized logistic and predictive analytics to improve efficiency. This chapter's goal is to examine various technologies in order to help supply chains become flexible and resilient so that sustainability targets can be met. Sustainable resource management, energy-efficient and Smart manufacturing are key components of Industry 4.0 that leads to a greener future.

Keywords Supply chain · Blockchain · Internet of Things · Industry 4.0 · Smart manufacturing

---

S. Khurana (✉) · V. Bhardwaj · J. Ahuja
Department of Engineering, Vivekananda Institute of Professional Studies, New Delhi, India
e-mail: sonal.khurana@gmail.com

B. Mannan
Department of Mechanical and Marine Engineering, University of Plymouth, Plymouth, UK

R. Singh et al. (eds.), *Zero Carbon Industry, Eco-Innovation and Environmental Sustainability*, World Sustainability Series,
https://doi.org/10.1007/978-3-031-80220-1_2

1 Introduction

Industry 4.0—a system powered by technology that uses Internet of Things (IoT) and internet enabled appliances, which track activities and manufacture commodities. Automation of industrial processes is intended to enhance the management of supply chains, efficiency, sustainability, and early detection of system bottlenecks (Qader et al. 2022). When analyzing industrial big data, manufacturers are able to obtain information that will help optimize the operations of specific pieces of equipment and of the manufacturing network as a whole. Artificial intelligence (AI) and machine learning (ML) are being applied for the building of supply chain resilience for zero carbon enterprises (El Baz and Ruel 2021). In order to build a cyber-physical environment with digital solutions and cutting-edge technologies like blockchain, additive manufacturing, and the IoT, among others, Industry 4.0 blends information technology and operational technology (Al-Talib et al. 2020). In the Industry 4.0 era (I4.0), advancements in machine learning (ML) and big data are transforming conventional supply chain into a resilient one (Qader et al. 2022). There are new opportunities arising from this paradigm shift. A network of physically linked electronic devices that facilitates data exchange and online communication is termed as the "Internet of Things" (IoT). The Internet of Things and the profusion of smart sensors are the main method through which huge quantities of industrial data-rich environments pertaining to every facet of supply chain and production can be curated and stored (Majeed and Rupasinghe 2017). Supply chain, sustainability, predictive maintenance, task scheduling, process optimisation, and quality improvement are just a few of the manufacturing and production applications where artificial intelligence (AI) subfield machine learning techniques have the capacity of becoming the primary force in identifying detailed and complex production and supply chain patterns and providing timely decision support (Dubey et al. 2019). Using Industry 4.0 technologies wisely can help in sustainable business growth. The leading global companies are extensively investing in the development of Fourth Industrial Revolution and utilizing it into their supply chains. These methods also automate data-driven learning, identify common underlying patterns, and facilitate well-informed decision-making. Despite all of these benefits, not enough study has been done to determine how supply chain resilience and industry 4.0 are related (Lee and Lee 2015). The aim of this research is to look at the potential outcome of Industry 4.0 on supply chain resiliency. Despite growing interest in this topic, not much research has been done on how Industry 4.0 can be employed for enhancing supply chain resilience (Dubey et al. 2020). Thus, the influence of Industry 4.0 technology on resilience of supply chains will be investigated by focusing on the essential components of supply chain resilience.

2 Literature Review

2.1 Industry 4.0

Industry 4.0, or The Fourth Industrial Revolution, emcompasses continuous automation and digitalization of production and industry. Building responsive and efficient industries is greatly aided by the blending of Industry 4.0 technologies (Fatorachian and Kazemi 2018). These advanced technologies help with collection of data, its analysis, interpretation, and in informed decision-making, which boosts productivity in manufacturing operations. Industry 4.0 encompasses the advancements in robots, big data, the Internet of Things (IoT), augmented reality, along with cloud manufacturing (Dastbaz 2019). The advancement of a more intelligent manufacturing process, encompassing autonomously communicating goods, machinery, gadgets, and production modules, depends on all of these technologies. Cyber-physical systems are used to merge the physical and digital worlds while embracing these next—generation technologies. Industry 4.0 has an immense amount of possibilities and will offer a variety of economic and social opportunities through its revolutionary shift in terms of work structure, business models and production technology (Ivanov et al. 2019). Cloud storage is tremendously packed with features and is stable, flexible.

2.1.1 Internet of Things (IoT)

Internet of Things (IoT) is an emerging model used to connect electronic devices through the internet to streamline our everyday lives (Zhou and Piramuthu 2018). IoT provides innovative solutions for our daily life challenges at home, business, school or office using smart devices and the internet. The term IoT implies a broad concept of the potential of sensing devices, data collection and communication over the internet (Pundir et al. 2019). It describes a framework of networked devices that exchange information and collaborate with one another via the internet. From industrial machinery to household appliances, these smart devices enable automation and remote monitoring, enhancing productivity, ease of use, and decision-making in a variety of spheres of life. Resilience in supply chain is improved by the Internet of Things (IoT), which offers real-time data and operational insights. IoT devices improve logistics, track inventory, and keep an eye on equipment, allowing for a quicker reaction to disruptions. Increased productivity, decreased downtime, and more adaptable and flexible supply chain management in dynamic environments are all made possible by this connectivity.

2.1.2 Machine Learning

Machine learning is one of Industry 4.0's useful technologies. One of the new technologies in Industry 4.0 that allows companies to access real-time data and modify patterns appropriately is machine learning (Queiroz and Wamba 2019). Applying machine learning (ML) to industrial processes is necessary to make promising algorithm-based decisions, and this takes a lot of data (Rejeb et al. 2019). Data often follows a certain life cycle, which includes the following stages: collection, source, processing, storage, transmission, visualisation, and application. Through the use of data-driven insights, machine learning improves operational efficiency, forecasting, and decision-making, thereby strengthening supply chain resilience. Machine learning analyzes enormous volumes of data from multiple sources, such as supplier performance, demand trends, and inventory levels, using sophisticated algorithms. This capacity lowers the possibility of stockouts and overstock scenarios by enabling more precise demand forecasting. Machine learning enables improved inventory management and optimized procurement strategies by predicting demand fluctuations and identifying trends, ensuring that supply chains can quickly adjust to changing conditions. Machine learning algorithms also improve risk management by spotting possible supply chain disruptions and weaknesses. These algorithms, for example, can use past data to forecast supply delays or equipment failures, enabling businesses to plan maintenance schedules and take preventative action. This proactive strategy reduces downtime and keeps things running continuously. Automation and process optimization are two more ways that machine learning increases operational efficiency. Machine learning can improve resource allocation, streamline logistics, and optimize routing by analyzing real-time data. As a result, operational expenses are decreased and service quality is raised. Furthermore, by continuously learning from fresh data and improving its models and predictions over time, machine learning helps adaptive supply chain strategies.

Noisy raw data can also be handled tactically because it is widely available and can contaminate accurate data, which can complicate decision-making. Machine learning's rapid data processing and visualization capabilities give firms visibility. As a result, machine learning provides a solid foundation for decision-making while successfully addressing the challenges of a quickly changing environment (Hohenstein et al. 2015). In light of the Fourth Industrial Revolution paradigm, the machine learning (ML) field has profoundly influenced the industrial sector. The industry 4.0 paradigm promotes the establishment of "smart factories," which employ sophisticated equipment, sensors, and devices to continuously gather production-related data (Qader et al. 2022). Machine learning strategies assist in producing implementable knowledge by analysing the gathered information to boost industrial productivity while minimally altering the required resources. Machine learning approaches have enabled a generation of smart decision support system in various manufacturing activities such as, real-time inspection, process control and improvement, quality assurance, condition-based maintenance, activities scheduling and supply chain (Majeed and Rupasinghe 2017).

2.1.3 Blockchain Technology

Blockchain technology is a digital record book that is decentralized and keeps track of transactions on several computers. By keeping an unchangeable record of data that is available to all participants, it guarantees security and transparency. This acts as another innovative solution that is centered around the simplest type of distributed ledger technology, which is bitcoin (Spieske and Birkel 2021). Johng et al. (2020) talked about various distributed ledger technology variations, including Ethereum, a worldwide platform for distributed applications powered by smart contracts, and the requirements that must be met for blockchain variations as well. They claimed that various blockchain scenarios, including public, private, and consortium (a mix of private and public), use different variations of the technology. In light of this, public Blockchain Technology is designed as a decentralized network in which every node or member logs transactions onto a public ledger (Ben-Daya et al. 2019). After every transaction has been collected, it is grouped or blocked. After that, an algorithm is applied to the data in order to produce a unique hash code of the specific data at a given moment. These codes are generated repeatedly, creating additional blocks, and ultimately accumulating the data to form a value chain. One company manages a private blockchain, allowing select parties restricted visibility rights. This ensures traceability and accountability at every step, reducing fraud and errors. By enabling real-time data sharing among stakeholders, blockchain improves coordination, speeds up decision formulation, and enhances the complete reliability of the supply chain.

2.2 *Industry 4.0 and Resiliency of Supply Chain for Zero Carbon Industry*

Supply Chain resilience is a demonstration of a multifaceted standard. Consequently, it's important formulate and comprehend the idea appropriately. The supply chain can adjust to the circumstances by being ready for unforeseen circumstances, reacting to disruptions while they are happening, and then, at the end, going through the recovery stage to keep connectivity, processes, and additional structural regulation in place.

There is an abundance of research available on the attributes of a resilient supply chain. The following elements of supply chain resilience have been discussed by Majeed and Rupasinghe (2017): sustainability, visibility, flexibility, adaptation, and cooperation. These characteristics facilitate competitive advantage and reduce disruptions to the supply chain. The challenges that conventional supply chains confront have a direct consequence on the qualities of supply chain resilience, which are crucial for it to handle changes and minimise interruptions. Increasing intricacy, growing expenses, mistakes, with increasing hazards are problems for traditional supply chains.

Industry 4.0 makes supply chains more robust systems. By facilitating live monitoring, proactive maintenance, and enhanced logistics, these technologies boost productivity and cut down on waste. By reducing energy use and emissions through resource optimization and intelligent energy management, this shift promotes sustainable practices for a zero-carbon industry. Industry 4.0 makes it easier to see and manage every part of the supply chain, which promotes the transition to circular economies that recycle and reuse materials. A more flexible and resilient supply chain netthat can withstand disruptions is fostered by this integration.

Prior to Industry 4.0's introduction into the supply chain, data and information were only conveyed with one actor, which decreased transparency (Zhou and Piramuthu 2018). In contrast, the smart supply chain allows for simultaneous collaboration with all necessary actors.

Visibility was further impeded across the supply chains by mistakes, inconsistencies, and information distortion brought on by a lack of real-time data and information. Due to their complexity and unpredictability, supply chains have faced a number of issues in the past, including delivery errors and new orders for pick-up that have cost and wasted time (Ali et al. 2017). The goal of implementing Industry 4.0 technologies is altering traditional supply chains' business models, which will necessitate redesigning earlier business models.

2.3 *Effect of Industry 4.0 on Resilience of Supply Chain Attributes*

Supply chain procedures are becoming more intelligent in tandem with the increasing speed at which Industry 4.0 is developing. According to Butner (2010), an intelligent supply chain must possess the following qualities in order to be considered innovative: it must be influential, intelligent, and consistent. Furthermore, as noted by Guha and Chakrabarti (2014), following its debut into business in the 1990s, information and communication technology (ICT) showed its importance and potential benefits through a variety of information systems.

Industry 4.0 integrates cutting-edge technologies that improve crucial operational dimensions, which has a significant impact on the resilience of supply chain attributes. The implementation of real-time data analytics and the Internet of Things (IoT) greatly enhances supply chain visibility, enabling thorough monitoring and prompt issue identification throughout the chain. Agility is enhanced by automation and artificial intelligence (AI), which make it possible to quickly adapt production and logistics in response to disruptions or changes in the market. While predictive maintenance ensures reliability by anticipating and addressing potential equipment failures before they impact operations, big data analytics improves efficiency by streamlining processes and cutting waste. Industry 4.0 also promotes improved cooperation by utilizing digital platforms that facilitate coordination and communication between supply chain participants. While advanced risk management tools offer insights to anticipate and mitigate potential disruptions, improved collaboration aligns objectives and optimizes resource utilization. When taken as a whole, these developments produce a supply chain that is more robust, flexible, and effective—capable of enduring and rebounding from a variety of difficulties.

According to Bhatti et al. (2021), corporate technologies from the manufacturing and services sectors are merged to generate digital transformation, which enables companies to adjust to evolving market dynamics, stable economic growth, and preferences of customers. In similar vein, digital supply chains give management the data they need to turn on their digital frameworks and standards, which link customer preferences for products and services. The management must understand how important IT capabilities are to enhancing their strategic management performance. To create effectiveness, reliability and specialization it is essential to create business relations based on the mutual comprehension and partnership.

Industry 4.0 has the potential to lower costs by making industries more effective, efficient, and sustainable. The qualities of decentralized management can be improved by blockchain technology's irreversible, immutable, and undeletable capabilities. Business process reengineering is a concept that describes how to change business operations to increase productivity and decrease implementation time and expense. Information technology-based machine design and development have been essential catalysts in industry 4.0 revolution (Kopp et al. 2016). Organizations must create criteria that strengthen their supply chains so they can better withstand supply chain disruptions and manage supply chain risks (Guha and Chakrabarti 2014). Industry 4.0's information technology might give the basis for decisions, allowing the supply chain industry to run more smoothly. Industry 4.0 may significantly affect the performance of supply chain performance as it improves them and provides organisations with real-time visibility into them. The supply chain resilience will be studied under the following attributes.

2.3.1 Visibility

Resilience of Supply chains depends on their visibility to the same extend as supply chain integration. According to Nothacker (2021), Supply Chain visibility boosts operational effectiveness and lowers Supply Chain risks. Additionally, Supply Chain Visibility facilitates access to important data for the seamless operation of the businesses for all parties involved. Additionally, Supply Chain Visibility facilitates access to important data for the seamless operation of the businesses for all parties involved. Additionally, labour hours may be cut down and communication can be improved to make better use of time. The definition of supply chain visibility was expanded to include demand, supply, and market awareness. Businesses may make required decisions by using supply visibility to get the information they want from all of their suppliers (McConaghy et al. 2017). Companies that have visibility into demand are better equipped to anticipate and plan for client demands. On the other hand, market visibility entails learning vital information about market trends, which aids in a company's ability to maintain its competitiveness (Kochan and Nowicki 2018). When supply chain insight is accessible, supply chain partners may more effectively manage risks of supply chain and take proactive measures to address problems. Industry 4.0 technologies for supply chain digitization give businesses access to real-time data that they may use to manage supply chain disruptions (Ivanov and Dolgui 2021). Furthermore, Industry 4.0 methods like blockchain and machine learning provide industries more visibility and transparency, which strengthens the robustness of the supply chain. By combining big data analytics, artificial intelligence, IoT, along with other technologies, Industry 4.0 dramatically improves supply chain visibility and resilience. These technologies allow businesses to track and analyze goods in real time, giving them the ability to more precisely monitor each step of the supply chain. Proactively managing disruptions, promptly identifying inefficiencies, and improving demand forecasting are all made possible by increased visibility. Predictive analytics and machine learning enable businesses to foresee future problems and modify operations accordingly. This enhanced responsiveness and transparency not only lowers risks but also fosters better stakeholder collaboration, making the supply chain more resilient and flexible.

2.3.2 Flexibility

This is important and related to SC resilience since it considers features of the product, system, and process like the flexibility to adjust the information architecture or reconfigure supply chains in response to customer demands (Closs et al. 2005). The process of gathering and sharing information between interested parties will inevitably become easier and more straightforward as supply chains become more integrated and connected (Lou et al. 2011). Real-time and Automated supply chains provide greater flexibility to respond swiftly, enabling the implementation of smart and self-reliant supply chains. But there is inadequate framework explaining the smart supply chain flexibility (Roberta Pereira et al. 2014). Digitizing supply chains influences supply chains flexibility directly because of communicating one machine with the other. The overall efficiency of the supply chain may profit from enhancing the flexibility component of these chains through connectivity. This has been validated from the study done by Majeed and Rupasinghe (2017) who studied the consequences of digitizing the supply chain on the performance of the supply chain by measuring the quality of the products, its response rate, delivery and productivity. Industry 4.0 uses technologies like IoT, AI, and advanced robotics to improve the resilience and flexibility of supply chains. Real-time data collection and analysis is made possible by these innovations, which enables businesses to react swiftly to disruptions and shifting market demands. Supply chains can be quickly rearranged to accommodate new product designs or changes in production volume thanks to automated procedures and smart manufacturing systems. AI-driven analytics enhance decision-making, enabling businesses to foresee problems and take proactive measures to address them. Because supply chains are now more adaptable, they can continue to function under a variety of circumstances, which lowers downtime, boosts productivity, and allows them to respond more quickly to customer demands.

2.3.3 Control

Control is the final component of resilience of supply chain. Supply chain information characteristics can be better controlled, planned, evaluated for quality, through the application of Industry 4.0 technologies and real-time methods (Verdouw et al. 2015). It may be possible to reduce the need for physical supply chain function monitoring by incoporating these technologies into control of supply chain. However, virtual monitoring surpasses human data collection in areas like sensor data and predicted and historical aspects of objects (Dweekat et al. 2017). Real-time control is essential to reduce expenses and boost output. Furthermore, because Industry 4.0 can provide information for inventory control, supply chain control costs can be reduced by using it. By combining technologies like automation, big data analytics, artificial intelligence (AI), and the Internet of Things (IoT), Industry 4.0 significantly improves the resilience of supply chain control. With the help of these technologies, data can be gathered and analyzed in real time, giving supply chain operations previously unheard-of visibility and control. IoT devices, which offer real-time asset and product monitoring, are vital to enhancing supply chain resilience. Goods can be tracked by companies along the supply chain in terms of location, condition, and status thanks to sensors and RFID tags attached to them. This degree of visibility makes the supply chain more responsive and efficient by assisting in the identification of bottlenecks, route optimization, and delay reduction. The analysis of large volumes of data produced by IoT devices and other sources is made possible by big data analytics and artificial intelligence (AI). Companies are able to make proactive adjustments to inventory levels and production schedules by using predictive analytics to forecast demand trends. Supply chains are better equipped to react swiftly to shifting market conditions thanks to AI-driven insights that assist in identifying possible risks and opportunities. This ability lessens the effect of disruptions and enhances overall control.

2.3.4 Collaboration

Developing a supply chain in collaboration is recognized as a critical step to fortify the resilience of supply chains (Calatayud 2017). In former times, an expanding number of producers have concentrated on cutting functional expenses to achieve maximum output. Owing to the presence of external partners, there is less information available at any stage of production in a globalized environment, exposing the firms to a variety of supply chain risks. The key to success in supply chain collaboration is rapid sharing of real-time information amongst partners and managers (Shahzad et al.

2021). Effective collaboration will decrease the likelihood of uncertainties and unexpected certainties. Lou et al. (2011) outlines the roles that supply chain collaboration with Industry 4.0 plays in production, transportation, as well as purchasing sectors. Purchasing domain seeks to obtain up-to-date product information. The fabricate sector prioritizes monitoring, while the transport section is primarily concerned with product location. By combining innovative technologies like blockchain, artificial intelligence (AI), cloud computing, and the Internet of Things (IoT), Industry 4.0 dramatically improves the resilience of supply chain collaboration. Supply chain partners can communicate and share data more easily thanks to these technologies, which promotes better cooperation and coordination.

Industry 4.0 empowers supply chain partners to instantly access vital information by facilitating real-time data exchange via cloud-based platforms and Internet of Things devices. By coordinating actions throughout the supply chain in real time, this communication helps cut down on delays and boost productivity. Partners can swiftly detect and resolve possible disruptions thanks to improved operational visibility, bolstering the supply chain's overall resilience. Supply chain partners' transparency and trust are greatly enhanced by blockchain technology. All stakeholders will have access to accurate and verified information because blockchain ensures an immutable and transparent ledger of transactions. By lowering the possibility of fraud and mistakes, this transparency encourages cooperation and trust among partners. Partners are more likely to work together productively and improve supply chain resilience when they have faith in the data they exchange.

3 Discussion on Findings

The research on using Industry 4.0 to improve resilience of supply chain has a revolutionary effect on reaching zero-carbon industry targets. Industry 4.0 technologies—which include blockchain, big data analytics, artificial intelligence (AI), and the Internet of Things (IoT)—offer vital resources for supply chain operations optimization that prioritize sustainability. IoT devices that collect data in real-time provide unmatched visibility into the operations of supply chain, making it possible to precisely monitor energy consumption and spot inefficiencies. A zero-carbon goal requires significant reductions in emissions and waste, which this capability supports. By averting unanticipated equipment failures, AI-driven predictive maintenance further strengthens resilience and ensures uninterrupted, effective operation over an extended period of time. Blockchain technology provides an immutable record of transactions, which guarantees sustainable practices and ethical sourcing, and thus improves transparency and accountability throughout the supply chain. Digital platforms facilitate improved communication and collaboration among stakeholders, allowing for more efficient resource allocation and waste reduction. Furthermore, supply chains can quickly adjust to shifts in demand and environmental challenges thanks to the flexibility offered by automation and robotics, maintaining resilience and reducing carbon footprints.

But embracing Industry 4.0 comes with a price tag, worries about data security, labor shortages, and regulatory compliance, among other issues. In order to successfully utilize Industry 4.0 technologies and create a robust, carbon-free supply chain, it is imperative that these issues are resolved. Overall, Industry 4.0 offers substantial opportunities to strengthen supply chains' resilience and sustainability, which will help the industry achieve its larger objective of being carbon neutral.

The problem is not with data availability in supply chains but the capability to address the data and turn them into insights for decision-making. (Khurana et al. 2024). Reconstructing new supply chain structures with Industry 4. 0 technologies enables them to overcome common management drawbacks like rigidity and decision making challenges (Qader et al. 2022). Furthermore, the utilisation of smart techniques in traditional approaches to supply chain management also increases sustainability and enhances product quality. When supply chains have the capabilities of self-learning, things such as sensors and machine learning techniques will give the necessary predictive abilities to reduce risks (McConaghy et al. 2017). For instance, they can issue alert notifications where there is an incidence of fire or production line accident, allowing swift management of the situation through real-time observation and efficient data transmission.

It is found that the deployment of intelligent technologies affects the major facets of the supply chain and strengthens its resilience. Firstly, Industry 4. 0 make them more visible due to the increased reliability and accuracy of the transmitted. Thus, the trust of the partners and customers enhances which contributes to effective decision-making. Secondly, Advanced technologies improve the supply chains by making it less rigid by enhancing stability along with increasing its ability to withstand disruptions through implementation of methods like smart scheduling. Thirdly, the overall purpose of collaboration of supply chain is the Machine to Human Interface (Roberta Pereira et al. 2014). Thirdly, primary objective of supply chain collaboration is machine-to-human collaboration (Roberta Pereira et al. 2014). Partners can synchronize efficiently using IoT technologies which enhance productivity and delivery through streamlined connectivity and automation. Fourthly, Real-time sensor data enables smart control, ensuring security across the supply chain and improving live tracking of resources. Accuracy in inventory management promotes the agility factor since the rate of information processing is increased. Ultimately, these advanced technologies of Fourth Industrial Revolution are said to boost resilience in supply chains, facilitate better control of errors and leverage big data analysis.

4 Conclusions

In summary, Industry 4.0 presents a significant chance to improve resilience of supply chain, thereby moving the industry closer to being carbon neutral. Supply chains can reach previously unheard-of levels of efficiency, transparency, and adaptability by leveraging technologies like blockchain, big data analytics, artificial intelligence (AI), and the Internet of Things (IoT). Predictive analytics and real-time information allow for exact operational monitoring and optimization, which significantly lowers waste and energy use—two factors that are essential for lowering carbon footprints. Blockchain technology ensures ethical and sustainable supply chain practices by fostering accountability and transparency. Digital platforms facilitate improved coordination among stakeholders through enhanced collaboration and communication, which in turn enhances resilience and resource management. Notwithstanding the significant advantages, Industry 4.0 technologies must be successfully implemented to overcome obstacles like high upfront costs, worries about data security, and the requirement for qualified staff. In order to fully realize Industry 4.0's potential for building robust and sustainable supply chains, it is imperative that these challenges be addressed. Businesses will be better equipped to meet the demands of a zero-carbon economy as they continue to adopt these technologies, accelerating the transition to a more resilient and sustainable future.

In this study, we have introduced an Industry 4.0-based resilient supply chain framework. By implementing this framework, they are engaged in short-term procedures that lead to increased robustness in the long run. Emphasis was given to four key components of resilient supply chains, which are flexibility, visibility, collaboration, and control. Owing to the shortage of studies examining the influence of technologies of Fourth Industrial Revolution on resilience of supply chains, our reconstruct conventional supply chains to ensure resilience across these four dimensions. Introducing smart technologies can significantly enhance supply chain functions like product tracking and tracing and maintaining a smooth information flow. These advanced technologies provide timely and accurate real-time data, ensuring that everyone in the supply chain has the information they need, when they need it.

References

Ali A, Mahfouz A, Arisha A (2017) Analysing supply chain resilience: integrating the constructs in a concept mapping framework via a systematic literature review. Supply Chain Manag Int J 22(1):16–39

Al-Talib M, Melhem WY, Anosike AI, Reyes JAG, Nadeem SP (2020) Achieving resilience in the supply chain by applying IoT technology. Procedia Cirp 91:752–757

Ben-Daya M, Hassini E, Bahroun Z (2019) Internet of things and supply chain management: a literature review. Int J Prod Res 57(15–16):4719–4742

Bhatti A, Malik H, Kamal AZ, Aamir A, Alaali LA, Ullah Z (2021) Much-needed business digital transformation through big data, internet of things and blockchain capabilities: implications for strategic performance in telecommunication sector. Bus Process Manag J 27(6):1854–1873

Butner K (2010) The smarter supply chain of the future. Strat & Lead 38(1):22–31

Calatayud A (2017) The connected supply chain: enhancing risk management in a changing world. Inter-American Development Bank. IDB-DP-508

Closs DJ, Swink M, Nair A (2005) The role of information connectivity in making flexible logistics programs successful. Int J Phys Distrib Logist Manag 35(4):258–277

Dastbaz M (2019) Industry 4.0 (I4.0): the hype, the reality, and the challenges ahead. In: Industry 4.0 and engineering for a sustainable future, pp 1–11. Springer, Cham

Dubey R, Gunasekaran A, Childe SJ, Blome C, Papadopoulos T (2019) Big data and predictive analytics and manufacturing performance: integrating institutional theory, resource-based view and big data culture. Br J Manag 30(2):341–361

Dubey R, Gunasekaran A, Bryde DJ, Dwivedi YK, Papadopoulos T (2020) Blockchain technology for enhancing swift-trust, collaboration and resilience within a humanitarian supply chain setting. Int J Prod Res 58(11):3381–3398

Dweekat AJ, Hwang G, Park J (2017) A supply chain performance measurement approach using the internet of things: toward more practical SCPMS. Ind Manag Data Syst 117(2):267–286

El Baz J, Ruel S (2021) Can supply chain risk management practices mitigate the disruption impacts on supply chains' resilience and robustness? Evidence from an empirical survey in a COVID-19 outbreak era. Int J Prod Econ 233:107972

Fatorachian H, Kazemi H (2018) A critical investigation of Industry 4.0 in manufacturing: theoretical operationalisation framework. Prod Plan & Control 29(8):633–644

Guha J, Chakrabarti B (2014) Making e-government work: adopting the network approach. Gov Inf Q 31(2):327–336

Hohenstein NO, Feisel E, Hartmann E, Giunipero L (2015) Research on the phenomenon of supply chain resilience: a systematic review and paths for further investigation. Int J Phys Distrib Logist Manag 45(1/2):90–117

Ivanov D, Dolgui A (2021) A digital supply chain twin for managing the disruption risks and resilience in the era of Industry 4.0. Prod Plan & Control 32(9):775–788

Ivanov D, Dolgui A, Sokolov B (2019) The impact of digital technology and Industry 4.0 on the ripple effect and supply chain risk analytics. Int J Prod Res 57(3):829–846

Johng H, Kim D, Park G, Hong JE, Hill T, Chung L (2020) Enhancing business processes with trustworthiness using blockchain: a goal-oriented approach. In: Proceedings of the 35th annual ACM symposium on applied computing, pp 61–68

Khuran S, Mannan B, Luthra S, Haleem A, Jain V (2024) Investigating barriers of sustainable supply chain in apparel & textile sector for realizing net zero economy: an AHP approach. In: Net zero economy, corporate social responsibility and sustainable value creation: exploring strategies, drivers, and challenges, pp 177–185. Springer Nature Switzerland, Cham

Kochan CG, Nowicki DR (2018) Supply chain resilience: a systematic literature review and typological framework. Int J Phys Distrib Logist Manag 48(8):842–865

Kopp R, Howaldt J, Schultze J (2016) Why Industry 4.0 needs workplace innovation: a critical look at the german debate on advanced manufacturing. Eur J Work Innov 2(1)

Lee I, Lee K (2015) The Internet of Things (IoT): applications, investments, and challenges for enterprises. Bus Horiz 58(4):431–440

Lou P, Liu Q, Zhou Z, Wang H (2011) Agile supply chain management over the internet of things. In: 2011 International conference on management and service science, pp 1–4. IEEE

Majeed AA, Rupasinghe TD (2017) Internet of things (IoT) embedded future supply chains for Industry 4.0: an assessment from an ERP-based fashion apparel and footwear industry. Int J Supply Chain Manag 6(1):25–40

McConaghy M, McMullen G, Parry G, McConaghy T, Holtzman D (2017) Visibility and digital art: blockchain as an ownership layer on the Internet. Strateg Change 26(5):461–470

Nothacker D (2021) Supply chain visibility and exception management. Disrupting logistics: startups, technologies, and investors building future supply chains, pp 51–62

Pundir AK, Jagannath JD, Ganapathy L (2019) Improving supply chain visibility using IoT-Internet of Things. In: 2019 IEEE 9th annual computing and communication workshop and conference (CCWC), pp 0156–0162. IEEE

Qader G, Junaid M, Abbas Q, Mubarik MS (2022) Industry 4.0 enables supply chain resilience and supply chain performance. Technol Forecast Soc Change 185:122026

Queiroz MM, Wamba SF (2019) Blockchain adoption challenges in supply chain: an empirical investigation of the main drivers in India and the USA. Int J Inf Manage 46:70–82

Rejeb A, Keogh JG, Treiblmaier H (2019) Leveraging the internet of things and blockchain technology in supply chain management. Future Internet 11(7):161

Roberta Pereira C, Christopher M, Lago Da Silva A (2014) Achieving supply chain resilience: the role of procurement. Supply Chain Manag Int J 19(5/6):626–642

Shahzad M, Qu Y, Zafar AU, Appolloni A (2021) Does the interaction between the knowledge management process and sustainable development practices boost corporate green innovation? Bus Strateg Environ 30(8):4206–4222

Spieske A, Birkel H (2021) Improving supply chain resilience through Industry 4.0: a systematic literature review under the impressions of the COVID-19 pandemic. Comput & Ind Eng 158:107452

Verdouw CN, Beulens AJ, Reijers HA, van der Vorst JG (2015) A control model for object virtualization in supply chain management. Comput Ind 68:116–131

Zhou W, Piramuthu S (2018) IoT security perspective of a flexible healthcare supply chain. Inf Technol Manage 19:141–153

Assessment of GHG Emission Towards Zero Carbon Industry in Rajasthan

Vineet Chouhan, Prashant Singh, and Shubham Goswami

Abstract The carbon footprint (CF) is a crucial metric used to evaluate the collective influence of governments, companies, and entities on climate change. Unlike traditional emissions accounting, which mostly focuses on emissions from local industrial activity, the CF include the emissions produced via use throughout universal supply networks for product and facilities. There has been a significant surge in interest in using CF assessment for municipalities. The reason for this is because municipalities play a substantial role in generating global carbon emissions, yet there is insufficient data available to adequately track their climate commitments. By integrating consumer surveys done at the household level with a comprehensive global supply chain database, it is possible to accurately quantify carbon footprints at both the district and household level in a spatially-detailed way. Thus far, our research has revealed previously undisclosed discrepancies in the carbon footprints of consumers in affluent countries. In this study, the carbon footprints of families in 12 Rajasthani districts are analysed and compared. Our estimations are derived from micro consumption data collected from 112 households. We analyse the variations in carbon footprints according to demographic, cultural, and economic variables. We establish that it is possible to simultaneously achieve the total eradication of extreme poverty in Rajasthan while implementing assertive strategies to address climate change. However, our research shows that in Rajasthan, efforts to reduce carbon footprint should prioritise high-expenditure families, since they are accountable for around 7 times the quantity of carbon emissions compared to low-expenditure households (who have a daily consumption of Rs. 192). The considerable disparities in carbon footprints among inhabitants of Rajasthan highlight the

V. Chouhan
Institute of Law, Nirma University, Ahmedabad, Gjuarat, India
e-mail: vineet.chouhan@nirmauni.ac.in

P. Singh
Hadarani Govt. P.G. College, Salumber, Rajasthan, India

S. Goswami (✉)
School of Management, Sir Padampat Singhania University, Udaipur, Rajasthan, India
e-mail: shubham.goswami@spsu.ac.in

R. Singh et al. (eds.), *Zero Carbon Industry, Eco-Innovation and Environmental Sustainability*, World Sustainability Series,
https://doi.org/10.1007/978-3-031-80220-1_3

need of discerning individual accountability for climate change in both domestic and international climate policy.

Keywords GHG emission · Zero carbon industry · Energy sector · Rajasthan

1 Introduction

A carbon footprint refers to the cumulative emissions of greenhouse gases, such as carbon dioxide and methane, resulting from human activity (Yaman 2024). The average per capita carbon footprint in the US is 16 tonnes, which is among the highest globally (Hertwich and Peters 2009). On a global scale, the average carbon footprint is around 4 tonnes (Kumar et al. 2014). In order to minimise the likelihood of a 2 °C increase in global temperatures, this are imperative that the global carbon footprint on average annual basis, be reduced to less than 2 tonnes by the year 2050 (Tetteh et al. 2021). GHG Platform India is a non-governmental effort that offers an impartial assessment and examination of India's GHG releases in important industries (Johnson 2023). The Platform seeks to eliminate data gaps and improve accessibility in order to maximize the benefit of ongoing GHG emissions estimating efforts (Yazdanie and Orehounig 2021). The objective is to enhance the level of analysis and policy discourse about India's GHG emissions sources, profile, and associated policies, extending beyond national inventories to encompass state inventories (Mahony 2014).

Urgent measures are required to restrict worldwide carbon releases for mitigating the magnitude and consequences of climate change. Modifications in habits of use are needed in order to restrict the carbon impact of civilization. In this particular setting, cities play a significant role in contributing to the release of greenhouse gases (GHG). Although cities only cover 2% of the world's surface area, they are responsible for 70–80% of worldwide carbon emissions. Only 100 cities contribute about 18% of the total global carbon footprint (Wei et al. 2021; Hoornweg et al. 2011). Currently, the number of people residing in urban regions surpasses those living in rural areas worldwide (Tassinari et al. 2024). "It is projected that over 68% of the world's population would live in cities by the year 2050" (O'Sullivan 2023). The growth of urban areas and population will have significant implications for the effects that cities experience and contribute to in terms of climate change (Laino and Iglesias 2023).

Over the last several years, cities have taken on a more prominent role in climate change governance. City-wide coalitions like the Local Governments for Sustainability, the "Covenant of Mayors for Climate & Energy, and C40 cities are examples of how this is being shown. More than 1,300 local governments across 26 nations have formally declared a climate emergency" since 2018 as a sign of their determination to decrease emissions (Martin et al. 2021). 87% of cities in the United States have issued such a proclamation. "The Climate and Environment Emergency Declaration in the United Kingdom, declared by the national government, was a response to pressure originating at the local level" (Harrington 2023). In light of this situation, there has been an increasing need to thoroughly assess, monitor, and reveal the carbon emissions of local regions (Arrieta and Davies 2024).

The "Carbon Disclosure Project" has identified 800 cities that are actively working towards this goal, with around 8000 planned urban sustainability initiatives taking place globally (Bashir et al. 2024). Nevertheless, there is a growing disparity between the pledges made at the district level for CF (Conservation Finance) and the amount of data available to support these obligations (Beer 2023). The available data for assessing the CF (carbon footprint) of districts is inconsistent, fragmented, and has limited coverage (Cai et al. 2023). Although still in its early stages, home carbon foot printing has shown significant promise in facilitating the discovery of previously unanticipated priorities for climate action, both at the national level and on a global scale (Liang et al. 2023). Recent research has shown the significance of wealth equality, social views, and religion as crucial factors influencing the carbon footprints of households (Lin and Yang 2023). An important challenge in this context is "determining the indirect demand for products and services, which links district-level consumption to production in national and international supply networks" (Mukhopadhyay et al. 2023). An in-depth investigation of major metropolitan centres revealed that the practice of determining boundaries for cities significantly underestimates their carbon footprint by around 60% (C40, 2018). However, some research on specific cities have reported even higher percentages.

However, recent studies linking personal consumption data to countrywide and international supply chain records present a viable method for gaining a thorough understanding of the variables affecting the carbon footprints of neighbourhood's and families (Madsen and Weidema 2023). District and household carbon foot-printing may reveal and clarify differences in carbon footprints across different socioeconomic and demographic groups within a nation, unlike national carbon foot-printing (Chen et al. 2023). The process of consumption and emissions-release patterns of people, organizations, and geographic areas varied significantly, according to preliminary study on lowering carbon foot-print calculation. Notable instances of spatially explicit carbon foot-printing may be found in Australia, Norway, and the United States. These examples demonstrate that a uniform strategy to reducing carbon emissions from consumption fails to consider the differences in factors that contribute to emissions within each country. Nevertheless, the lack of sufficient data hindered the ability to uncover the connection between consumption, emissions, and variations in spending habits, cultural background, and social views across populations (Winterich et al. 2023). Recent data indicates that these variables have a significant impact on the carbon footprints of households. However, the study of carbon footprints, which takes into account both geographical and social factors, is still constrained by the lack of data from developing nations and China (Li et al. 2024a, b). Assessments of emerging nations are necessary due to the intricate and swiftly evolving factors that drive their consumption and their progressively substantial role in climate change (Chipangamate and Nwaila 2023). In this research, we assess and contrast the carbon footprint of districts and families in Rajasthan.

2 Literature Review

A carbon footprint is a quantitative assessment that measures the total quantity of greenhouse gases released into the atmosphere, which are accountable for the retention and release of heat, hence contributing to climate change. The measurement accounts for the emission of many greenhouse gases. The main factors that contribute to the carbon footprint of a building are operational carbon emissions, embedded carbon emissions, and construction processes. Operational carbon emissions are the CO_2 emissions that occur when a structure is in operation, whereas embedded carbon emissions are the emissions associated with the manufacturing and construction stages of a building. Construction activities, including the production and transportation of materials, have the potential to add to the carbon footprint. Energy-saving technology implemented in buildings can decrease the amount of energy consumed and the emissions produced. However, it is important to note that these technologies may also contribute to higher rates of building renovation, which might result in a rise in carbon pollution. The carbon footprint of a structure may be greatly influenced by factors such as the utilization of cement-based materials, metal, and insulation in construction, as well as the consumption of high-carbon materials like concrete. This section contains the studies that have evaluated greenhouse gas (GHG) emissions with the goal of achieving a zero-carbon industry.

Lee et al. (2021) showed that the carbon footprint (CF) has become a crucial metric for determining how much each nation, industry, and individual has contributed to global warming. Thus far, this method has revealed previously unnoticed variations in the carbon footprints of consumers in industrialized nations. A total of 623 Indian districts had their household carbon footprints calculated and analysed using micro consumption data from 203,313 residences. They also explain the difference in these footprints by factors related to economics, culture, and demography. They demonstrate that aggressive climate change mitigation in India does not conflict with the elimination of severe poverty. Our research indicates that high-expenditure families—who consume \$1.9 per day—are accountable for over seven times as much carbon emissions as low-expenditure households, which should be the focus of India's CF reduction programs. These stark differences in the carbon footprints of Indian residents demonstrate the necessity of distinguishing between personal and planetary responsibility for atmosphere alteration in country and at international climate policy.

Pathak et al. (2010) demonstrated that carbon emissions happen at different points in the food product's life cycle. Methane (CH_4) emissions were mostly caused by rice agriculture and animal food products (meat and milk), whereas nitrous oxide (N_2O) emissions were primarily caused by food products derived from crops. These findings were based on the GHG released from 24 different Indian food items. CO_2 was released during agricultural activities, input production, transportation, food processing, and food preparation. The life cycle of cooked rice produced 2.8 times as much greenhouse gas emissions as "the life cycle of chapatti, a wheat flour product". Rice, chapatti, salmon, and mutton each released 12.9 times, 12.9 times, and 11.9

times more greenhouse gas emissions than milk. Since most Indians eat fresh food that is grown nearby, food production accounted for 87% of emissions, with preparation (10%), processing (2%) and transportation (1%), following in order of importance. An adult Indian guy followed a healthy diet (vegetarian) and consumed 1165 g of food, emitting 723.7 g of CO_2 equivalents each day. 1.8 times the greenhouse gas emissions from a non-vegetarian lunch containing mutton, 1.5 times from non-vegetarian meals containing chicken and ovo-vegetarian meals, and 1.4 times from lacto-vegetarian meals were produced. Thus, changing eating habits may provide a chance to reduce greenhouse gas emissions. Kulkarni (2019) disclosed that the amount of greenhouse gases (GHGs) released by every organization or activity is measured by its carbon footprint (CF). One step toward adopting sustainable teaching methods is to report the amount of carbon footprints (CFs) from educational campuses in units of CO_2e. Using a bottom-up methodology, the current study evaluates the CFs resulting from the resource consumption of each academic department independently, and the net is produced by aggregating the contributions from each department. One benefit of this strategy is that it avoids possible computation overlaps. Thus, this method of calculating carbon footprints can result in an accurate CF evaluation of them.

Ramachandra (2012) said that the production of power, transportation, home energy use, industry, and agriculture account for the majority of India's carbon emissions. Soil and biomass found in forests store most carbon. Using state-specific carbon sequestration capability and region-specific emission parameters, this article focuses on state-level carbon emissions (CO_2, CO, and CH_4). According to the estimate, India emits "965.9 Tg of CO_2, 22.5 Tg of CO, and 16.9 Tg of CH_4 annually". The production of electricity accounts for 35.5% of all CO_2 emissions; transportation accounts for the remaining portion. Only the transportation of vehicles accounts for 25.5% of all emissions (Dwivedi and Soni 2023). According to the data, Maharashtra is the state with the most CO_2 emissions, followed by "West Bengal, Andhra Pradesh, Uttar Pradesh, Gujarat, and Tamil Nadu". For every federal state, the carbon status is calculated as the yearly carbon storage divided by the annual carbon emission. This indicates that smaller states and union territories (UTs) with higher carbon sequestration rates due to their rich forest cover and have a carbon status greater than 1 (such as "Arunachal Pradesh, Mizoram, and the Andaman and Nicobar Islands"). Arunachal Pradesh, Madhya Pradesh, Chhattisgarh, and Orissa account for 34% of the 7.35% of annual carbon emissions that are stored in soil or forest biomass.

Sreedhar et al. (2016) attempted to create a toolset known as the "Carbon Footprint Calculator" in order to measure the carbon footprints of the various pavement technologies that are utilized in the building of highways. The study's tool took into account the primary sources of GHG emissions, such as the design of the pavement, the production and transportation of materials to the site, the construction techniques employed in the various pavement systems, and the anticipated vehicle usage during the pavement's design life. As part of this investigation, a mathematical model was also created to calculate the entire quantity of GHG emissions in terms of total $kgCO_2$ equivalent ($kgCO_2e$). A life cycle assessment was utilized to evaluate greenhouse gas emissions in terms of carbon footprints for the various pavement technologies used in the building of Indian roads. It is anticipated that design engineers would find good use for this tool in optimizing pavement design approach and construction methods with regard to the creation of a more sustainable and greener environment.

Jaiswal and Agrawal (2020) examines the large contribution of the agriculture sector in the overall amount of GHG emissions. Along with population growth comes an increase in the amount of GHG emissions from the agriculture sector. The term "carbon footprint of agriculture" refers to the overall quantity of GHGs released by the agricultural sector's operations, expressed in units of carbon equivalent (C-eq). Agriculture-related operations include tilling, plowing, manuring, irrigation, growing a range of crops, raising animals, and operating associated equipment release a substantial quantity of greenhouse gases (GHGs), which are divided into three carbon footprinting tiers and kept apart by fictitious borders. "Energy input from fossil fuels, electricity, machinery, and livestock management accounts for a significant amount of agriculture's carbon emissions" (Mor et al. 2021). More greenhouse gases are produced by crop cultivation systems, which primarily grow grains, than by agricultural systems that grow fruits and vegetables. In addition, changes in land use such as the conversion of natural ecosystems to agricultural uses, deforestation, and burning of crop residues after harvest all greatly increase carbon emissions (Kumar and Aravindakshan 2022). The carbon footprint of agriculture, including energy-related inputs, fertilizers, pesticides, organic manure, and processes that influence carbon emissions from agriculture, will be the main topic of this review article. We'll also examine the mitigation strategies that work well in lowering the carbon footprint left by different types of agriculture. We'll talk about strategies that could help lower greenhouse gas emissions from the agriculture sector, like crop rotation, improved soil carbon sequestration through plantations and straw returns, diversified cropping systems, crop rotation systems, and limiting deforestation (Sah and Devakumar 2018).

An evaluation of carbon reduction activities, including the separate but interconnected effects on emissions and the economy, can offer a practical comprehension of a company's actions and objectives when seen through the lens of the "Natural Resource-Based View (NRBV)". Previous research on carbon reduction projects may be classified into two categories of investigations: analytical studies and empirical exploration/validation studies. Although there is a wide range of analytical studies available (such as those conducted by Jiang et al. 2022; Raihan et al. 2022; Li et al. 2022a, b; Zhang et al. 2022; Fang et al. 2022), there is a limited number of empirical research that focus on the carbon reduction efforts of enterprises. The limited number of empirical research have mostly focused on the methods and effectiveness of carbon footprint-based initiatives. Research has mostly examined topics such as precise carbon footprint calculation (Wang and Zhao 2023), life cycle assessment (LCA) (e.g., Kiehle et al. 2023; Rinne et al. 2022), and the impact of stakeholder pressure (Wang and Zhao 2023; Eleftheriadis and Anagnostopoulou 2024). The success of carbon reduction initiatives relies on the utilization of management tools, the dedication of the company, and the alignment of internal processes (M'baye 2022). When examining the architecture of green supply chain networks, researchers regularly evaluated various policies such as carbon cap, carbon offset, cap-and-trade, and carbon tax to evaluate attempts to reduce carbon emissions (Li et al. 2024a, b; Linghu et al. 2022).

Multiple studies have examined the challenge of obtaining a precise assessment of a company's carbon footprint (Liu et al. 2022; Mariette et al. 2022) because of differences in definitions and assumptions on emission standards, technology, and inputs. The stylized analytical literature has offered valuable information on the cost-effectiveness of several carbon reduction initiatives. Key factors that contribute to the problem are the influence of product attributes (Taufique et al. 2022), cooperation within the supply chain (Xia et al. 2023; Lintukangas et al. 2023), incentives and financial support driven by market forces (Godinho et al. 2022), environmentally friendly technology (Jumaev and Tursunov 2022), and a means to establish a connection between the environmental performance of the supply chain and its economic outcomes (Amjad et al. 2022).

Karthick and Uthayakumar (2022) examined the impact of production, inventory replenishment, and routing decisions on greenhouse gas emissions in a supply chain. The GHG supply chain models were integrated with factors like as energy consumption and flawed processes across various coordination choices (Xue et al. 2022). Mechouar et al. (2022) observed the impact of production technology on the connection between carbon tax and sustainable facility placement. Huang et al. (2023) conducted a comprehensive CO_2 emission reduction in shipping of container across many trades. Furthermore, Chekoubi et al. (2022) analysed the compromises between economic and environmental efficiency in creation and inventory routing difficulties. On the other hand, Abbasian et al. (2022) formulated a model for resilient optimisation for effective and environmentally friendly supply chain planning by employing a delay technique. Collectively, these analytical studies encompass a wide range of topics related to methods aimed at reducing greenhouse gas emissions.

The existing research on evaluating decarbonization efforts primarily focuses on the significance of carbon disclosure. Carbon disclosure assists managers in comprehending the origins of climate change risks and opportunities, which subsequently aids them in determining the type of projects aimed at reducing carbon emissions (Li et al. 2022a, b). These disclosures also aid in recognizing any prejudices against economically efficient methods of reducing emissions within their companies and throughout external supply chains (She 2022). Consequently, corporations' benefit from reduced equity costs as a result of heightened investor confidence, despite the fact that these firms disclose a substantial level of carbon emissions (Li et al. 2022a, b). "In general, there has been a rise in the number of firms that choose to reveal their efforts to reduce carbon emissions" (Lee and Hussain 2022; Liu et al. 2023). As a result, there is an increase in the reporting of companies' efforts to reduce carbon emissions (Hidayat et al. 2022; Haque and Ntim 2022). Another emerging factor of significance is the need to examine the complete supply chain network when evaluating greenhouse gas (GHG) emissions. Sharma et al. (2022) conducted a study to investigate the correlation between the complexity of buyer–supplier networks and the greenhouse gas (GHG) emissions of enterprises. The researchers measured complexity in terms of horizontal, vertical, and geographic complexity, and also explored how the reach of the buyer–supplier network influenced this connection. Due to the lack of extensive empirical research in this field, our knowledge of the dynamics related to carbon reduction measures and their effects on emissions and finances is limited. Our study aims to fill this gap by examining the effects of commonly implemented initiatives that focus on reducing emissions, promoting responsible product management, and fostering sustainable development with assessment of the GHG emission towards zero carbon industry.

3 Methodology of the Paper

The paper analyses the magnitude and factors influencing carbon footprints in Rajasthan, for this purpose first it presents the analysis of macro and microeconomic data to evaluate the financial condition at the district level in Fig. 1 as under.

A schematic diagram (Fig. 3.1) showing "the integration of household expenditure data from the National Sample Survey Organisation (NSSO) Consumer Expenditure Survey (CES) of Indian households, administered by Ministry of Statistics & Programme Implementation, with the Eora database."

Using data from "Consumption Expenditure Surveys" and the "Eora (2020) database", a Choropleth map is created to illustrate the regional distribution of per capita carbon footprints for Rajasthan and Indian families (Fig. 2). (Source: *Authors*).

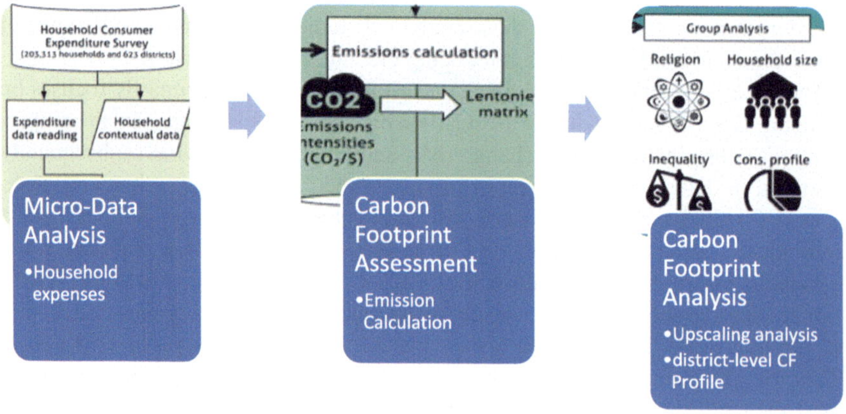

Fig. 1 Analysing macro and microeconomic data to evaluate the financial condition at the district level. *Source* Authors creation

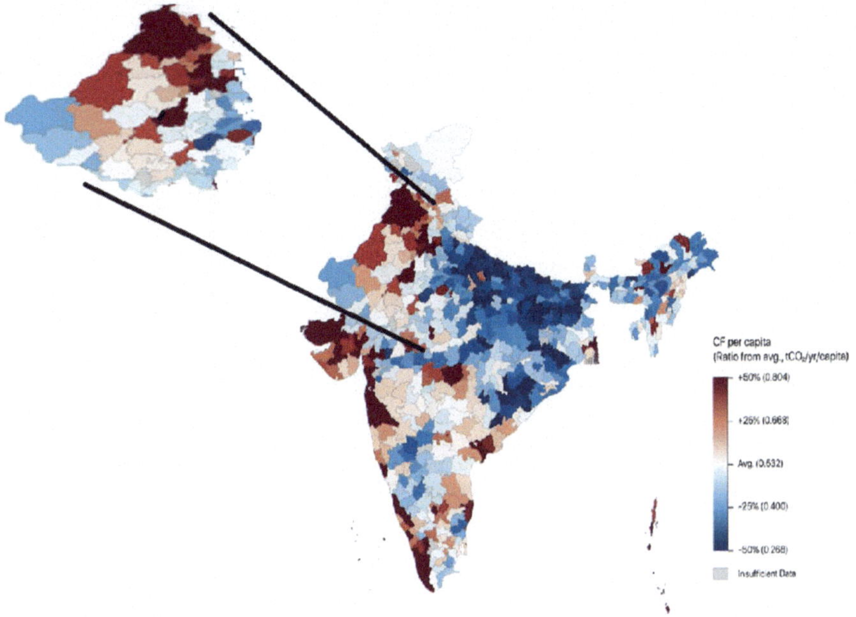

Fig. 2 Carbon footprint per capita in various regions in Rajasthan and India

3.1 Data Analysis

An objective assessment and study of India's GHG emissions across major industries is provided by the civil society project GHG Platform India. By expanding beyond national inventories to include state inventories, addressing data gaps and accessibility issues, and the Platform intends to strengthen the numerous ongoing Efforts to estimate GHG emissions are being intensified by expanding the scope of research and policy discourse on the origins, characteristics, and related policies of India's GHG emissions. The program calculates and evaluates GHG removals and emissions from the following industries: Energy-related waste has been recorded under the Energy sector, including fuel burned for direct fuel combustion (industrial energy) and fuel combusted for captive power production (auto-producers) (see Fig. 3).

Figure 3.3 shows that Rajasthan's emissions increased at a CAGR of 5.90%, from 72.21 Mt CO_2e in 2005 to 152.12 Mt CO_2e in 2018. In all of the reference years, the energy sector contributed significantly to Rajasthan's overall GHG emissions. From 64% in 2005 to 69% in 2018, the energy sector accounted for a larger portion of emissions. However, from 14% in 2005 to 12% in 2018, the Agriculture, Forestry, and Other Land-Use (AFOLU) sector's share of the economy's emissions decreased. The share of the Waste sector decreased from 5% in 2005 to 3% in 2018, while the IPPU (Industrial Process and Product Uses) sector's share decreased from 17% in 2005 to 16% in 2018 (see Fig. 4).

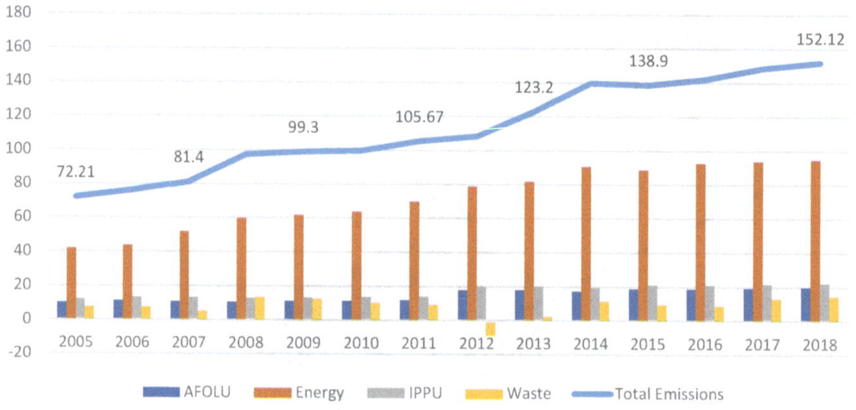

Fig. 3 GHG emission estimates of Rajasthan (2005–2018). *Source* Chakraborty et al. (2022). Major emitting sources from the livestock sub-sector: trends in India and mitigation opportunities

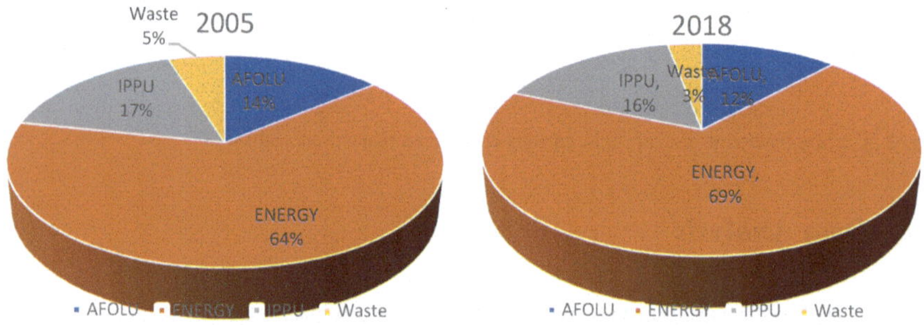

Fig. 4 Share of emissions of various sectors

Rajasthan's per-capita emissions were somewhat less than those of India, continuous during study period (see Fig. 5). But from 1.18 t CO_2e/capita in 2005 to 1.99 t CO_2e/capita in 2018, they grew at an estimated CAGR of 4.11%, which was greater than India's CAGR (3.41%). Fuel combustion and fugitive emissions are the two types of emissions from the energy industry. Emissions from transportation, industries, agriculture, commercial, residential, and captive power plants are all included in the category of fuel combustion. Producing fuel is the cause of fugitive emissions. In 2018, 69% of all emissions in the Rajasthani economy came from the energy sector. From 46.26 Mt CO_2e in 2005 to 104.82 Mt CO_2e in 2018, the energy sector's emissions grew at a CAGR of 6.49%.

"Livestock, Land and Aggregate Sources, and Non-CO_2 Emissions Sources on Land are the three primary sub-sectors that contribute to the AFOLU sector's emissions." In 2018, the AFOLU industry accounted for 12% of Rajasthan's overall economic emissions. Throughout the reference period, the Land sub-sector was a net sink of greenhouse gases, in contrast to the above-mentioned Emissions Sources

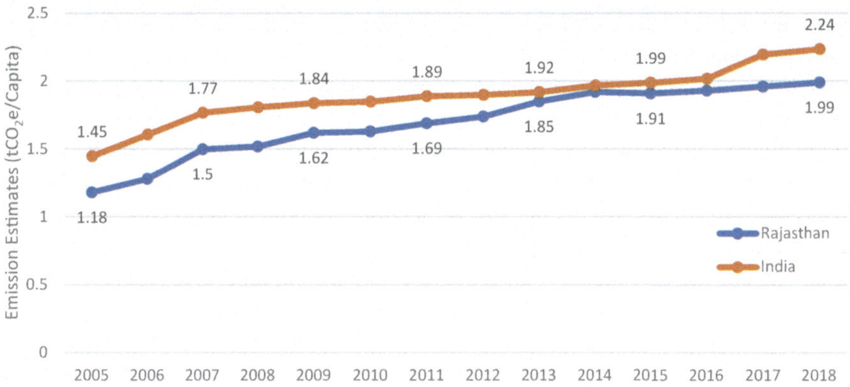

Fig. 5 Per capita GHG emission of Rajasthan and India (2005–2018)

on Land sub-sectors, which were net emitters. Total AFOLU emissions rose from 10.35 Mt CO_2e in 2005 to 18.17 Mt CO_2e in 2018, a CAGR of 4.43%. Post 2011, a decrease was registered in the Land sub-sector removals. This can be attributed to the decreasing sink of Agricultural Land and Other Land categories. During the reference period, the Land sub-sector in Rajasthan removed 8.33 Mt CO_2e annually on average, which equates to about 35.87% of the average annual gross AFOLU emissions. The primary sources of greenhouse gas emissions in the waste industry are "additional solid waste disposal, domestic wastewater, and industrial wastewater". In Rajasthan, the waste industry accounted for over 3% of all emissions in the state in 2018. The waste industry's GHG emissions rose from 3.32 million metric tons CO_2e in 2005 to 4.83 million metric tons CO_2e in 2018, at an estimated CAGR of 2.93%. The overall Waste sector emissions increased in 2011 due to growth in Domestic Wastewater emissions of Rajasthan.

4 Conclusion

An accurate comprehension of the individuals, locations, and factors that impact carbon emissions within a country is crucial for the implementation of effective national climate policy. Conducting a detailed analysis of carbon emissions that takes into account specific locations is crucial for achieving this goal. While the use of this approach to determine the causes and patterns of carbon emissions is common in affluent nations, it is not well used in poor countries. It is crucial to address this lack of awareness in order to reduce carbon emissions in developing countries and find ways to do so without hindering poverty reduction efforts. Aligning these goals is a significant priority in the next ten years, as outlined in the UN SDG's and "national policy agendas as per Hubacek et al. (2017)". Nowhere is this more crucial than in Rajasthan, where a quarter of India's impoverished population resides and a growing portion of the world's carbon emissions are produced. The computation of household and individual carbon footprints used an assumption that specific goods in the household survey had the similar carbon emissions strength. This assumption was made since detailed information on the carbon emissions of similar products was not available for the majority of the Eora (2020) nation's databases. In the future, when MRIO databases improve in sectoral resolution, it will be possible to loosen this assumption. This will enable a more accurate alignment between spending and carbon emissions predictions. Nevertheless, these enhancements were beyond the parameters of our investigation. In order to create a comprehensive carbon footprint profile for both the district and the whole country, additional assumptions had to be made based on the census data collected from homes. In order to achieve this scaling, it is assumed that the household expenditure profile in each district is the same for both the unsampled and the sampled population.

5 Implications

Industrial deep decarbonisation for achieving a more than 60% reduction in industrial emission by 2050 is a distinctly different and much bigger global climate and industrial policy challenge. It is these aspects that must be further stressed, mainly the zero-emission solutions such as electrification and hydrogen energy, as well as hydrogen carriers and transports, and the implications of industrial transition on the region. In some places, some form of carbon pricing has been used as one of the key weapons in combating climate change. But it has not been done on a scale that has provided the much-needed targeted cuts in emissions from energy and carbon-intensive industries. However, there are several business entities that can be said to have been untouched by the working of the price mechanism. Therefore, the authors highlighted that the Rajasthan state needs to develop and implement an industrial strategy that targets the emission-intensive subsectors of the industry for dealing with an issue of zero emissions. Wang et al. (2022) reviewed the proportion of the AFOLU sector's greenhouse gas emissions in the economy using various research disciplines, including engineering, economics, policy, governance, and innovation. There are ways suggested to reduce these emissions and the following is our recommendation. Governments must take explicit, and ambitious initiatives to map out strategies and laid down polices that would chart the way forward. This includes remapping of markets, extension of research and innovation frameworks to capture zero emission technologies for industries, capacity building for themselves and understanding of industrial transformation, advocacy for this issue at climate and trade fora. An efficient strategy should also be flexible so as to accommodate changes along the technological and systems' development and the socio-cultural changes that occur over time. Forcing industrial policy plans maybe more effective if they are especially tailored to small sub sectors. They may also be altered to suit the needs of different national and regional conditions that is present in the Rajasthan state whereby it is beneficial to acquire certain cleaner inputs and source of energy. Therefore, it is essential to come up with, plan and implement strategies, approaches and instruments drawing strict attention to the context. These elements include the level of economic development, governing tradition, availability and distribution of resources and the industrial structure.

Furthermore, it is equally pertinent to note that analysing and formulating the strategy about how the carbon emissions in the industries can be cut down, would require number of measures to sort out the issue. Such measures should encompass many sub-sectors of trade and also contain the trading of green products. Each government, business, as well as the research sector, needs capacity building to improve and enhance the models used for decarbonisation. Another factor demonstrated is the rather high level of differentiation concerning the maturity and application of impact minimization methods. There exists a need for a better understanding of the available options to minimize the negative consequences of the industrial sector transformation processes, and the policy implications of such decisions. How to apply them strategically and what priority they should be implemented on is the next level of knowledge. More than that it has to learn from the past policy experiences and modify the current ones.

References

Abbasian M, Sazvar Z, Mohammadisiahroudi M (2023) A hybrid optimization method to design a sustainable resilient supply chain in a perishable food industry. Environ Sci Pollut Res 30(3):6080–6103. https://doi.org/10.1007/s11356-022-22115-8

Amjad A, Abbass K, Hussain Y, Khan F, Sadiq S (2022) Effects of the green supply chain management practices on firm performance and sustainable development. Environ Sci Pollut Res 29(44):66622–66639. https://doi.org/10.1007/s11356-022-19954-w

Arrieta T, Davies JS (2024) Crisis management in English local government: the limits of resilience. Policy & Polit 1–19. https://doi.org/10.1332/03055736Y2024D000000029

Bashir N, Donti P, Cuff J, Sroka S, Ilic M, Sze V, Delimitrou C, Olivetti E (2024) The climate and sustainability implications of generative AI

Beer CM (2023) Bankrolling biodiversity: the politics of philanthropic conservation finance in Chile. Environ Plan E Nat Space 6(2):1191–1213. https://doi.org/10.1177/25148486221108171

Cai M, Ren C, Shi Y, Chen G, Xie J, Ng E (2023) Modeling spatiotemporal carbon emissions for two mega-urban regions in China using urban form and panel data analysis. Sci Total Environ 857:159612. https://doi.org/10.1016/j.scitotenv.2022.159612

Chakraborty M, Solanki S, Dutt R (2022) Major emitting sources from the livestock sub-sector: trends in India and mitigation opportunities

Chekoubi Z, Trabelsi W, Sauer N, Majdouline I (2022) The integrated production-inventory-routing problem with reverse logistics and remanufacturing: a two-phase decomposition heuristic. Sustainability 14(20):13563. https://doi.org/10.3390/su142013563

Chen X, Zhen Y, Chen Z (2023) Household carbon footprint characteristics and driving factors: a global comparison based on a dynamic input–output model. Energies 16(9):3884. https://doi.org/10.3390/en16093884

Chipangamate NS, Nwaila GT (2023) Assessment of challenges and strategies for driving energy transitions in emerging markets: a socio-technological systems perspective. Energy Geosci 100257. https://doi.org/10.1016/j.engeos.2023.100257

Dwivedi AK, Soni A (2023) The carbon footprint of India and its drivers: a structural decomposition analysis of global value chains. Sustain Energy Technol Assess 56:103109. https://doi.org/10.1016/j.seta.2023.103109

Eleftheriadis I, Anagnostopoulou E (2024) Developing a tool for calculating the carbon footprint in SMEs. Sustainability 16(5):1905. https://doi.org/10.3390/su16051905

Fang G, Gao Z, Tian L, Fu M (2022) What drives urban carbon emission efficiency?–Spatial analysis based on nighttime light data. Appl Energy 312:118772. https://doi.org/10.1016/j.apenergy.2022.118772

Godinho R, Lacerda de Oliveira MI, Rodrigues L, Moreira M, EA Seabra J (2022). Carbon credits and the bioethanol industry: governmental programs and incentives. In: Liquid biofuels: bioethanol, pp 363–384. Springer International Publishing, Cham. https://doi.org/10.1007/978-3-031-01241-9_16

Haque F, Ntim CG (2022) Do corporate sustainability initiatives improve corporate carbon performance? Evidence from European firms. Bus Strateg Environ 31(7):3318–3334. https://doi.org/10.1002/bse.3078

Harrington C (2023) Climate change as a "Threat multiplier": the construction of climate security by the United Kingdom—2007–2020. In: Climate security in the anthropocene: exploring the approaches of United Nations security council member-states, pp 297–318. Springer International Publishing, Cham. https://doi.org/10.1007/978-3-031-26014-8_14

Hertwich EG, Peters GP (2009) Carbon footprint of nations: a global, trade-linked analysis. Environ Sci Technol 43(16):6414–6420. https://doi.org/10.1021/es803496a

Hidayat I, Ismail T, Taqi M, Yulianto AS (2022) Investigating in disclosure of carbon emissions: influencing the elements using panel data. Jurnal Reviu Akuntansi dan Keuangan 12(3):721–732. https://doi.org/10.22219/jrak.v12i3.23072

Hoornweg D, Sugar L, Trejos Gómez CL (2011) Cities and greenhouse gas emissions: moving forward. Environ Urban 23(1):207–227. https://doi.org/10.1177/0956247810392270

Huang D, Wang Y, Yin C (2023) Selection of CO_2 emission reduction measures affecting the maximum annual income of a container ship. J Mar Sci Eng 11(3):534. https://doi.org/10.3390/jmse11030534

Jaiswal B, Agrawal M (2020) Carbon footprints of agriculture sector. Carbon Footpr Case Stud Build Househ Agric Sect 81–99. https://doi.org/10.1007/978-981-13-7916-1_4

Jiang T, Yu Y, Jahanger A, Balsalobre-Lorente D (2022) Structural emissions reduction of China's power and heating industry under the goal of "double carbon": a perspective from input-output analysis. Sustain Prod Consum 31:346–356. https://doi.org/10.1016/j.spc.2022.03.003

Johnson B (2023) An assessment of clean development mechanism projects in the energy sector in India. Doctoral dissertation, Department of Commerce and Management Studies, University of Calicut. https://hdl.handle.net/20.500.12818/1500

Jumaev K, Tursunov B (2022). Environmentally friendly technology for obtaining fuel briquettes from oil waste. IOP Conf Ser Earth Environ Sci 1112(1):012005. (IOP Publishing)

Karthick B, Uthayakumar R (2022) Impact of carbon emission reduction on supply chain model with manufacturing decisions and dynamic lead time under uncertain demand. Clean Logist Supply Chain 4:100037. https://doi.org/10.1016/j.clscn.2022.100037

Kiehle J, Kopsakangas-Savolainen M, Hilli M, Pongrácz E (2023) Carbon footprint at institutions of higher education: the case of the University of Oulu. J Environ Manage 329:117056. https://doi.org/10.1016/j.jenvman.2022.117056

Kulkarni SD (2019) A bottom up approach to evaluate the carbon footprints of a higher educational institute in India for sustainable existence. J Clean Prod 231:633–641. https://doi.org/10.1016/j.jclepro.2019.05.194

Kumar M, Sharma L, Vashista PK (2014) Study on carbon footprint. Int J Emerg Technol Adv Eng 4(1):345–355

Kumar BM, Aravindakshan S (2022) Carbon footprints of the Indian AFOLU (Agriculture, forestry, and other land use) sector: a review. Carbon Footpr 1(7):2–20. https://doi.org/10.20517/cf.2022.04

Laino E, Iglesias G (2023) Extreme climate change hazards and impacts on European coastal cities: a review. Renew Sustain Energy Rev 184:113587. https://doi.org/10.1016/j.rser.2023.113587

Lee CC, Hussain J (2022) Optimal behavior of environmental regulations to reduce carbon emissions: a simulation-based dual green gaming model. Environ Sci Pollut Res 29(37):56037–56054. https://doi.org/10.1007/s11356-022-19710-0

Lee J, Taherzadeh O, Kanemoto K (2021) The scale and drivers of carbon footprints in households, cities and regions across India. Glob Environ Chang 66:102205. https://doi.org/10.1016/j.gloenvcha.2020.102205

Li K, Ma M, Xiang X, Feng W, Ma Z, Cai W, Ma X (2022a) Carbon reduction in commercial building operations: a provincial retrospection in China. Appl Energy 306:118098. https://doi.org/10.1016/j.apenergy.2021.118098

Li N, Shi B, Wu L, Kang R, Gao Q (2022b) Climate-related development finance, energy structure transformation and carbon emissions reduction: an analysis from the perspective of developing countries. Front Environ Sci 9:778254. https://doi.org/10.3389/fenvs.2021.778254

Li P, Wang W, Xia X (2024) Assessing the effectiveness of carbon cap-and-trade and hybrid subsidy policies: The perspective of production, environment, and consumer. J Clean Prod 143074. https://doi.org/10.1016/j.jclepro.2024.143074

Li R, Yu Y, Cai W, Liu Y, Li Y (2024) Exploring the gradient impact of climate and economic geographical factors on city-level building carbon emissions in China: characteristics and enlightenments. Sustain Cities Soc 105637. https://doi.org/10.1016/j.scs.2024.105637

Liang L, Chen M, Zhang X (2023) Measuring inequality of household carbon footprints between income groups and across consumption categories in China. J Clean Prod 418:138075. https://doi.org/10.1016/j.jclepro.2023.138075

Lin B, Yang M (2023) Choosing the right policy: Factors influencing the preferences of consumption-side personal carbon reduction policies. J Environ Manage 326:116706. https://doi.org/10.1016/j.jenvman.2022.116706

Linghu D, Wu X, Lai KH, Ye F, Kumar A, Tan KH (2022) Implementation strategy and emission reduction effectiveness of carbon cap-and-trade in heterogeneous enterprises. Int J Prod Econ 248:108501. https://doi.org/10.1016/j.ijpe.2022.108501

Lintukangas K, Arminen H, Kähkönen AK, Karttunen E (2023) Determinants of supply chain engagement in carbon management. J Bus Ethics 186(1):87–104. https://doi.org/10.1007/s10551-022-05199-7

Liu G, Liu J, Zhao J, Qiu J, Mao Y, Wu Z, Wen F (2022) Real-time corporate carbon footprint estimation methodology based on appliance identification. IEEE Trans Industr Inf 19(2):1401–1412. https://doi.org/10.1109/TII.2022.3154467

Liu X, Cifuentes-Faura J, Zhao S, Wang L (2023) Government environmental attention and carbon emissions governance: firm-level evidence from China. Econ Anal Policy 80:121–142. https://doi.org/10.1016/j.eap.2023.07.016

M'baye A (2022) Energy performance management in an industrial site: definition and application of a specific methodology for carbon emissions reduction. Low Carbon Econ 13(03):148–162. https://doi.org/10.4236/lce.2022.133008

Madsen ST, Weidema BP (2023) Input–output modelling for household activity-level environmental footprints: a systematic literature review. Environ Res Lett 18(4):043003. https://doi.org/10.1088/1748-9326/acc59c

Mahony M (2014) The predictive state: science, territory and the future of the Indian climate. Soc Stud Sci 44(1):109–133. https://doi.org/10.1177/0306312713501407

Mariette J, Blanchard O, Berné O, Aumont O, Carrey J, Ligozat A, Lellouch E, Roche PE, Guennebaud G, Thanwerdas J, Bardou P, Ben-Ari T (2022) An open-source tool to assess the carbon footprint of research. Environ Res Infrastruct Sustain 2(3):035008. https://doi.org/10.1088/2634-4505/ac84a4

Martin MA, Sendra OA, Bastos A, Bauer N, Bertram C, Blenckner T, Woodcock J (2021) Ten new insights in climate science 2021: a horizon scan. Glob Sustain 4:e25. https://doi.org/10.1017/sus.2021.25

Mechouar Y, Hovelaque V, Gaigne C (2022) Effect of raw material substitution on the facility location decision under a carbon tax policy. EURO J Transp Logist 11:100061. https://doi.org/10.1016/j.ejtl.2021.100061
Mor S, Madan S, Prasad KD (2021) Artificial intelligence and carbon footprints: roadmap for Indian agriculture. Strateg Chang 30(3):269–280. https://doi.org/10.1002/jsc.2409
Mukhopadhyay K, Prabhu VS, Shrivastava S, Ajatasatru A, Klauer B (2023) Measuring food-energy-water nexus footprint using a systematic input–output approach: a case study of Pune district. Nat Resour Forum 47(4):632–671. (Oxford, UK: Blackwell Publishing Ltd.). https://doi.org/10.1111/1477-8947.12288
O'Sullivan JN (2023) Demographic delusions: world population growth is exceeding most projections and jeopardising scenarios for sustainable futures. World 4(3):545–568. https://doi.org/10.3390/world4030034
Pathak H, Jain N, Bhatia A, Patel J, Aggarwal PK (2010) Carbon footprints of Indian food items. Agr Ecosyst Environ 139(1–2):66–73. https://doi.org/10.1016/j.agee.2010.07.002
Raihan A, Muhtasim DA, Pavel MI, Faruk O, Rahman M (2022) An econometric analysis of the potential emission reduction components in Indonesia. Clean Prod Lett 3:100008. https://doi.org/10.1016/j.clpl.2022.100008
Ramachandra TV (2012) Decentralised carbon footprint analysis for opting climate change mitigation strategies in India. Renew Sustain Energy Rev 16(8):5820–5833. https://doi.org/10.1016/j.rser.2012.05.035
Rinne R, Ilgın HE, Karjalainen M (2022) Comparative study on life-cycle assessment and carbon footprint of hybrid, concrete and timber apartment buildings in Finland. Int J Environ Res Public Health 19(2):774. https://doi.org/10.3390/ijerph19020774
Sah D, Devakumar AS (2018) The carbon footprint of agricultural crop cultivation in India. Carbon Manag 9(3):213–225. https://doi.org/10.1080/17583004.2018.1457908
Sharma A, Pathak S, Borah SB, Adhikary A (2022) Collaboration strategies in buyer-supplier relational (BSR) networks and sustainable firm performance: a trade-off story. Int J Prod Econ 253:108558. https://doi.org/10.1016/j.ijpe.2022.108558
She G (2022) The real effects of mandatory nonfinancial disclosure: evidence from supply chain transparency. Account Rev 97(5):399–425. https://doi.org/10.2308/TAR-2020-0178
Sreedhar S, Jichkar P, Biligiri KP (2016) Investigation of carbon footprints of highway construction materials in India. Transp Res Procedia 17:291–300. https://doi.org/10.1016/j.trpro.2016.11.095
Tassinari F, Kleine-Rueschkamp L, Veneri AP (2024) Life satisfaction along the urban-rural continuum: a global assessment. Appl Econ 1–17. https://doi.org/10.1080/00036846.2024.2303617
Taufique KM, Nielsen KS, Dietz T, Shwom R, Stern PC, Vandenbergh MP (2022) Revisiting the promise of carbon labelling. Nat Clim Chang 12(2):132–140. https://doi.org/10.1038/s41558-021-01271-8
Tetteh EK, Amankwa MO, Yeboah C (2021) Emerging carbon abatement technologies to mitigate energy-carbon footprint-a review. Clean Mater 2:100020. https://doi.org/10.1016/j.clema.2021.100020
Wang J, Zhao C (2023) Reducing carbon footprint in a resilient supply chain: examining the critical influencing factors of process integration. Int J Prod Res 61(18):6197–6214. https://doi.org/10.1080/00207543.2022.2063088
Wang Q, Li S, Li R, Jiang F (2022) Underestimated impact of the COVID-19 on carbon emission reduction in developing countries–a novel assessment based on scenario analysis. Environ Res 204:111990. https://doi.org/10.1016/j.envres.2021.111990
Wei T, Wu J, Chen S (2021) Keeping track of greenhouse gas emission reduction progress and targets in 167 cities worldwide. Front Sustain Cities 3:696381. https://doi.org/10.3389/frsc.2021.696381
Winterich KP, Reczek RW, Makov T (2023) How lack of knowledge on emissions and psychological biases deter consumers from taking effective action to mitigate climate change. J Acad Mark Sci 1–20. https://doi.org/10.1007/s11747-023-00981-z

Xia T, Wang Y, Lv L, Shen L, Cheng TCE (2023) Financing decisions of low-carbon supply chain under chain-to-chain competition. Int J Prod Res 61(18):6153–6176. https://doi.org/10.1080/00207543.2021.2023833

Xue L, Li H, Xu C, Zhao X, Zheng Z, Li Y, Liu W (2022) Impacts of industrial structure adjustment, upgrade and coordination on energy efficiency: empirical research based on the extended STIRPAT model. Energ Strat Rev 43:100911. https://doi.org/10.1016/j.esr.2022.100911

Yaman C (2024) A review on the process of greenhouse gas inventory preparation and proposed mitigation measures for reducing carbon footprint. Gases 4(1):18–40. https://doi.org/10.3390/gases4010002

Yazdanie M, Orehounig K (2021) Advancing urban energy system planning and modeling approaches: gaps and solutions in perspective. Renew Sustain Energy Rev 137:110607. https://doi.org/10.1016/j.rser.2020.110607

Zhang Z, Hu G, Mu X, Kong L (2022) From low carbon to carbon neutrality: a bibliometric analysis of the status, evolution and development trend. J Environ Manage 322:116087. https://doi.org/10.1016/j.jenvman.2022.116087

A Conceptual Study on Utilizing Technology for Attaining Net Zero Through Industry 4.0

Rubee Singh, Walter Leal Filho, Raghuveer Katragadda, and Shahbaz Khan

Abstract Industry 4.0 as a concept which was originated in the continental Europe is on the process or converting into reality for the traditional corporate houses through acceptance and implementation of the advancing technology. The changes and consequences which this adoption has in a number of aspects are certain to promote fare long-lasting effect in the corporate spectrum, and are also noteworthy for the academia. This research chapter's purpose is to analyze the association between the development of products that are more environmentally dangerous and the emergence of Industry 4.0. In the current century's business environment, Industry 4.0. There is no study that positively depicts the environmental sustainability effects of 0 in manufacturing and its spin-off technologies including IoT and CPS. These are actions which result in the release of polluting substances in the environment, the wastage disposal problem, and the exponential use of natural resources, knowledge- and energy-information. Hence, incorporating Industry 4.0 obtain a result of 0 with sustainable development initiatives can enhance environmental sustainability. Thus, this chapter seeks to find out the consequences of Industry 4.0 are estimated to be in changes in the economy/businesses, and thus in the pathway, SCD achieves the net zero target and balances sustainability.

Keywords Industry 4.0 · Net zero · Technology

R. Singh (✉)
Institute of Business Management, GLA University, Mathura, Uttar Pradesh, India
e-mail: dr.rubeerajput@gmail.com

W. L. Filho
Faculty of Life Sciences, Humburg University of Applied Sciences, Humburg, Germany

R. Katragadda
School of Management, V.R. Siddhartha Engineering College (Deemed to Be University), Kanuru, Vijayawada, India

S. Khan
Department of Industrial Engineering, University of Tabuk, Tabuk, Saudi Arabia

R. Singh et al. (eds.), *Zero Carbon Industry, Eco-Innovation and Environmental Sustainability*, World Sustainability Series,
https://doi.org/10.1007/978-3-031-80220-1_4

1 Introduction

The IPCC has raised a globe alarm in 2018 regarding the potential consequences of current trends in climate change. The main sources of greenhouse gas (GH) emissions are economic activities, particularly in industries, transport, and production. The IPCC warns that any continued interference with climate will lead to severe, irreversible impacts, such as rising sea levels, sinking land, shifting agricultural characteristics, lowered productivity, and the spread of epidemics and natural disasters. The alarm is based on extensive research into the factors contributing to climate change. (Hamilton and Tol 2005; Roson and Dominique 2012).

The IPCC and IEA report optimistically suggest that human activities can limit climate change's potential outcomes and provide a better future. To achieve temperatures below 1 °C, proper insulation and ventilation control are necessary. To achieve a carbon neutral society by 2050, a systemic shift in the economy, starting with the industry, which accounts for 24% of carbon emissions, is needed. The IEA reports that the industrial sector contributed 37% of total energy consumption in 2017, with a carbon consumption rating of 80% of global ratings up to 42. Concrete has set a target of reducing energy use by 5% of 2014, while reducing overall emissions of 45 Caucasian people. Key changes for the industry in 2030 include drastic changes in energy use and a focus on reducing carbon emissions.

To address the climate change challenge, transformation of technology at scale is needed, which means that new forms of digital technologies will disrupt most industries. These include the IoT, cloud computing, augmented reality and virtual reality, drones, and other such related technologies. Technologies that continue to make the connection between the Internet and tangible systems open various possibilities for new uses for the applications and future products. The name given to this development in technology is Industry 4.0, commonly referred as smart industry or the 4th industrial revolution. Industry 4.0 main goal is to increase economic performance and maximise industrial outcomes. Fritzche et al. (2018), Stock and Seliger (2016), Ciarli and Savona (2019) studies have attempted to explore the potential of Industry 4.0 in mitigating climate change impacts. However, the results are often mixed, with some arguing for a positive impact and others suggesting it may have negative or no effect. Most articles focus on energy conservation and reduction features of Industry 4.0, excluding potential increases or ideal strategies. This qualitative study aims to explore processes that enhance Industry 4.0 to support sustainability and limit carbon emissions in the global economy. Theoretical analysis of these changes is discussed, and directions for Industry 4.0's progress to zero are presented. The focus is on analyzing industrialization in developing countries and their ability to bridge the gap with developed counterparts. This work is aimed at meeting the goals of the Paris Agreement, which requires reducing the greenhouse effect and maintaining average temperature increases below 2 °C. The impact on socio-economic conditions is not covered in this work.

2 The Need for Zero Carbon Emission (ZCE)

Every year, the emission of Green House Gases (GHGs) as a result of human activities and their discharge to the atmosphere continues to rise. This results in a condition where the levels of these gases in the atmosphere are progressively being on the rise. Since 2010, the global human society has been annually emitting over 300 billion tons of GHGs into the atmosphere. Some of these gases escape to the atmospheric naturally, while the others are likely to stay in the atmosphere for hundreds of years.

The general trend of the increase in the level of anthropogenic greenhouse gas emissions in the atmosphere is described in the scientific literature. Human activities that lead to emission of greenhouse gases are the major cause of increase in global mean surface air temperature both in the past and in the future. According to assessment of IPCC's Fifth Assessment Report of 2014, there is a linear relationship between the volume of CO_2 emissions caused by human activities and increase in the global temperature. To get to the state where no anthropogenic factors are contributing to Global Warming it means getting to a point where anthropogenic GHG emissions are no longer being added to the atmosphere. This state is popularly referred to as the net-zero emissions. As stated by IPCC net zero emissions means that the volume of Green House Gases released into the atmosphere by human activities is countered by the volume of Green House Gases that are being 'drawn down' in the same period.

The global net-zero emissions require the absolute cut of all anthropogenic GHG emissions which is now estimated at 55 $GtCO_2e$ annually. If it is impossible to decrease emissions because of the certain technical or economic factors, it will be necessary to apply an equivalent amount of the CDR to neutralize the emissions. The main goal of climate change mitigation is the total decarbonization of economies and removal of net emissions of greenhouse gases. This is one of the major goals of the measures described in the Paris Agreement which calls to achieve the peak of the greenhouse gas emissions as soon as possible and to balance, in the second half of the century, emissions caused by human activities and emissions removed by natural processes.

3 Corporates and NCZ

The 'United Nations Framework Convention on Climate Change' (UNFCCC) has reported that both state and non-state actors have committed to achieving net-zero carbon emissions by mid-century, accounting for almost half of the global GDP. This commitment is a significant part of the global GDP and a quarter of total CO_2 emissions worldwide. As companies set net-zero targets, it is clear that more actions are being taken to address climate change in the economy. However, the concept of these commitments is ambiguous, with some companies setting targets for further emission reductions and shifting their business models to suit the net-zero economy, while others focus on cutting emissions and increasing offset use. This lack of synchronized net-zero ambitions hinders the evaluation and comparison of targets and impacts by various stakeholders.

The Paris Agreement, a global climate change treaty, has prompted many companies to develop climate change strategies aiming to achieve zero emissions by 2050. The goal is to limit the average global temperature rise to below 2 °C, preferably to 1 °C. This requires governments and businesses to work together to achieve this goal. The Science Based Targets initiative (SBTi) has recently introduced the Net-Zero Standard, a rigorous requirement for setting corporate net-zero targets. This certification ensures that corporations' net-zero goals align with the scientific criteria of the Paris Agreement, a protocol limiting global temperature increase at 1.5 °C. The SBTi is part of the We Mean Business Coalition, a group of seven non-governmental organizations focusing on climate change. The United Nations Global Compact encourages businesses worldwide to adopt sustainable and socially responsible business policies. CDP operates a global reporting system for investors, companies, cities, states, and regions to assess their environmental impact. The World Resources Institute collaborates with stakeholders to address global issues, including climate change, and the World Wide Fund for Nature focuses on environmental protection.

SBTi and the UN Global Compact have partnered to set goals for over 1,000 companies to achieve a 1.5 °C trajectory in corporate climate action. The Business Ambition for 1.5 °C initiative, initiated in 2019, is led by SBTi due to the IPCC's special report calling for a 1.5 °C increase in global temperature. This can be achieved by cutting emissions by at least half by 2030 and reaching net-zero emissions by 2050. Over 1,000 enterprises have agreed to set science-based targets linked to a 1.5 °C trajectory through the Corporate Ambition for 1.5 initiative. The UN Global Compact, a set of principles with over 9,500 business participants, includes over 600 companies who have signed the Climate Pledge, pledging to reach net-zero emissions status by 2050. SBTi is now endorsing companies' net-zero targets starting in 2022.

Organizations that adopt the SBTi's Net-Zero Standard must set science-based intermediate and long-term goals for all aspects of their business, including operational activities, energy and power purchases, and supply chain emissions. This holistic approach ensures the elimination of emissions without excluding any part of a company's emissions. The SBTi's net-zero emissions Standard is a crucial tool in combating climate change, as it defines the principles for establishing evidence-based goals for the near and distant future.

The Net-Zero Standard, a global standard for greenhouse gas emissions reporting, aims to achieve a 90–95% reduction in carbon emissions by 2050. This standard requires business organizations to achieve this by utilizing carbon removal measures, such as direct air capture with geological storage and afforestation. The GHG Protocol, a widely adopted method for measuring greenhouse gas emissions, is used by 92% of Fortune 500 companies. To comply with the Net-Zero Standard, organizations must calculate total emissions in Scopes 1 and 2 and assess Scope 3 categories deemed to be material. Tools like the GHG Protocol Scope 3 Guidance, Scope 3 Evaluator, and CDP Supply Chain Program can assist in this process. Most corporations aim to cut emissions by 90% to reach net zero by 2050, while the power sector aims to achieve this by 2040. The standard is crucial for businesses to meet intermediate and long-term science-based targets, ensuring they meet their emissions reduction goals.

Industry 4.0 It is for this reason that technologies could be said to pose the highest potential of altering the world, positively affecting the lives of people, and creating more opportunities for business and, therefore, sustainable development. These technologies increase productivity and production through the use of new manufacturing strategies and better supply chain management. For instance, organizations that use sensors in their production lines stand to gain a lot on the economic sustainability by managing their inventory and warehousing system. IoT applications enhance supply chain through decreasing stock discrepancies and the amount of time it takes to get products to the market. This helps in the economic viability of Industry 4.0 (Stock and Seliger 2016; García-Muiña et al. 2020; Hawkins 2021). This chapter primarily centers around the subsequent key research objectives: This chapter primarily centers around the subsequent key research objectives:

- Research Objective 1. In this paper, an attempt has been made to briefly discuss about the Industry 4. 0 and its minimum specification.
- Research Objective 2. In this paper the main focus is on the identification of the basic technologies that underpin Industry 4.0 This paper has identified zero strategies that organizations need to adopt to attain **sustainability** and net zero.
- Research Objective 3. Components that will help in identifying and analysing the key application areas of Industry 4.0.

4 Industry 4.0

Industry 4.0 refers to the vast and growing family of technologies, tools, and processes that are used in an integrated manner during the entire product realization system. These technologies facilitate the self-reliant production, integrated management, decentralized control and initiation, and low human interaction (Castelo-Branco et al. 2019). This world of technology is rich in many innovations that have surfaced over the years, and these include the IoTs, CPS, robots, and visualization technologies such as Virtual Reality (VR) and Augmented Reality (AR). Other emerging developments are cloud computing, the integration of block chain, big data analytics, additive manufacturing, and twin technology. A lot of work has been done to develop these technologies in the recent past and these technologies are expected to bring in changes across many industries. There is a need to gain knowledge on these advancements thus the information herein is critical, according to Culot et al. (2020), Liao et al. (2017), Tao et al. (2019). However, gradually integrating the concept of Industry 4. Thus, making Industry 4.0 a reality, manufacturer must consider the shift from the traditional Industry 3.0 technology. Four design principles of Industry 4. No technology should be disregarded as they can greatly help organizations. The leaders need to consider these principles to realize the benefits of using this new production paradigm.

This is according to Hermann et al. (2015) on their analysis of Industry 4. Some key characteristics of 0 include interoperability, virtualization, de-centralization, real time, service oriented, and modularity. Interoperability is one of the most important features of Industry 4. It is zero because it allows systems and devices developed by different companies to inter-operate or work in harmony (Qin et al. 2016). Wang and Xu introduced ICMS in 2013 that is a type of manufacturing system that operates on the cloud computing environment and combines the physical equipment and the software. This system helps in the adaptation by controlling the services and by integrating the resources that leads to the reduction in the cost and improvement in the production and quality of products (Zeid et al. 2019).

This chapter conceptually seeks to argue that decentralization enables timely and flexible decision making giving entities, operational workers, and machines an opportunity to arrive at their own decisions. This is especially the case as the demand for certain items increases in that the management of control systems becomes a rather complex process. Decentralized control systems are important in organizing the complexity of planning, controlling and monitoring fluid forms of organizations (Schumacher and Hummel 2016). Cyber-Physical systems (CPS) produced by various manufacturers are integrated through Interoperability; while virtualization allows CPS to view the physical process as a result of the interaction between machines and surveillance.

Also, it is possible to highlight virtualization, real-time data gathering, and analysis, and introduction of technologies related to Industry 4.0, all of them are being used in the enhancement of decision-making among the organizations. Kamble et al. (2018) focuses on the possibility of virtualization in the improvement of safety and solving of system issues. They also present an ontology based virtual factory model for production planning. Also, they stress on the need for the improvement of the archaic systems and the incorporation of advanced systems for real time handling. Moreover, business organizations tap into big data from various sources including social media, point of sale, and suppliers to support many aspects of the business and its customers.

Modular components are thus systems that give a company the ability to make changes or additions to the production of goods and services through changing or adding modules. This facilitates the consideration of changes in the demand as well as changes in the specific product needs. Modularity enables the modeling of Products, Planning of production and Engineering of production within the manufacturing facilities. Industry 4.0 is based on the provision of services and supports a system where products and services are combined. It allows for faster and more flexible reaction to the changes in the environment of the market. These concepts allow for bridging the systems and the equipment, enable sound decision making, create a digital twin of the real world, acquire real time data, and bridge the physical and cyber space (Carvalho et al. 2018; Hermann et al. 2014; Kamble et al. 2018).

5 Technologies Fundamental for Industry 4.0 Practices

Focus on the processes and methods that are used in the achievement of sustainability 4.0 successfully, such as cloud computing, Internet of Things (IoT), additive manufacturing, big data analytics, cyber-physical systems, artificial intelligence, and many more as well as the related issues. These components have become important in the achievement of the sustainable development 4.0. The advantages of adopting sustainability in an organization include; 4. The eight core competencies include; Real-time computing, Immediate response, Reliability and security, Image processing, Adaptive manufacturing and Digital manufacturing. It, therefore, goes to show that sustainability 4.0 has some positive implications deployment. This helps the reader to consider these strategies in his own business (Ghobakhloo 2020; Jamwal et al. 2021a, b).

Industry 4.0 can be defined as the fourth revolution in the world of Industrialization. 0 Presents many opportunities for companies across the globe to enhance the effectiveness of their production systems and goods. Thus, this makes it possible for firms to play an active role in the enhancement of the sustainable development goals and the cultivation of green business opportunities. Due to the increased population awareness of the impact of pollution and other vices, global warming and its causes, natural disasters, and increased scientific findings over the years regarding the adverse effects of non-sustainable practices, the importance of sustainability has been on the rise. The cut of emissions, the utilization of efficient energy sources is beneficial not only for people and the environment but also for companies. At the present time, the concept of sustainability is highly relevant in industrial processes, which makes it possible to fulfil social and environmental requirements and, at the same time, achieve tangible results such as increased sales and reduced costs (Waibel et al. 2017, Çınar et al. 2020).

Thus, to enhance and modernize the current procedures, new processes must be developed that use fewer hazardous ingredients and produce less pollution in order to be deemed eco-friendly. The application of sophisticated automated machinery has eliminated some of the most dangerous employment sectors and reduced the incidence of employees' accidents (Ramirez-Peña et al. 2019). The application of automation makes the company conserve the environment since there are fewer rejected components since the process is more accurate. There are other advantages of waste reduction not only the environmental aspect; but also it will affect the community in a positive way by decreasing the formation of pollutants that could be a threat to the health of the community. According to Gregori et al. (2017), the fourth industrial revolution affects the environment positively and the people engaged in green businesses. As a result of the COVID-19 outbreak, investors and consumers' focus has shifted towards sustainability and the efficient use of resources. All sectors' leaders should promote the operational efficiency, use the renewable energy sources, adopt a holistic view to their offerings and make sure that supply chain information is easily accessible for the full realization of the potential of Industry 4.0 goals (Jin et al. 2017 and Klymenko et al. 2021).

Manufacturers should review and improve on the energy efficient measures and activities through monitoring and evaluation of energy usage of process units, lines, and equipment and then managing the energy consumption, cost, and emissions. The application of the digital technologies for energy management in combination with the decentralized energy resources implies two-way control and communication, and opens up new possibilities for demand and energy saving.

This underlines the importance of the two in the most recent developments in the management of energy. Industry 4.0 promotes the development of connections that allow for the use of one process's by-product as an asset in another. Sustainability, through Industry 4.0 should be a strategic response that organizations of all sizes can use to create a competitive advantage by designing smart and efficient approaches to production as well as income generation. The fourth industrial revolution aims at reducing or eradicating the wastage in the production and processing through adopting the eco-friendly methods and the new technology that is friendly to the environment.

To date, the focus on the manufacturing sector in competitiveness has been rather shifted to the sustainability of the manufacturing process. The challenge of integrating sustainability into the competitiveness of the manufacturing companies has not been easily solved. The adoption of a feasible and sustainable recycling plan has the possibility of achieving the goal of reducing on waste while at the same time contributing to the enhancement of the environment. It is possible to study and apply efficient ways of modifying the production process in order to isolate reusable, repairable, and recyclable materials. The opportunities for the expansion of manufacturing industries in the developing countries are immense but this has the disadvantage of subjecting the environment to a lot of pollution. Due to the growing concern by the consumers and business people on the negative impacts of pollution, many firms have/are shifting from the conventional methods of production and disposal of wastes. They are not interested in the linear model of consume, dispose, and repeat; instead, they are going for the circular economy. Beside the positive impact on the environment, this approach seems to provide numerous economic opportunities. To reduce the usage of resources in the production cycle and minimize the disposal of items to landfills, one has to adopt the reuse and recycling of resources, which in turn can lead to cutting down on costs and improving the returns.

6 Prominence of Industry 4.0 in Attaining Net Zero Carbon Emissions

In 2019, there was a significant increase in worldwide carbon dioxide emissions, reaching a peak of 36.700 billion tons, primarily due to the expansion of the economy. This was an increase of 60% over the emissions of 1990, at the same time the need for efficient management and reduction of these emissions grew. This is because the carbon dioxide emitted in the course of manufacturing is one the global pollutants of the environment. The following is therefore a plan as to how the sector can take a leading role in solving this problem. Concrete production only accounting for a third of the industry's total output produces around 8% of the yearly CO_2 emissions. The industry stands in a very vulnerable position to address these pollutants and therefore they should have to do it properly.

Many business associations have pledged to reduce emissions to zero in the middle of this century or even earlier. This fits the concept of net-zero, which means that the industry will attempt to eliminate the production of GHG emissions while still allowing for the use of some emissions. To this end, these groups are cooperation and endorsing their collective responsibility to analyze a range of possibilities in order to achieve this goal, including the use of processes which do not release greenhouse gases or to find ways of offsetting emissions within the remit of their activities. This may include the use of sophisticated technologies for the capture and storage of carbon dioxide and other emissions of the greenhouse gases or purchasing carbon credits from organizations that are engaged in the business of carbon capture and storage.

6.1 Exploring Energy Consumption: From Production to Distribution

The process of getting different industries to adopt zero carbon emission will not be an easy one as all the segments of the supply chain will be involved. Such collaboration is essential for companies to claim responsibility for greenhouse emissions that result from different activities within the supply chain and begin to reduce or eliminate such emissions. An area with great opportunity to cut cost is in the energy section, which is a key factor in adding to the greenhouse gases in the supply chain of manufacturing. Energy will always be needed but switching to clean energy sources can indeed help in the process of attaining net-zero carbon emissions.

Of importance, the JIT supply management has been one of the most discussed issues in the industrial production systems in the recent past. However, this pursuit of the zero net balance may affect the typical notion of the most efficient plan of action. For instance, if a manufacturing plant has invested in wind or solar power to produce electricity that is used within the plant, the plant has an option of using all the electricity produced or even supply some of it to the public grid. However, renewable power sources seem to possess a major drawback in that power generation cannot be easily synchronized with the demand. Nevertheless, this is a problem, it also presents an opportunity. Excess energy which cannot be fed back to the grid can be used to decrease the carbon footprint thus being a ray of light in the race to sustainability.

6.2 The Relevance of Cloud Computing

A very large feedback loop that can exploit the computational resources of cloud and edge computing is necessary for the efficient control of the so-called controlling systems that are critical to the functioning of the system. Thus, cloud servers allow AI models to adapt to changes in power consumption and respond to various conditions in the environment. Sometimes, the temperature requirement of Carbon Intensive processes may reduce the rate at which a system can contribute to a reduction of global emissions. However, this underpins the need to always assess the energy blend that is being used in the society. This monitoring responsibility is quite useful for the strategic planning of carbon credits and carbon capture solutions. Thus, it enables industrial control integrators and operators to monitor their processes in a preventive manner.

6.3 Significance of Process Control in Applications

This is crucial in order to increase efficiency, guarantee a good standard of the final product and reduce on wastage. Applying process control can result in enhancing the productivity and reducing the costs. Such processes need to be controlled in order to reduce carbon dioxide emissions in areas like cement production, for instance. Cement production process is responsible for emitting various greenhouse gases of which the first stage chemical reaction accounts for almost half of the emissions. Global cement production has tried to improve energy efficiency in its operations, however, according to the International Energy Agency's 2021 report, some cement manufacturers in specific areas still had work to do. The issue as identified by researchers is that there is need to improve the control of reaction conditions in the production of some major cement products. Thus, improving process control, it is possible to decrease carbon dioxide emissions, which opens up a large potential for faster net-zero emissions goals.

"It is possible to state that the effective process regulation can lead to the increase of the overall energy consumption efficiency in numerous other industries at different levels. This can help in avoiding generation of unwanted heat and other unwanted by-products if the process is better controlled. It is possible to reduce energy consumption on transportation if material movements are well predicted. The improvement of controls is actually limited by the actuators and motors used individually. Motor systems can use up to 70% of the total energy consumption in a manufacturing plant. Consequently, conventional applications of rather ineffective asynchronous AC motors have been expected in industrial production processes because they are cheaper and easier to maintain as compared to synchronous motors."

6.4 Communication Process

Understanding the right time and the right way to engage the use of machinery in the process of work is quite important towards ensuring that less energy and materials are used in the production process. Industry 4. None of the technologies are useful in this regard. An important factor of the Industry 4.0 is Thus, 0 architecture involves both short and long-range communications to enable information exchange between the local control systems. These systems can assure that conveyor belts are in use only at the time when the product has to be transferred from one location to another. Machine tools can be turned off when a component is not to be processed and turned on again when a new component is to be introduced to your production cell. Sensors and distributed computing platforms are of great importance in the acquisition of data from various zones of the production system and decision making. Such decisions may be made with the help of the remote servers, relying on the platforms' awareness of the environment.

There is the increasing use of wireless protocols such as 'Wireless Fidelity', Bluetooth, and others and the 'Long Range Wide Area Network facilitates' the installation of sensors at strategic points, thus improving the instrumentation of present machine tools and their management systems. Thus, the integration of these sensors and cloud support can significantly change the control systems to some extent. The control systems of PLCs that directs each machine tool can remain in utilization for a longer time. However, industrial computers can be incorporated in the DIN rail to enable them fit well in the factory settings hence increasing efficiency. For better control algorithms, better PLCs like those made by Industrials Shields and Kunbus can be used as upgrades to the current PLCs to enhance their performance. Edge servers are capable of Ethernet and other high-speed connections to gather data from several PLCs and control systems. The information can then be used in complex models to ensure proper co-ordination all over the entire factory floor. The systems can also be switched on and off to avoid power extravagant use.

6.5 The Critical Component of Industry 4.0: Smart Sensors

Combining sensors with monitoring systems that are built to the highest standards, makes it possible to prevent inefficiency of industrial equipment and reduce waste. Through frequent checks and observations of the operational parameters like temperature, pressure, and vibration among others, it is easy to detect any variations from the normal behavior during the normal check-ups and tests. Proper management of the company's schedules for maintenance and repairs will assist in avoiding costly processes such as scrapping and reconditioning which are hazardous to the environment and the business.

Besides, the manufacturers can also use other maintenance strategies like the predictive maintenance where the performance of the machine tools and other mechanical systems are managed in a more proactive manner. Through the analysis of data, it is possible to plan maintenance works to coincide with other environmental goals for instance; low carbon maintenance. Preparations for maintenance were made in the past mainly with regard to operational practices and schedules regardless of their effect on the environment. But with the help of analysis programs and AI models, it is possible to adjust the schedule of maintenance works to the time when the consumption of low-carbon energy sources is most effective which impacts positively on the cost and the environmental footprint of the maintenance process. These AI models that are capable of making decisions and defining optimal maintenance schedules based on current and past events assure the system's effectiveness.

6.6 Integration

Finally, many experts have agreed that changing or updating only one thing will not greatly help in decreasing the greenhouse gas emission from the industrial sector. Several measures can be put in place by manufacturers to ensure that there is reduced carbon emissions and energy utilization especially in the course of implementing Industry 4.0 principles. All these strategies are vital in the overall process of trying to attain the desired net-zero operation. An example of this can be 'Farnell', a distributor of specialist electronic components who are aiding this transition. Thus, by partnering with companies that provide technical support as a re-seller opportunity, companies are able to assess the existing investment and gain guidance on the implementation of the various kinds of technologies such as the advanced PLCs, sensor modules, industrial computers, and many other subsystems. Worth mentioning is that these technologies should be applied to prepare an effective strategy for reaching the net-zero emissions.

7 Foundations of Industry 4.0

Therefore, the strategies for Industry 4. No developments have offered manufacturers a competitive advantage as much as flexible and effective automation powered by data. These include; competitiveness, performance, flexibility, resilience, profitability, and customer service. Besides, Smart Factory technology shows the importance of Industry 4. Impact no negative effects, enabling manufacturers to create better products with higher efficiency (Müller et al. 2018; Jamwal et al. 2021a, b). The Industry 4.0 vision is the integration of the plants and the devices and goes beyond the factories, bringing data to all and incorporating enhanced application software. The significance of capacity building and cultural change of the employees through training in analytics and new technologies is well understood from the fact that data has become a critical asset for organisations (Miśkiewicz and Wolniak 2020; Vrchota et al. 2020).

The current evolution commonly known as Industry 4. The term 4.0 or the fourth industrial revolution is changing industrial processes and defining the future of business. Exploring the application of the current advanced technologies such as IoT, AI, and big data analytics (Atzeni et al. 2023). Data-Driven Insights through Industrial Retrofitting: An Anonymized Data Set and the Machine Learning Applications, also, the use of machine learning has the potential of providing several opportunities for the enhancement of sustainability as well as realization of the net-zero emissions target. This revolution is expected to revolutionize the business world in all industries, thus leading to organization improvement and better future.

8 Challenges and Opportunities

The achievement of a NZE future is largely dependent on technological advancement which needs to be stepped up. This entails large scale use of sustainable energy technologies, stressing on CCS technologies, and focusing on the financial aspect. Some initial costs are needed to shift to an NZE, but its advantages include lower environmental and health costs and the generation of new jobs. Sectors that produce a lot of GHGs like steel and cement must change to meet the net-zero target. This calls for spending on the green options and the enhancement of innovation in the production of eco-friendly products. Realizing net-zero emissions requires changes in people's behavior and decisions and upgrades or replacements of the existing systems. Long term and stable policies and international cooperation and compliance with mitigation measures are needed (Shafik 2024 and Singh et al. 2024).

NZE initiatives come with many benefits like emission reduction and better climate quality. This shift results in the creation of business opportunities, technological advancements, reduced emissions, higher energy security, and increased preparedness to adverse impacts of climate change. Also, going to net-zero emissions also supports the move of circular economy, which focuses on minimizing waste and reuse of materials in a way that does not cause harm to the environment (Shafik 2024).

9 The Bottom Impact

The new industrial revolution that focuses on the development in the environmental, social and economic aspects of the business plays an important role in enhancing sustainability (Braccini and Margherita 2018). It does not involve the utilization of hazardous substances and energy, maximized the reuse of materials and services, and practices the circular economy system (Braccini and Margherita 2018; Glavič and Lukman 2007; Ruggieri et al. 2016). But its impact goes further: has led to the emergence of what is known as industry 4.0 improves energy efficiency, reduces carbon footprint and offers possibilities for sustainable production. Industry 4.0 is characterized by the application of the following technology amongst others. One such technology that is very relevant in this area is additive manufacturing which reduces on carbon footprint through reducing waste and logistics (Oettmeier and Hofmann 2017).

While assessing the social aspect, it is crucial to focus on the adequate working conditions of employees (Braccini and Margherita 2018) and the development of teamwork and the enhancement of people's potential (Herrmann et al. 2014). Industry 4.0 offers advantages to employees in various settings and, hence, suggests that workers must be introduced to new technologies through training (Müller et al. 2018). The main competencies that are crucial for the Industry 4.0 context defines the personal, social/interpersonal, action-related, and domain-related competencies (Erol et al. 2016). However, it has also been argued that this innovation has the potential of creating job losses and increased automation which has adverse effects on the society. It also presents possible opportunities for technologically related jobs in the future as well. Alternatively, the economic view is related to the financial and monetary condition of the organization. The arrival of Industry 4.0 presents numerous economic advantages for businesses: 0 presents numerous economic advantages for businesses:

- CPS also results in cutting down the expenses on materials, improves the production processes and productivity of the entire manufacturing system.
- Industry 4. This minimises the time that machines and changeovers are down hence reducing the costs and increasing the efficiency in production.
- Industry 4.0 improves the labor productivity by integrating the concept of human–robot coexistence.
- It helps in inventory management, demand forecasting, and quality of the product.
- At the same time, the application of advanced analytics decreases the amount of waste (Hasan et al. 2021) and increases the speed of the R&D processes.
- New services give an option of fixing the items and, therefore, using them for a longer time.

9.1 Conclusion

This chapter presents an overview of the innovations which have taken place and the necessity to focus on the climate change issue. The chapter also focuses on critical needs of Industry 4.0 as it aims at supporting the decarbonization of the society and numerous endeavors in this regard. This chapter stipulates the main technologies of Industry 4.0. It also focuses on the relevance of these diseases to climate change, and the capability of the diseases to enhance growth that is sustainable. About the goals stated in the Paris Agreement, climate change challenges, and sustainable development it is crucial to adopt technology of Industry 4.0 and move towards the opposite of the conventional behaviour of organizational entities. Thus, the circular economy that implies the use of additive manufacturing and IoT will reduce material waste and bring about economical, ecological and social value. It shall help in transforming the garbage into valuable commodities. This paper also sought to establish how the concept of sustainability can be integrated with the principles of Industry 4.0. This paper has identified that the implementation of 0 will lead to better resource management and reduction in wastes within the closed loop supply chain. Industry 4. Of course, the SDG 0 opens the chances to fight climate change, but the least developed nations need the technology transfer from the developed countries to ensure the sustainable development and decarbonisation.

In the recent pandemic, less human contact and measures that involved social distancing saw the global CO_2 emissions reduced by half which is good for the environment. Industry 4. The management of vital industries and farming using a limited number of people could be possible with the adoption of Industry 4.0 technologies post-pandemic. Also, the shift to Industry 4. However, where we stand in the fight against climate change with the use of 0 has some limitations that have to be considered. Data centers are known to release the same amount of CO_2 into the atmosphere as the whole aviation industry. Therefore, cloud computing services must change their energy source to renewable energy. Many large corporations have confirmed their commitment to the process of achieving a state when the carbon output of the organization would be fully balanced with the reduction and when the usage of electricity in hosting facilities would be minimal. Moreover, it is equally important to accurately evaluate the socio-political implications of this shift as well as the possibilities of dramatic unemployment rates. However, it is crucial that we establish good governance and ethical frameworks for the fourth industrial revolution to ensure that the new technologies do not harm people's well-being and, on the contrary, help build better societies.

References

Atzeni D, Ramjattan R, Figliè R, Baldi G, Mazzei D (2023) Data-driven insights through industrial retrofitting: an anonymized dataset with machine learning use cases. Sensors 23(13):6078. https://doi.org/10.3390/s23136078

Braccini A, Margherita E (2018) Exploring organizational sustainability of Industry 4.0 under the triple bottom line: the case of a manufacturing company. Sustainability 11(1):36. https://doi.org/10.3390/su11010036

Carvalho N, Chaim O, Cazarini E, Gerolamo M (2018) Manufacturing in the fourth industrial revolution: a positive prospect in sustainable manufacturing. Procedia Manuf 21:671–678. https://doi.org/10.1016/j.promfg.2018.02.170

Castelo-Branco I, Cruz-Jesus F, Oliveira T (2019) Assessing Industry 4.0 readiness in manufacturing: evidence for the European Union. Comput Ind 107:22–32. https://doi.org/10.1016/j.compind.2019.01.007

Ciarli T, Savona M (2019) Modelling the evolution of economic structure and climate change: a review. Ecol Econ 158:51–64. https://doi.org/10.1016/j.ecolecon.2018.12.008

Çınar ZM, Nuhu AA, Zeeshan Q, Korhan O, Asmael M, Safaei B (2020) Machine learning in predictive maintenance towards sustainable smart manufacturing in Industry 4.0. Sustainability 12(19):8211. https://doi.org/10.3390/su12198211

Culot G, Nassimbeni G, Orzes G, Sartor M (2020) Behind the definition of Industry 4.0: analysis and open questions. Int J Prod Econ 226:107617. https://doi.org/10.1016/j.ijpe.2020.107617
Erol S, Jäger A, Hold P, Ott K, Sihn W (2016) Tangible Industry 4.0: a scenario-based approach to learning for the future of production. Procedia CIRP 54:13–18. https://doi.org/10.1016/j.procir.2016.03.162
Fritsche I, Barth M, Jugert P, Masson T, Reese G (2018) A social identity model of pro-environmental action (SIMPEA). Psychol Rev 125(2):245–269. https://doi.org/10.1037/rev0000090
García-Muiña FE, Medina-Salgado MS, Ferrari AM, Cucchi M (2020) Sustainability transition in Industry 4.0 and smart manufacturing with the triple-layered business model canvas. Sustainability 12(6):2364. https://doi.org/10.3390/su12062364
Ghobakhloo M (2020) Industry 4.0, digitization, and opportunities for sustainability. J Clean Prod 252:119869. https://doi.org/10.1016/j.jclepro.2019.119869
Glavič P, Lukman R (2007) Review of sustainability terms and their definitions. J Clean Prod 15(18):1875–1885. https://doi.org/10.1016/j.jclepro.2006.12.006
Gregori F, Papetti A, Pandolfi M, Peruzzini M, Germani M (2017) Digital manufacturing systems: a framework to improve social sustainability of a production site. Procedia CIRP 63:436–442. https://doi.org/10.1016/j.procir.2017.03.113
Hamilton JM, Tol RSJ (2007) The impact of climate change on tourism in Germany, the UK and Ireland: a simulation study. Reg Environ Change 7(3):161–172. https://doi.org/10.1007/s10113-007-0036-2
Hasan MA, Raghuveer K, Pandey PS, Kumar A, Bora A, Jose D, Kshirsagar PR, Hung BT, Chakrabarti P, Khanapurkar MM (2021) Internet of Things and its application in Industry 4.0 for smart waste management. J Environ Prot Ecol 22(6):2368–2378
Hawkins M (2021) Cyber-physical production networks, Internet of Things-enabled sustainability, and smart factory performance in Industry 4.0-based manufacturing systems. Econ Manag Financ Mark 16(2):73. https://doi.org/10.22381/emfm16220214
Hermann M, Pentek T, Otto B (2015) Design principles for Industrie 4.0 scenarios: a literature review. https://doi.org/10.13140/RG.2.2.29269.22248.
Herrmann L, Chotte JL, Thuita M, Lesueur D (2014) Effects of cropping systems, maize residues application and N fertilization on promiscuous soybean yields and diversity of native rhizobia in Central Kenya. Pedobiologia 57(2):75–85. https://doi.org/10.1016/j.pedobi.2013.12.004
Jamwal A, Agrawal R, Sharma M, Kumar V, Kumar S (2021a) Developing a sustainability framework for Industry 4.0. Procedia CIRP 98:430–435. https://doi.org/10.1016/j.procir.2021.01.129
Jamwal A, Agrawal R, Sharma M, Giallanza A (2021b) Industry 4.0 technologies for manufacturing sustainability: a systematic review and future research directions. Appl Sci 11(12):5725. https://doi.org/10.3390/app11125725
Jin M, Tang R, Ji Y, Liu F, Gao L, Huisingh D (2017) Impact of advanced manufacturing on sustainability: an overview of the special volume on advanced manufacturing for sustainability and low fossil carbon emissions. J Clean Prod 161:69–74. https://doi.org/10.1016/j.jclepro.2017.05.101
Klymenko O, Halse LL, Jæger B (2021) The enabling role of digital technologies in sustainability accounting: findings from Norwegian manufacturing companies. Systems 9(2):33. https://doi.org/10.3390/systems9020033
Liao SH, Chen CC, Hu DC, Chung YC, Liu CL (2017) Assessing the influence of leadership style, organizational learning and organizational innovation. Leadersh Org Dev J 38(5):590–609. https://doi.org/10.1108/lodj-11-2015-0261
Miśkiewicz R, Wolniak R (2020) Practical application of the Industry 4.0 concept in a steel company. Sustainability 12(14):5776. https://doi.org/10.3390/su12145776
Müller JM, Kiel D, Voigt KI (2018) What drives the implementation of Industry 4.0? The role of opportunities and challenges in the context of sustainability. Sustainability 10(1):247. https://doi.org/10.3390/su10010247

Oettmeier K, Hofmann E (2016) Additive manufacturing technology adoption: an empirical analysis of general and supply chain-related determinants. J Bus Econ (Zeitschrift Für Betriebswirtschaft) 87(1):97–124. https://doi.org/10.1007/s11573-016-0806-8

Qin J, Liu Y, Grosvenor R (2016) A categorical framework of manufacturing for Industry 4.0 and beyond. Procedia CIRP 52:173–178. https://doi.org/10.1016/j.procir.2016.08.005

Ramirez-Peña M, Sotano AJS, Pérez-Fernandez V, Abad FJ, Batista M (2020) Achieving a sustainable shipbuilding supply chain under I4.0 perspective. J Clean Prod 244:118789. https://doi.org/10.1016/j.jclepro.2019.118789

Roson R, Dominique VDM (2010) Climate change and economic growth: impacts and interactions. Soc Sci Res Netw. https://doi.org/10.2139/ssrn.1594708

Ruggieri A, Braccini A, Poponi S, Mosconi E (2016) A meta-model of inter-organisational cooperation for the transition to a circular economy. Sustainability 8(11):1153. https://doi.org/10.3390/su8111153

Schuhmacher J, Hummel V (2016) Decentralized control of logistic processes in cyber-physical production systems at the example of ESB logistics learning factory. Procedia CIRP 54:19–24. https://doi.org/10.1016/j.procir.2016.04.095

Shafik W (2024) Industry 4.0 technologies' opportunities and challenges for realising net-zero economy. In: CSR, sustainability, ethics & governance, pp 19–41. https://doi.org/10.1007/978-3-031-55779-8_2

Singh R, Khan S, Kumar A, Luthra S, Chokshi H (2024) Net zero economy, corporate social responsibility and sustainable value creation. In: CSR, sustainability, ethics & governance. https://doi.org/10.1007/978-3-031-55779-8

Stock T, Seliger G (2016) Opportunities of sustainable manufacturing in Industry 4.0. Procedia CIRP 40:536–541. https://doi.org/10.1016/j.procir.2016.01.129

Tao F, Zhang H, Liu A, Nee AYC (2019) Digital Twin in Industry: State-of-the-Art. IEEE Trans Industr Inf 15(4):2405–2415. https://doi.org/10.1109/tii.2018.2873186

Vrchota J, Pech M, Rolínek L, Bednář J (2020) Sustainability outcomes of green processes in relation to Industry 4.0 in manufacturing: systematic review. Sustainability 12(15):5968. https://doi.org/10.3390/su12155968

Waibel M, Steenkamp L, Moloko N, Oosthuizen G (2017) Investigating the effects of smart production systems on sustainability elements. Procedia Manuf 8:731–737. https://doi.org/10.1016/j.promfg.2017.02.094

Wang XV, Xu XW (2013) ICMS: a cloud-based manufacturing system. In: Springer series in advanced manufacturing, pp 1–22. https://doi.org/10.1007/978-1-4471-4935-4_1

Zeid A, Sundaram S, Moghaddam M, Kamarthi S, Marion T (2019) Interoperability in smart manufacturing: research challenges. Machines 7(2):21. https://doi.org/10.3390/machines7020021

Innovative Solutions: Zero Carbon Production Practices for a Sustainable Future

Samriti Devi and Kawal Nain Singh

Abstract The chapter dive into the critical transition towards sustainable industrial practices which are aimed at minimizing the climate change occurring because of industrial practices. As the world have witnessed the environmental repercussions of previous industrial revolutions this new wave of industrialisation focusses primarily on zero carbon production as the focus strategy. The chapter starts with the definition of zero carbon production and underscore its significance in contemporary environmental disclosures. As a part of the analysis, it traces the historical evolution of industrial practices highlighting the urgent requirement for a shift towards sustainability in the industrial practices. Exploring the current state of zero carbon initiative the chapter highlights Global advancement and identifies leading Nations and industries progressing in these in initiative. Further it highlights that how the government and the national and international authorities are approaching towards zero carbon industry and repairing policies for zero carbon production. The key concepts which are helping to transform the industry are renewable energy carbon capture and green manufacturing. Along with their chapter address the economic and social impact of zero carbon production emphasizing its potential to create job foster skill development and enhance community well being and review the existing policy and regulatory framework discussing the future direction and necessary changes. All the barriers and challenges including the technical and financial hurdles are critically analysed to provide a comprehensive understanding of obstacles and future trends and predictions contemplating the long term economic and environmental impacts of zero carbon industry.

Keywords Zero carbon · Policies · Framework · Renewable energy · Hurdles · Trends · Sustainability

S. Devi (✉)
CT University, Ludhiana, India
e-mail: devisamriti014@gmail.com

K. N. Singh
Department of Management, Chandigarh School of Business, CGC Jhanjeri, Chandigarh, India

R. Singh et al. (eds.), *Zero Carbon Industry, Eco-Innovation and Environmental Sustainability*, World Sustainability Series,
https://doi.org/10.1007/978-3-031-80220-1_5

1 Introduction

Zero carbon production refers to the process of manufacturing goods and services without emitting any carbon dioxide or other greenhouse gases into the atmosphere (Samadzadegan et al. 2021). This concept was established with the principle of achieving a balance between the amount of greenhouse gases emitted from the industries and the amount removed from the atmosphere. The day today industrial practices usually lead to large amount of carbon dioxide emitted in the environment leading to environmental damage. In practice zero carbon production involve the adoption of clean and renewable energy resources such as wind solar and hydroelectric power as well as the implementation of advanced technology and innovative processes that minimise or eliminate carbon emission at every stage of production (Zhao et al. 2023). Zero carbon production involve implementing such technology is with the help of the company can operate at the same efficiency but by reducing the amount of carbon dioxide produced from those operation. The concept of zero carbon production also encompasses strategies like carbon capture and storage which capture carbon dioxide emission from industrial processes and store them underground and carbon of setting where companies invest in environmental project that reduce carbon emission exhibit to compensate for their own emission. This concept can be considered as indirect way of reducing the carbon production from the industrial operation where from the daily operation the company will EMIT the same amount of carbon dioxide but through its Innovation and investment it can reduce the carbon emission from other industries so that its own carbon emission can be compensated.

1.1 Importance of Zero Carbon Production in the Context of Climate Change

In the context of climate change the importance of zero carbon production cannot be overstated because climate change is driven by the increase in greenhouse gases in the atmosphere due to human activities which possess and existential set to ecosystem human health and economies worldwide. One of the largest sources of greenhouse gas emission contributing significantly to the global warming is the industrial activities which include production of various boards and services on day today basis (Zhang et al. 2015). The transition to zero carbon production is essential to mitigate the adverse impact of climate change. As it is stipulated by the Paris agreement reducing the industrial emission is crucial for limiting Global temperature rise to below 1.5 °C above preindustrial levels (Zhang et al. 2016). The catastrophic climate impacts can be reduced and prevented by the achievement of this target. The impact on the climate can be extreme whether events sea level rise loss of biodiversity and description of food and water supplies. Moreover, zero carbon production is vital for ensuring long term economic resilience and sustainability and bear reducing dependence on fossil fuels and enhancing energy efficiency industry can lower operational cost, improve the energy security, and foster Innovation and competitiveness in the market.

Another major importance of the zero-carbon production is that it can create new economic opportunities such as green jobs and industries thereby contributing to economic growth and social development (Wu et al. 2014). When new source of energy is being used then individuals with more skills in this area can get employment in the industry and contribute to the success of these innovation. For managing the green sources new departments can be created and further Innovation and research department can employ new employees with innovative ideas to develop modern technology which can contribute to the process of zero carbon production.

Also, the zero carbon production technique address the growing consumer demand for environmentally responsible products and practices which can help the business to enhance its brand reputation and Market positioning. There is increasing awareness in the consumer group about the environmentally sustainable practices which should be used by the companies to run their operation and by following this stand the company can get the attraction of customer group.

1.2 Overview of the New Wave of Industrialisation

The new wave of industrialisation often referred to as the industry 4.0 is characterized by the integration of digital technologies, automation, and sustainability into the industrial processes. This Era marks a significant departure from the traditional market practices which were previously driven by mechanization maths production and fossil fuel consumption a new wave of industrialisation which is defined by several key trends and innovation that collectively aim to enhance productivity efficiency and sustainability (Samadzadegan et al. 2021). The industry 4.0 is based on the concept of using less resources to produce more outcome and using such process and technique through which the negative impact on the environment can be reduced. One of the key aspects of the new industrial revolution is the adoption of smart manufacturing technology such as internet of things artificial intelligence and Big Data analytics where this technology is enablingreal time monitoring control and optimisation of industrial processes leading to significant improvement in operation and resource utilization by the company (Wimbadi and Djalante 2020).

With the help of advanced analytics, the company can use the data generated from daily operations to get valuable insights into performance maintenance needs and energy consumption. With the help of the artificial intelligence in algorithms the company can leverage the data to predict equipment failure and optimum production schedule and the advantage of the opportunities in the market in terms of the innovative technology and innovation can also be taken. Another key aspect of the new wave of industrialisation is the emphasis on sustainable production practices where the goal is to develop and adopt clean technology such as renewable energy and energy efficient equipment and low carbon materials (Wang et al. 2022). The new industrial revolution focusses on the combination of technology and sustainable practices like the use of renewable sources and new equipment through which less energy can be consumed for the same level of operation. Another major focus is on the promotion of circular economy principle where it AIIMS to minimize waste and maximize the use of resources through recycling reusing and re purposing material.

This involve restructuring the waste into the resources. This practice of circular economy helps significantly to reduce the environmental footprint of industrial activities because more value can be generated from waste also (Omer 2008). Significant reason for the promotion of circular economy and sustainable resources is the growing focus on corporate social responsibility and environmental social and governance criteria which highlights duties of the companies to return the favour to the society. With the help of these principles and practices focused upon in the industrial practices the company now integrate sustainability into their business strategy and operations not only to comply with regulatory requirement but also to meet the expectation of investor customers and other stakeholder. This has now increased the transparency in the business operation and accountability of the businesses in their reporting to manage the environmental impact and to as well as promote proactive efforts to address social and environmental challenges.

Importantly it can be said that the new wave of industrialisation is underpinning collaborative approach which involve partnership between Government industry academy and civil society where government player crucial role in setting the policy and regulatory framework which incentives the transition to zero carbon production.

2 Historical Context

2.1 Evolution of Industrialisation

Industrialisation has been a defining force in shipping modern civilization transforming economy societies and environment. The history of industrialisation went back to 18th century from agrarian economy to those dominated by industry and manufacturing. In every new wave of industrialisation there was one common characteristics of production capacity enhancement and significant social and economic change in the civilization due to adoption of innovative technology and processes to execute the same operation. The first industrial revolution occurring from the late 18th century to mid 19th century originated in Britain and was marked by the transition from hand production methods to machine. The previous process which was done by the labourers and was manual were now started to be performed with the help of machines where a single human can operate multiple machines and can achieve higher efficiency. Some of the key innovation included the steam engine mechanised spinning and weaving machine and advancement in iron production which drastically increase the productivity and efficiency leading to rise in factory and mass production of goods and services. During this phase of industrialisation people started moving from rural areas to urban areas in the search of employment opportunities in the new industrial centres.

The second revolution was built upon the advancement of the first wave and introduced innovative technology is that further accelerated industrial growth (Wang et al. 2022). This era of industrialisation is also known as technology revolution where the industry witness to the advent of electricity the internal combustion Indian and innovation in steel production which led to advancement and development of new industry such as automotive chemical and electrical industries and various existing industries also revolutionised. Business is started to establish their large-scale manufacturing and assembly lines and one of the major examples can be seen by Henry force introduction of mass production technique in the automobile industry.

The third industrial revolution was known to as the digital revolution which begin in the mid-20thcentury, and it is characterized by the shift from mechanical and analogue electronic technology to digital electronics. This industry revolution is known for the development of computers internet and advanced telecommunications which revolutionised the information processing storage and communication (Vimal et al. 2022). Automation and robotics become integral part of manufacturing processes which enhanced precision and efficiency in the industrial processes. Major development which can be seen in this industrial revolution was globalisation where the industrial do not remained limited to a single region but started to expand where operation International also beyond their domestic boundaries which was facilitated by the advancement in transportation and the communication technologies.

2.2 Previous Waves of Industrialisation and Their Environment Impact

The previous waves of industrialization had a significant impact on the economic and social life of the individual and it brought several benefits. All the industrial waves lead to significant amount of growth, technology, and business which provided employment opportunities and opportunity for development to the economy and several other boost to the existing Innovation and development but and the same time there was significant environmental impact of all the industrial revolutions. The first industrial revolution marked the beginning of large-scale exploitation of fossil fuels particularly coal with power steam engine and industrial machinery and the widespread use of coal resulted in substantial air pollution characterised by six mog and suit that enveloped industrial cities. The rapid urbanization and the industrial growth also led to various other environmental damages which include deforestation habitat destruction and water pollution as factory is discharged untreated waste into the rivers and streams (Mulugetta and Urban 2010).

The environmental consequences of the second industrial revolution were even more pronounced. The increased use of fossil fuels such as oil and natural gas for energy and transportation significantly elevated greenhouse gas emission which further lead to deterioration of environment. Industrial activities result in enormous quantities of pollutant released into the environment including Sulphur dioxide nitrogen oxide and particulate matters contributing to acid rain smog and respiratory disease. The intensive exploitation of natural resources including minerals and metals lead to white spread land degradation soil erosion and loss of biodiversity. More and more land were converted from forest areas into industries in order to serve the rising demand for the industrial produce. As a result of this Era also witness the emergence of large-scale chemical production resulting in release of hazardous substances and contamination of air water and soil.

Digital revolution despite its advancement in technology and efficiency was not able to prevent itself from damaging the environment. The production of electronics and extraction of rare Earth metals required for manufacturing resulted in significant environmental degradation and pollution (Dominkovic et al. 2016). Also, considerable number of digital resources were consumed it increased the demand for energy consumption and electronic waste also skyrocket in this period of industrial revolution. Additionally, the globalise nature of production and supply chain led to increase the carbon emission from transportation and logistic activities from various routes of road transportation and water transportation. The cumulative effect of these industrial activities has been the significant increase in greenhouse gas concentration in the environment which further contributed to global warming and climate change.

2.3 *Shifting Focus Towards Sustainable Practices*

There has been a growing shift towards sustainable practices in the recent decades after recognising the detrimental impact of industrialization on the environment. The transition from the traditional industrial practices where high amount of fossil fuels and other energy resources were utilised to a more sustainable practice where more of the renewable energy resources and other elements are used to spend the energy requirement of the industry is driven by the urgent need to mitigate climate change and conserve natural resources and ensure long term sustainable economic and social well being. Earlier the entire focus was on the development of the industrial practices, but the focus has now shifted, and the major aim nowadays is balance economy growth with environment protection and social equality through which we can faster A resilience and inclusive industrial future.

One of the key strategies which is being utilised in the process of shift to the renewable and sustainable practices is the use of clean and renewable energy resources and transition from fossil fuels to natural resources like solar wind and hydroelectric power which is essential for reducing greenhouse gas emission and mitigating the climate change. The renewable energy resources technology has advanced significantly becoming more efficient and cost effective and along with this government industry and community is also increasingly investing in the renewable energy projects including large scale solar farm wind turbines and hydro power plants.

Traditionally the integration of smart grades and energy storage system enhance the reliability and stability of renewable energy supply.

Moreover the concept of circular economy is Central to sustainable industrialisation where unlike traditional economy where the waste which was generated but disposed of into the nature is now shifted to a more sustainable practice where the waste generated is further considered as a resource and utilised in another industry so that it can be converted into value through further utilisation. This approach reduces the waste and conserve resource and minimize the environmental impact and it is because of these industries are increasingly adopting circular economy principle by designing product durability replicability and recyclability. For example, companies are developing modular products that can be easily disassembled and repaired extending their lifespan and reduce waste and additionally some of the industries are implementing closed loop system where waste material from one process can be utilised as a resource in another which create a self-sustaining cycle. Sustainable industrial practices also emphasize the development and adoption of green technology and innovation. Various industries including the building materials petroleum-based products and plastic industry also have started to utilise green technology in the production of their product and services which help them to reduce the environmental footprint of construction and building operation. Advancement in green chemistry enable the production of chemicals and material using environmentally friendly process and renewable feed stocks. Biodegradable plastics produce from renewable sources such as corn-starch and sugarcane offer and alternative to traditional petroleum-based plastic.

The evolution of industrialization has brought significant economic and social benefit, but it also resulted in profound environmental impact. The definition of environmental impacts has led to the growing shift toward sustainable industrial practices which aims to balance economic growth with environmental protection and social equity fostering an inclusive industrial future.

3 Current Stage of Zero Carbon Production

3.1 Current Global Status of Zero Carbon Production

The Global post towards zero carbon production has gained significant momentum in the recent year due to the growing recognition of economic and social benefits of sustainable industrial practices. Carbon production aims to eliminate net greenhouse gas emission from industrial activities is now a key component of climate strategies worldwide where countries industries and companies are increasingly committing ambitious target and implementing innovative solutions to achieve zero carbon production. These commitments are reflected in the national policies regulatory framework and the investment strategies designed to promote the transition to zero carbon production. Various authorities and even at the lower scale most corporate bodies are focusing extensively on the shift towards a zero-carbon production.

Despite the efforts from various macro and micro authorities the status of zero carbon production varies significantly across reasons and sectors where developed countries are region such as European Union States Japan are at the four fronts of zero carbon initiative which is given by strong regulatory framework substantial financial resources and advancement in the technology. All these regions are investing heavily in renewable energy, energy efficiency and sustainable industrial practices and due to the heavy investment there witnessing a gradual decoupling of economic growth from greenhouse gas emission. This contrast in the developed and developing country is happening due to Limited financial resources technology constraints and competing development priorities. Due to lack of proper financial support from the authorities and meeting other priorities like improving the standard of living and investment in other areas of the economy the developing countries are lacking behind in the process of promoting zero carbon production. However, in the recent times many developing countries are making tried in this direction often with the support of international organisations climate finance mechanism and bilateral cooperation with the Global authorities. Country such as China and India which have rapidly growing economies and significant industrial bases are increasingly focusing on clean energy and sustainable practices as part of their development strategies. Various partnerships have been executed with the Global authorities to move forward in this direction.

3.2 Leading Countries and Industries in Zero Carbon Initiative

The growth in the zero-carbon initiative across various countries is quite imbalanced. For instance, the European Union is a notable leader having committed to becoming climate neutral by the year 2050 (Li et al. 2011). The European green deal is a comprehensive policy of framework which has the objective to make the European Union economy sustainable by turning climate an environmental challenge into opportunity. Some of the key initiative under the green deal include increasing the share of renewable energy enhancing energy efficiency, promoting circular economy practices, and supporting green technology. With the help of replacing the traditional source of energy with the new and more sustainable resources the company is targeting to achieve carbon neutrality.

With the help of Energiewende policy driving significant investment in renewable energy and energy efficiency the country is target into achieve a 65% share of renewable energy in electricity consumption by 2030. Also, the company is targeting to reduce greenhouse gas emission by at least 55% as compared to the 1990 level. The industrial sector in Germany is also making tribes towards zero carbon production with major companies like Siemens and BASF who are implementing comprehensive sustainability strategies in their day-to-day operations and have integrated the sustainable strategies with their overall corporate objectives (Omer 2008).

Japan and China are another key player in zero carbon production. China being one of the largest emitters of greenhouse gases is targeting to achieve carbon utility by 2060, with the help of heavy investment in renewable energy particularly in solar and wind power. With the help of the investment companies focusing on improving the energy efficiency and developing green technology. Companies like Alibaba and TENCENT in China are increasingly integrating sustainability into their operations and supply chain leading to reduction in the greenhouse gas emissions at the ground level also rather than just a policy created by the government at the macro level (Fujii et al. 2016). Countries like United States under the Biden administration as renewed its commitment to addressing climate change with a goal of achieving A 52% reduction in greenhouse gas emissions by 2030.

3.3 Recent Advancements and Milestones

Recent years have witnessed significant advancement and milestone in zero carbon production which are primarily driven by technology innovation policy support from the government and increasing awareness of the need for sustainability at the micro and macro level. One of the most notable advancements is the rapid growth of renewable energy resources which primarily include the solar and wind power which have now become most cost competitive in nature leading to a search in installations worldwide. According to International renewable energy agency the renewable energy capacity grew by 260 gigawatts in 2020 which is the largest annual increase on the record. This growth is supported by technological improvement economies of scale and favourable policy framework (Dominkovic et al. 2016). With the advancement of the technology more amount of energy can now be created and stored for later use, and it is because of this high amount of energy production can be witnessed in the year 2020 (Dinga and Wen 2022). Along with this the energy consumption has also reduced significantly due to the innovations such as high efficiency motors advanced heat pumps and industrial heat recovery system which are helping to reduce not only the carbon emission but also the reduction in energy consumption. Carbon capture utilisation and storage technology is making stride. In the private sector corporate commitments to sustainability are playing a significant role in the recent development of the sustainable practices across distinct reasons.

Organisation such as the United Nation and the international energy agency along with the world economic forum are facilitating global corporation on climate action and sustainability. Initiative like science-based target initiative (SBTI) and the RE100 CAMPAIGN which are encouraging companies to set ambitious emission reduction targets and transition to 100% renewable energy by a specific target period (Maktabifard et al. 2023). It can be said that current stage of zero carbon production is marked by noteworthy progress across several regions and growing momentum where leading countries and industries are setting ambitious target and implementing innovative solutions to achieve notable milestone. The rapid growth of renewable energy advancement in energy storage and energy efficiency are driving the transition towards zero carbon production. Further the policy support and the regulatory framework along with the corporate commitments are essential in sustaining this progress and achieving the Global climate goals.

3.4 Specific Government Initiative Send Program

Government all around The World are implementing by range of initiative and program to support zero carbon production. These initiatives often include financial incentive regulatory measures and research and development support which are provided to the company having good role in implementation of sustainable industry practices. Financial incentive sir one of the most critical incentives which help in driving investment in clean technology and sustainable practices. For example, many countries offer subsidies tax credits and grants for renewable energy project energy efficient technology and low carbon innovation. In the United States the investment tax credit and the production tax credit provide financial incentive for solar and wind energy project, respectively. In the European union also horizon you are a program allocate significant pending for a search and innovation in clean energy and climate technology with help the business to contribute a certain percentage of the total expenditure and the rest is funded by the government reducing the risk for the business and the financial burden.

Research and development support is also important from the government which helps the business to focus on innovation focused upon the development of new and more efficient energy sources. With the help of research and development support provided by the government the company can invest in RND department to develop innovative technology that can reduce emission and enhance sustainability. For example, United States department of energy funds research in renewable energy, energy storage and carbon capture and storage technology. In the European Union also clean energy transition partnership supports collaborative research and development project to accelerate the development and deployment of clean energy technology across various sector (McJunkins and Reilly 2021).

Specific government programs highlight the diverse approaches to promote zero carbon production where certain countries are sharing the burden of the corporate by providing finance or certain government provide rebate to the companies who are investing in low carbon production. For example, in Germany the development Bank provide low interest loan for energy efficient and renewable energy project through its renewable energy program and energy efficiency program (Meegahapola et al. 2021). In United Kingdom also green industrial revolution plan includes initiative to support of shore winds hydrogen production and carbon capture and storage and electric vehicle. Chinese government has also launched the National green development fund to finance project that promote sustainable development and environmental protection (Li et al. 2011).

3.5 Role of International Agreement and Collaboration

Because the resources available to a specific country is limited in the international agreement and collaborations are vital in addressing the Global nature of climate change and frosting collective action towards zero carbon production. The Paris agreement is the corner stone of international climate effort setting the framework for global corporation on emissions reduction and climate residence. By submitting their consent, the countries commit to specific climate action targets and measures creating a shared vision for limiting Global temperature rise (Wang et al. 2022). For instance, the conference of party's conferences held annually under the United Nation framework convention on climate change provide a platform for countries to negotiate in advance climate action. Conferences bring together government business civil society and other take holder to discuss program, share best practices, and enhance ambition. With the help of the conferences various countries can participate into the decision making and share their valuable inside which can help the other countries also to grow and developed more advanced measures and implement the same in their industrial practices to promote a sustainable practice.

The conference of people twenty-six conference held in Glasgow in 2021 resulted in Glasgow climate pact which emphasize the need for stronger emission reduction commitment and increase support for adoption and resilience by the Global authorities (Zhang et al. 2015). Other international collaboration and initiative complement the efforts of the Paris agreement. The climate and clean air collision focus are reducing short lived climate pollutant such as methane and black carbon which have large amount of climate and health impact. Accordingly, the international solar alliance promotes the development and deployment of solar energy in member countries particularly in tropical region with high solar potential. The mission innovation initiative brings together countries to accelerate innovation in clean energy technology through increased research and development investment and collaboration. The various initiative targets the sharing of resources and knowledge among the member countries to develop and deploy new measures of energy production and promoting sustainable industry.

4 Economic and Social Impacts of Zero Carbon Production

4.1 Economic Benefits of Zero Carbon Production

The transition to zero carbon production present substantial economic benefit including transformation of industries promotion of innovation and festering sustainable growth in the economy. As countries and companies commit to reducing greenhouse gas emissions the economic landscape is being re-shaped offering numerous opportunities for economic prosperity. The development of zero carbon production technique creates massive opportunities for the business and the economy because it opens the door for new market and industries to be created and this is one of the primary economic benefits of the zero-carbon production. The shift towards renewable energy source such as solar wind and hydroelectric power is driving significant investment and generating economic activity. According to the international renewable energy agency the renewable energy sector could attract up to 131 trillion dollars in investment by the year 2050 (Zhang et al. 2015). The attraction of such amount of investment will support economic growth and stability and it will also and compass the development of renewable energy infrastructure manufacturing of components and provision of related services to the other industries.

4.2 Job Creation and Skill Development

As mentioned according to International renewable energy agency the renewable energy sector employed over twelve million people globally in the year 2020 and in the recent future this number is expected to grow significantly. The transition to zero carbon production is a powerful engine for job creation and skill development and it is likely to grow exponentially as industries adopt substantial equipment and process focused on sustainable practices. Jobs in renewable energy compasses wide range of role including manufacturing installations operations and maintenance of renewable energy system. For example, the workers employed in solar industry can work on the production of solar panel installation of solar system and maintenance of solar farms.

4.3 Social Benefit Including Health and Community Well Being

There are various social benefits also of contributing to improved health and community well being. The transition to zero carbon production will reduce the gas emissions and environment pollutants which can help in enhancement of public health and create healthier community and promote social well being. One of the most significant social benefits which is recognised as of now is the reduction in the air pollutant due to which the air quality has improved significantly. The main reason for the increase in the air pollutant was the extensive use of fossil fuels in the power plant industrial facilities and transportation which has now reduced due to adoption of sustainable practices by certain industries and as this concept will be accepted by more industries the social benefits of improved health and quality of life will also enhance. The pollutants were linked to respiratory and cardiovascular diseases including asthma and heart attack and by transiting to renewable energy source an implementation of energy efficiency measures their can be seen production in air pollution and improved health outcomes.

5 Challenge and Barrier

5.1 Technical and Financial Challenges

Transition to zero carbon production involve lot of technical and financial challenges which can lead to hindrance in the white spread adoption of clean technology and sustainable practices making it difficult for industries to achieve zero carbon goal. Despite prominent level of awareness among various individual about the importance of renewal energy and clean technology the main reason that the adoption is still Limited across various regions and industryis the lack of proper technical and financial resources available through which the innovation can be continued further.

Not only the technical challenges but financial challenges are equally significant in the non adoption of zero carbon production technique. Zero carbon production technique and technology are highly capital intensive and it require the business to invest substantial amount in the innovative technology infrastructure and process. This can be possible for the industry established in the developed countries but many industries especially in the developing countries flag the financial resources to make these investments (Chen et al. 2013). Access to the finance is a critical barrier and this is quite high in the medium size enterprises and small-scale enterprises which find the difficult to secure loans or attract investors to put money helping them to move further towards the adoption of zero carbon production technique and clean technology.

5.2 Market and Consumer Acceptance Issue

Market and consumer expectation is another issue in the transition towards zero carbon production technique. Overcoming resistance from market stakeholders and gaining consumer by in our essential for white spread adoption of sustainable practices in technology in the industry. As mentioned the industries who are dependent on fossil fuel for the operation are capital intensive in nature and it is because of this they have heavenly invested in the technology which require fossil fuels for their operational task. It is because of this the market stay holder do not favour the adoption of sustainable and clean technology because it will lead to the waste of their investment in the technology requiring fossil fuels.

Government initiative place a crucial role in this transition because such policy should be created through which the industries do not have to wear the loss alone for the investment already made in the technology requiring fossil fuels. Such policy and framework can be introduced through which the industries can be compensated for their loss, and they can be motivated for the adoption of sustainable techniques and technology implementation.

6 Future Trends and Prediction

We cannot operate for a longer time by continuously using the fossil fuels in the industries mentioned above and there is stringent requirement to control pollution and other negative aspects of use of fossil fuels to address current challenges and open new opportunities and possibilities for sustainable industrialisation. One of the most promising emerging technologies is green hydrogen which is produced using renewable energy and can serve as a clean energy source for various industrial processes particularly in steel and chemical industries. Further carbon capture utilisation and storage technology can also play a significant role in the future of zero carbon production where it can capture the carbon emission from various industries which can further utilised as input in another industrial process. This concept can be further applied in other industries which can create new industries and job and enhance the Global competitiveness. As per the report of international energy agency they have estimated the achievement of net zero emission by 2050 and 14 million new jobs in energy sector while other sixteen million jobs created in efficiency and electric vehicle manufacturing. Further it is estimated that with the help of proper implementation of zero carbon production techniques can lead to the economic benefits of more than five trillion by the year 2030 (Caiazzo et al. 2022).

7 Conclusion

The new wave of industrialisation in terms of the transition to zero carbon production reflect critical shift in how industries operate which is highly driven by the urgent need to address climate change and its associated impact. Since the evolution of industrialisation through multiple waves each contributing to significant economic growth but also substantial environmental degradation, the current industrial wave is focused highly towards the adoption of sustainable practices to mitigate the environmental impact and also to create positive economic and social impacts through job creation and improving the public health by reducing the carbon emission and pollutants in the nature.

There are various challenges in the adoption of the zero-carbon emission across various regions and industries as studied which are of technical and financial nature but with the proper guidance and participation of various stakeholder and authorities all the barriers can be overcome. The importance of zero carbon production cannot be overstated in terms of reduction in the environment pollution. The scientific consensus is clear to avoid the worst impact of climate change Global carbon emission must be drastically reduced and it must be achieved by mid century. It is only proved this we can ensure that the future generation will also have access to the natural resources which we are exploiting as of now. All the stakeholders including industries Government and consumer have significant role in the adoption of zero carbon production where industries should invest heavily to reduce their carbon footprint and enhance their competitiveness through offering significant opportunities for Innovation and economic growth. Government should be setting ambitious climate target and should support policies and financial incentives through which zero carbon initiative can be promoted. Regulatory framework should make it easy for the industries to invest and developed the resources rather than creating hindrance in the adoption of the technology and practices in their daily operations.

References

Arevalo-Gallegos A, Ahmad Z, Asgher M, Parra-Saldivar R, Iqbal HM (2017) Lignocellulose: a sustainable material to produce value-added products with a zero-waste approach—a review. Int J Biol Macromol 99:308–318. https://doi.org/10.1016/j.ijbiomac.2017.02.097

Aziz S, Ahmed I, Khan K, Khalid M (2024) Emerging trends and approaches for designing net-zero low-carbon integrated energy networks: a review of current practices. Arab J Sci Eng 49(5):6163–6185. https://doi.org/10.1007/s13369-023-08336-0

Caiazzo B, Di Nardo M, Murino T, Petrillo A, Piccirillo G, Santini S (2022) Towards zero defect manufacturing paradigm: a review of the state-of-the-art methods and open challenges. Comput Ind 134:103548. https://doi.org/10.1016/j.compind.2021.103548

Chen C, Li YP, Huang GH (2013) An inexact robust optimization method for supporting carbon dioxide emissions management in regional electric-power systems. Energy Econ 40:441–456. https://doi.org/10.1016/j.eneco.2013.07.022

Dinga CD, Wen Z (2022) China's green deal: can China's cement industry achieve carbon neutral emissions by 2060? Renew Sustain Energy Rev 155:111931. https://doi.org/10.1016/j.rser.2021.111931

Dominković DF, Bačeković I, Ćosić B, Krajačić G, Pukšec T, Duić N, Markovska N (2016) Zero carbon energy system of South East Europe in 2050. Appl Energy 184:1517–1528. https://doi.org/10.1016/j.apenergy.2016.03.046

Feng JC, Yan J, Yu Z, Zeng X, Xu W (2018) Case study of an industrial park toward zero carbon emission. Appl Energy 209:65–78. https://doi.org/10.1016/j.apenergy.2017.10.069

Fujii M, Fujita T, Dong L, Lu C, Geng Y, Behera SK, Park HS, Chiu ASF (2016) Possibility of developing low-carbon industries through urban symbiosis in Asian cities. J Clean Prod 114:376–386. https://doi.org/10.1016/j.jclepro.2015.04.027

Guo D, Li B, Yu W, Baroutian S, Young BR (2024) A system engineering perspective for net zero carbon emission in wastewater and sludge treatment industry: a review. Sustain Prod Consum 46:369–381. https://doi.org/10.1016/j.spc.2024.02.033

Kumar D, Soni G, Jabeen F, Tiwari NK, Sariyer G, Ramtiyal B (2024) A hybrid Bayesian approach for assessment of Industry 4.0 technologies towards achieving decarbonization in manufacturing industry. Comput & Ind Eng 190:110057. https://doi.org/10.1016/j.cie.2024.110057

Li YP, Huang GH, Chen X (2011) Planning regional energy system in association with greenhouse gas mitigation under uncertainty. Appl Energy 88(3):599–611. https://doi.org/10.1016/j.apenergy.2010.07.037

Maktabifard M, Al-Hazmi HE, Szulc P, Mousavizadegan M, Xu X, Zaborowska E, Li X, Mąkinia J (2023) Net-zero carbon condition in wastewater treatment plants: a systematic review of mitigation strategies and challenges. Renew Sustain Energy Rev 185:113638. https://doi.org/10.1016/j.rser.2023.113638

McJunkin TR, Reilly JT (2021) Net-zero carbon microgrids (No. INL/EXT-21-65125). Idaho National Lab. (INL), Idaho Falls, ID (United States). https://doi.org/10.2172/1831061

Meegahapola L, Mancarella P, Flynn D, Moreno R (2021) Power system stability in the transition to a low carbon grid: a techno-economic perspective on challenges and opportunities. Wiley Interdiscip Rev Energy Environ 10(5):e399. https://doi.org/10.1002/wene.399

Mulugetta Y, Urban F (2010) Deliberating on low carbon development. Energy Policy 38(12):7546–7549. https://doi.org/10.1016/j.enpol.2010.05.049

Muradov N (2017) Low to near-zero CO_2 production of hydrogen from fossil fuels: status and perspectives. Int J Hydrogen Energy 42(20):14058–14088. https://doi.org/10.1016/j.ijhydene.2017.04.101

Omer AM (2008) Focus on low carbon technologies: the positive solution. Renew Sustain Energy Rev 12(9):2331–2357. https://doi.org/10.1016/j.rser.2007.04.015

Pan W (2014) System boundaries of zero carbon buildings. Renew Sustain Energy Rev 37:424–434. https://doi.org/10.1016/j.rser.2014.05.015

Samadzadegan B, Samareh Abolhassani S, Dabirian S, Ranjbar S, Rasoulian H, Sanei A, Eicker U (2021) Novel energy system design workflow for zero-carbon energy district development. Front Sustain Cities 3:662822. https://doi.org/10.3389/frsc.2021.662822

Vimal KEK, Kumar A, Sunil SM, Suresh G, Sanjeev N, Kandasamy J (2022) Analysing the challenges in building resilient net zero carbon supply chains using influential network relationship mapping. J Clean Prod 379:134635. https://doi.org/10.1016/j.jclepro.2022.134635

Wang P, Wang J, Jin R, Li G, Zhou M, Xia Q (2022) Integrating biogas in regional energy systems to achieve near-zero carbon emissions. Appl Energy 322:119515. https://doi.org/10.1016/j.apenergy.2022.119515

Wimbadi RW, Djalante R (2020) From decarbonization to low carbon development and transition: a systematic literature review of the conceptualization of moving toward net-zero carbon dioxide emission (1995–2019). J Clean Prod 256:120307. https://doi.org/10.1016/j.jclepro.2020.120307

Wu W, Yang HT, Hwang JJ (2014) Conceptual design of syngas production systems with almost net-zero carbon dioxide emissions. Energy 74:753–761. https://doi.org/10.1016/j.energy.2014.07.042

Zhang Q, Mclellan BC, Li H (2015) An integrated scenario analysis for future zero-carbon energy system. Int J Energy Res 39(7):993–1010. https://doi.org/10.1002/er.3313

Zhang J, Terrones M, Park CR, Mukherjee R, Monthioux M, Koratkar N, Kim YS, Hurt R, Frackowiak E, Enoki T, Chen Y, Bianco A (2016) Carbon science in 2016: status, challenges and perspectives. Carbon 98:708–732. https://doi.org/10.1016/j.carbon.2015.11.060

Zhao X, Chen H, Zheng Q, Liu J, Pan P, Xu G, Zhao Q, Jiang X (2023) Thermo-economic analysis of a novel hydrogen production system using medical waste and biogas with zero carbon emission. Energy 265:126333. https://doi.org/10.1016/j.energy.2022.126333

Sustainable Supply Chains in the Zero-Carbon Era

A. Arun Kumar, Rakesh Suryadevara, K. Rakesh, and Veera Swamy Marri

Abstract A key tactic for businesses looking to balance social responsibility, environmental stewardship, and economic viability in their supply chains and operations is sustainable supply chain management (SSCM). The importance of SSCM in tackling modern issues like resource scarcity, climate change, and ethical sourcing practices is examined in this study. In addition to offering helpful suggestions for companies and legislators, it highlights essential obstacles to adopting SSCM, such as supplier preparedness problems, regulatory uncertainties, and cost concerns. The recommendations emphasize supply chain collaboration, technological innovation, leadership commitment, and open reporting of sustainability metrics. Enhancing operational effectiveness, controlling risks, fostering better relationships with stakeholders, and advancing long-term sustainability objectives are all implications of SSCM. In the future, research will focus on developing theoretical frameworks, broadening the scope of empirical studies to include a variety of industries and geographical areas, and investigating how emerging technologies can improve the efficacy of SSCM. Organizations can develop shared value across international supply chains, promote innovation, increase resilience, and lessen risks and environmental effects by incorporating SSCM principles. The research outcomes of this study will not only enhance scholarly comprehension but also furnish companies with feasible approaches to accomplish their carbon neutral objectives.

Keywords Sustainable · Supply chain management · Zero · Carbon era

A. A. Kumar (✉) · R. Suryadevara
ICFAI Law School, ICFAI Foundation for Higher Education (Deemed to Be University Under Sec 3 of UGC Act, 1956), Hyderabad 501203, Telangana, India
e-mail: arunkumar@ifheindia.org

K. Rakesh
Department of BBA, Anish Degree College, Osmania University, Hyderabad 500007, Telangana, India

V. S. Marri
Loyola Academy (Degree and PG College), Old Alwal, Secunderabad 500010, Telangana, India

R. Singh et al. (eds.), *Zero Carbon Industry, Eco-Innovation and Environmental Sustainability*, World Sustainability Series,
https://doi.org/10.1007/978-3-031-80220-1_6

1 Introduction

Significant shifts toward greater sustainability are driven by the global climate crisis, particularly in the corporate and industrial sectors. Research on Sustainable Supply Chain Management (SSCM), has become increasingly focused as industries work to lessen their carbon impact. This strategy reduces the adverse effects on the environment while enhancing the competitiveness and resilience of the company in a market that is changing quickly. Given industrial activities' substantial environmental impact, supply chain management must incorporate sustainability. Businesses can save more money by implementing SSCM techniques because they will improve resource management, waste reduction, and energy efficiency.

In addition, SSCM helps with more general societal advantages like pollution reduction and resource conservation. Global climate change is one of the most enduring issues of our day, mainly due to human activity. These activities contribute to global warming and the greenhouse effect by releasing significant amounts of CO_2. Large amounts of CO_2 and other greenhouse gases are released during industrial processes like steel, cement, and chemicals. Strong greenhouse gases like nitrous oxide and methane are released in agriculture by rice paddies, livestock, and the digestion of synthetic fertilizers. The greenhouse effect is exacerbated by deforestation caused by logging, urban development, and agriculture because fewer trees are left to absorb CO_2 from the atmosphere. A natural landscape's ability to store carbon can be destroyed, and emissions can be increased by conversion to agriculture. Methane is produced during the anaerobic breakdown of organic waste in landfills and waste treatment facilities. One significant contributing factor to greenhouse gas emissions is poor waste management practices (Acquaye et al. 2014).

Extreme weather events cause enormous financial losses that impact livelihoods, agriculture, and infrastructure. Coastal communities are seriously threatened by rising sea levels, which can result in erosion, flooding, and a more significant number of displaced people. Increasing sea levels and temperatures threaten coastal and marine ecosystems, including mangroves and coral reefs. The destruction and fragmentation of habitats brought on by climate change will endanger biodiversity and cause the extinction of entire species. Ecosystems will be disrupted, and changes in temperature and precipitation patterns will impact food webs and species interactions. Heat-related illnesses and fatalities will increase with rising temperatures, particularly for vulnerable groups (Zhu and Sarkis 2004).

Due to climate change, new health risks may be associated with the spread of vector-borne illnesses like dengue fever and malaria. Emissions and energy consumption are decreased by increasing energy efficiency in transportation, buildings, and industrial processes (Ageron et al. 2012). Water resources will always be available if sustainable water management techniques are implemented, even in shifting precipitation patterns. In the face of a changing climate, food security can be preserved by cultivating climate resilient crops and implementing sustainable farming methods.

International accords such as the Paris Agreement seek to unite nations to curb global warming and lower greenhouse gas emissions. Governments must enact laws and policies that encourage environmentally friendly behavior, provide incentives for renewable energy sources, and lower emissions. There are many different aspects to global climate change, and each one calls for swift and concerted action from people, corporations, governments, and international organizations. Using technological advancements, policy interventions, and community involvement, it is feasible to alleviate the detrimental consequences of climate change and safeguard the Earth for posterity (Yang et al. 2009).

Initiatives with a zero-carbon footprint are essential for combating climate change and promoting sustainable development. By reducing emissions of greenhouse gases (GHGs) like carbon dioxide (CO_2), these programs seek to lessen the effects of global warming. The main cause of global warming, greenhouse gas emissions, is significantly reduced by zero-carbon projects. By reducing emissions from industry, transportation, and energy production, among other sectors, these initiatives aid in maintaining a stable global temperature and reducing carbon emissions, shielding ecosystems and biodiversity from additional harm. Many species' survival depends on the health of the oceans, forests, and other natural habitats maintained by lower pollution levels. In addition to releasing harmful air pollutants, burning fossil fuels is one of the many sources of carbon emissions. Zero-carbon initiatives can improve air quality by drastically reducing contaminants like sulphur dioxide (SO_2), nitrogen oxides (NOx), and particulate matter. Reduced respiratory and cardiovascular illnesses are two immediate benefits of improved air quality, reduced healthcare expenses, and enhanced general health (Naslund and Williamson 2010).

Zero-carbon projects support resource efficiency and waste reduction and encourage recycling and reuse. Reforestation and afforestation projects, which restore natural habitats and biodiversity and sequester CO_2, are frequently included in initiatives. Investment and innovation in green technologies, including sustainable transportation, energy efficiency, and renewable energy, are stimulated by zero-carbon initiatives. This promotes technological development and opens new markets and business opportunities. New job opportunities are created by the shift to a zero-carbon economy in several industries, such as sustainable agriculture, green building, and renewable energy. This can lower unemployment rates and promote economic growth. A common component of zero-carbon initiatives is efficiency improvement, which can save homes and businesses significant money. Lower energy use results in lower operating and utility costs. Zero-carbon projects help prevent the financial consequences of extreme weather events like floods, hurricanes, and droughts by reducing climate change. This can prevent damages and disaster recovery costs in the billions of dollars (Pullman et al. 2009).

Businesses that implement zero-carbon practices can obtain a competitive advantage by satisfying consumer demand for goods and services that are ecologically friendly. Customer loyalty and brand reputation may both benefit from this. Businesses can avoid potential fines and penalties by actively participating in zero-carbon initiatives that guarantee compliance with current and future environmental regulations. By lowering pollution levels, zero-carbon initiatives lessen the health risks of air and water pollution. Communities become healthier, and people's quality of life increases. Increasing access to renewable energy sources can help underserved populations receive dependable, reasonably priced electricity, raise living standards, and create job opportunities, particularly in developing nations. Initiatives aiming at achieving zero carbon can advance social justice by widely distributing the advantages of the shift to a sustainable economy. This includes providing retraining and job placement assistance to employees in carbon-intensive industries. Involving local communities in zero-carbon initiatives promotes sustainable practices at the regional level by creating a sense of ownership and accountability (Olga 2012).

Education components that increase public awareness of sustainability and climate change are frequently included in zero-carbon initiatives. People are now better equipped to advocate for environmental protection and make educated decisions. To prepare the workforce for upcoming challenges, the green economy's growth offers opportunities for education and skill development in new technologies and sustainable practices. Zero-carbon initiatives are essential to address the global climate crisis and advance sustainable development. Lowering greenhouse gas emissions, enhancing air quality, and protecting natural resources have several positive environmental effects. From an economic standpoint, these endeavors stimulate innovation, generate employment, reduce expenses, and improve the competitiveness of businesses. Socially, they foster social justice, enhance public health, and offer educational opportunities (Montoya-Torres and Ortiz-Vargas 2014).

Shifting to a zero-carbon economy is essential for safeguarding the environment and for a more prosperous, just, and resilient future. A comprehensive strategy for controlling the flow of goods and services, SSCM, ensures that social, environmental, and economic sustainability are incorporated into every part of the supply chain. To help lay the foundation for understanding SSCM's role in achieving a zero-carbon economy, this introduction examines the idea, significance, guiding principles, and changing practices of SSCM. SSCM plans, coordinates, and manages supply chain operations to reduce adverse environmental effects, advance social justice, and preserve financial sustainability. Integrating sustainable practices goes beyond conventional supply chain management (Lambert et al. 1998).

They prevent pollution throughout the supply chain, minimize waste, conserve resources, and lower carbon emissions. Maintaining fair labour practices, fostering innovation and competitiveness, promoting worker safety and well-being, giving back to the communities where supply chain activities occur, and achieving cost efficiency are all critical goals. SSCM aids in preserving biodiversity and natural habitats by encouraging resource efficiency. Businesses can guarantee that environmental regulations are followed and stay out of trouble by implementing SSCM. By implementing sustainable best practices, they stay ahead of regulation changes (Frohlich and Westbrook 2001).

Reducing waste and using less energy are examples of sustainable practices that can save money. Investors and consumers increasingly value strong sustainability credentials, improving a company's market positioning. In addition to promoting human rights and ensuring that workers are treated fairly, SSCM also advances social equity. Businesses can forge strong bonds with their surrounding communities by taking part in sustainable and socially conscious initiatives.

They ensure supply chain procedures produce both financial gains and favorable environmental and social effects, and they consider products' environmental and social impacts. They include suppliers, customers, and communities in decision-making processes and uphold clear and transparent information about supply chain operations, such as sourcing, manufacturing, and distribution. Assess and refine supply chain operations regularly to boost sustainability outcomes. They set specific goals for sustainability and use Key Performance Indicators (KPIs) to monitor their progress (Harris et al. 2011).

Choosing materials with less environmental impact and giving preference to suppliers who follow social and environmental standards are also crucial in reducing environmental impact by using recyclable, biodegradable, and renewable materials and putting energy-saving techniques and technologies into manufacturing, shipping, and storage and switching supply chain operations over to renewable energies. Transportation involves the use of electric, hybrid, and other low-emission vehicles. They streamline distribution and logistics systems to reduce emissions and fuel use (Frohlich and Westbrook 2001).

Guaranteeing that every employee in the supply chain receives fair compensation, is treated fairly, and works in safe conditions. SSCM will continue to develop, spurring innovation and resulting in more sustainable business practices globally as awareness of climate change and sustainability issues spreads across the globe. Even though the significance of SSCM is becoming increasingly apparent, many companies still struggle to put into practice practical plans that lead to supply chains with zero carbon emissions. These difficulties include organizational resistance to change, a lack of regulatory support, and the high cost of sustainable technologies. A more thorough study is also required to comprehend the innovations, best practices, and effects of SSCM on environmental sustainability and business performance (Feng et al. 2007).

SSCM still has unanswered questions. These gaps restrict the efficacy of SSCM strategies and impede the complete realization of sustainable supply chains. Achieving the objectives of a zero-carbon economy and progressing in the field depends on recognizing and comprehending these gaps. Instead of taking an integrated approach, most current research concentrates on specific facets of sustainability, such as the environmental, social, or economic dimensions. Research must determine how these factors interact and how supply chains can successfully balance them. Research in particular fields is frequently conducted alone. To create thorough SSCM frameworks, more interdisciplinary research combining knowledge from the social sciences, economics, management, and environmental science is needed (Walsh et al. 2000).

Standardized metrics and indicators are needed to thoroughly evaluate supply chains' sustainability performance. Most of the time, current instruments are either industry-specific or fail to account for sustainability impacts fully. More sophisticated and user-friendly lifecycle assessment tools are required to assess how processes and products affect the environment and society at large along the supply chain. Emerging technologies such as the Internet of Things (IoT) and other Artificial Intelligence (AI) are being largely implemented. Still, their practical adoption and implementation in SSCM are receiving less attention. Research frequently looks at technology solutions in regulated settings or pilot projects with little room for growth and valuable application (Iraldo et al. 2011).

There is a knowledge vacuum regarding how employee behavior and organizational culture affect the adoption of sustainable practices. Further investigation is required regarding managing change and cultivating a sustainable culture within an organization. Sustainability is driven by consumer demand. However, additional study is needed to comprehend how consumer behavior affects supply chain procedures and how businesses can interact with customers to encourage sustainable decisions. Further empirical studies are required to determine how well-received SSCM is by current laws and regulations. This involves assessing how policies affect things locally, nationally, and globally. Particularly in a globalized economy, research is required to support sustainable supply chains by integrating and harmonizing various policy instruments (Elgazzar et al. 2012).

Robust monitoring and evaluation systems are necessary for SSCM to improve continuously. Many businesses lack the procedures and systems to evaluate and enhance their sustainability performance routinely. The ongoing adoption of cutting-edge methods and technologies is interrupted. To remain ahead of the sustainability curve, businesses must cultivate a culture of creativity and flexibility. Many companies struggle to effectively inform stakeholders and customers about their sustainability initiatives and accomplishments. Building trust and engagement requires believable, transparent, and clear communication. More input from stakeholders such as customers, staff members, and communities is necessary when creating and implementing SSCM strategies (Frohlich and Westbrook 2001).

Achieving the objectives of a zero-carbon economy and advancing sustainable supply chain management depends on filling in these gaps in research and practice. Researchers and practitioners can increase the effectiveness of SSCM through the integration of sustainability dimensions, the development of solid assessment tools, the utilization of technological innovations, the comprehension of behavioral factors, and the promotion of efficient policies. To fully realize the potential of sustainable supply chains, businesses must also increase transparency, engage suppliers, balance economic considerations, commit to continuous improvement, and effectively communicate with stakeholders. By offering thorough insights into SSCM practices and their effects, as well as valuable suggestions for businesses and policymakers, this study seeks to close a gap in the literature.

2 Objectives of the Study

This study's primary goals are to

1. Examine how SSCM contributes to the achievement of zero-carbon goals.
2. To determine the most important SSCM technologies and strategies.

3. To evaluate how SSCM affects environmental sustainability and business performance.

3 Results and Discussion

Globally reaching zero-carbon targets depends on SSCM. Businesses can drastically lower their carbon footprint, improve operational effectiveness, and support more general environmental goals by incorporating sustainable practices into every part of the supply chain. This analysis looks at the role that SSCM plays in reaching zero-carbon goals. It highlights critical tactics, technological advancements, and these strategies' effects on environmental sustainability and business performance. Raw material suppliers can lessen their carbon footprint by favoring those who follow social and environmental standards. It is crucial to source recyclable, eco-friendly, and renewable materials. Green procurement requires suppliers to be routinely assessed for their environmental performance and assisted in strengthening their sustainable practices. Energy consumption and related emissions are decreased in manufacturing, warehousing, and transportation by implementing energy-efficient technologies and procedures.

Reducing waste and carbon emissions can be achieved by implementing circular economy principles, which include recycling, reusing, and repurposing materials. Achieving zero-carbon goals requires lean manufacturing techniques to reduce waste, increase efficiency, and use less energy. Logistics emissions can be reduced by using electric, hybrid, and other low-emission vehicles for transportation. Fuel consumption and emissions can be decreased by optimizing logistics and distribution networks to reduce travel times and increase load efficiency. Utilizing the IoT and blockchain technologies to improve traceability and transparency across the supply chain guarantees that all procedures follow sustainable standards. Companies can pinpoint areas for improvement and swiftly address them by putting in place real-time monitoring systems to track emissions and resource consumption throughout the supply chain (Harris et al. 2011).

Blockchain technology guarantees transparent and safe product tracking from manufacturer to customer, making it possible to confirm sustainable practices at every turn. Intelligent automated contracts make it easier to follow sustainability guidelines and expedite transactions, lowering administrative costs and guaranteeing zero-carbon pledges. IoT devices give businesses access to real-time data on emissions, resource consumption, and energy usage, which they can use to monitor and optimize their operations for sustainability. Predictive maintenance enabled by the IoT lowers energy waste and equipment downtime, resulting in more sustainable and effective supply chain operations (Iraldo et al. 2011).

AI and Machine Learning (ML) algorithms can reduce waste and increase efficiency in supply chain operations, from inventory management to demand forecasting. AI-driven analytics allow businesses to adopt focused emission reduction strategies by helping spot trends and forecast emissions. Lower carbon emissions are a result of additive manufacturing, also known as 3D printing, which minimizes material waste and permits the use of sustainable materials. Achieving zero-carbon targets requires cutting-edge manufacturing technologies that optimize production processes for minimal waste and energy efficiency (Harris et al. 2011).

Effective resource use and decreased waste result in lower overall costs for operations, which is why sustainable practices like energy efficiency and waste reduction are so important. Businesses that have a solid reputation for sustainability can become more competitive. Sustainable practices guarantee resource efficiency, lessen environmental deterioration, and protect natural ecosystems. Improving air quality and public health outcomes are facilitated by lowering emissions and pollution. Businesses dedicated to sustainability and zero-carbon goals frequently have a stronger reputation and brand image, increasing customer loyalty and trust. Respecting sustainability guidelines and environmental laws keeps one out of trouble with the law and improves corporate governance (GreenBiz 2018).

Reaching zero-carbon goals requires sustainable supply chain management. Businesses can drastically reduce their carbon footprint by implementing sustainable transportation, cutting waste, improving energy efficiency, adopting green procurement practices, and utilizing technology advancements. In addition to providing social and economic advantages, these practices support environmental sustainability and enhance overall business performance and reputation. Sustainability will continue to be a top priority worldwide. SSCM will be an essential tactic for companies looking to achieve zero-carbon targets and contribute to international efforts to tackle climate change (Schmidt et al. 2017).

Adopting circular economy concepts, emphasizing material recycling, reuse, and repurposing to reduce waste. Use lean manufacturing techniques to cut waste, eliminate extra inventory, and boost productivity. Driving low-emission automobiles like hybrids and electric cars decreases carbon emissions. To cut down on travel time, increase load efficiency, and save fuel, optimize distribution and logistics networks. Use real-time monitoring systems to track emissions and resource usage throughout the supply chain to enable quick corrective action. Use technology that tracks products in-depth from the manufacturer to the final customer, guaranteeing that sustainability guidelines are followed. Involve suppliers, clients, and local communities to synchronize sustainability objectives and procedures. To improve sustainability initiatives and utilize specialized knowledge, collaborate with NGOs, government agencies, and other organizations. Supply chain processes should be continuously assessed and improved to enhance sustainability performance (Bhattacharya et al. 2014).

Blockchain ensures that products are tracked safely and transparently from the manufacturer to the customer, confirming sustainable practices at every turn. Automated intelligent contracts simplify transactions and help comply with sustainability standards while cutting down on administrative work. IoT devices help businesses monitor and optimize operations for sustainability by providing real-time data on energy usage, emissions, and resource consumption. AI-driven analytics allow businesses to adopt focused emission reduction strategies by helping spot trends and forecast emissions. Lower carbon emissions are a result of additive manufacturing, also known as 3D printing, which minimizes material waste and permits the use of sustainable materials. Achieving zero-carbon targets requires cutting-edge manufacturing technologies that optimize production processes for minimal waste and energy efficiency. Carbon emissions significantly decrease when solar and wind energy systems are used in power manufacturing facilities and supply chain activities (Oxborrow and Brindley 2014).

A steady and dependable energy supply is ensured by managing and storing renewable energy using sophisticated energy storage technologies like batteries. Analyzing waste generation and resource usage data facilitates finding room to improve and gain efficiency. Putting material recovery and recycling technologies into practice decreases waste and the need for new raw materials because resources are reused. Technology that monitors and controls a product's whole lifecycle helps businesses to design with sustainability and end-of-life recovery in mind (Croxton et al. 2001).

Global efforts to combat climate change benefit from significant reductions in resource consumption, waste generation, and carbon emissions. Cost savings through improved operations, decreased waste, enhanced competitiveness, market uniqueness, and energy efficiency. More significant social equity through ethical labour practices and community involvement, better public health and well-being, and strengthening community ties (Laeequddin et al. 2010).

Key technologies and strategies in sustainable supply chain management are required to meet zero-carbon goals. Businesses can significantly improve the sustainability of their supply chains by incorporating green procurement, energy efficiency, waste reduction, sustainable transportation, transparency, collaboration, and continuous improvement. These efforts are strengthened by utilizing cutting-edge technologies like big data, blockchain, IoT, AI, advanced manufacturing, renewable energy, and circular economy technologies. Companies are positioned for long-term success in a sustainable future by implementing these strategies and technologies, which also offer economic and social benefits and lessen environmental impact (Iraldo et al. 2011).

Sustainability in the environment and business performance are strongly impacted by sustainable supply chain management or SSCM. This section thoroughly evaluates the effects of SSCM practices and technologies on these two critical areas. Energy-efficient technology can lower operating costs and utility bills by reducing energy consumption. For instance, costs can be significantly reduced by utilizing LED lighting, energy-efficient HVAC systems, and production process optimization. Lean manufacturing and circular economy practices increase material efficiency, cut waste, and lower disposal costs. Reusing and recycling materials instead of buying new raw materials helps businesses save money. SSCM promotes product and process innovation, allowing businesses to stay one step ahead of rivals and adjust to shifting consumer needs. New environmentally friendly goods and services can be developed due to sustainable practices (Schmidt et al. 2017).

Compliance with environmental regulations and standards aids in the prevention of fines and legal penalties for companies. Adherence to sustainability regulations reduces the potential for harm to one's reputation. Risks related to resource scarcity, supply chain disruptions, and climate change are decreased by SSCM. Businesses can guarantee continuity and stability by implementing resilient practices and diversifying their supply sources. Selling recycled materials, providing sustainable products, and breaking into green markets are just a few examples of how sustainable practices can create new revenue streams. The long-term financial gains from SSCM, such as lower operating costs and higher sales, offset the potentially significant initial investments and increase overall profitability (Koh et al. 2013).

Reducing resource depletion and environmental degradation is ensured through implementing a circular economy and sustainable sourcing practices. For instance, materials are conserved when products are made to be long-lasting and recyclable. SSCM techniques include minimizing waste, cutting consumption, and optimizing energy and water use. Adoption of renewable energy sources and water-efficient procedures support conservation initiatives. Waste generation is reduced by applying lean manufacturing and waste reduction techniques. By ensuring that waste is recycled, reused, or repurposed, circular economy practices lessen the need to dispose of waste in landfills. Minimizing packaging waste and utilizing eco-friendly materials help lessen the environmental impact (Cheng and Fu 2013).

Pollution of the air, water, and soil is decreased by using sustainable manufacturing, transportation, and waste management techniques. Pollution of the environment is reduced by employing non-toxic materials and cleaner production methods. The environmental impact of supply chain activities is significantly reduced by implementing technologies like electric vehicles and renewable energy systems that minimize emissions and pollutants (Schmidt et al. 2017).

Unilever's dedication to sustainability has produced cost savings, higher sales, and improved brand recognition through its Sustainable Living Plan. Unilever has decreased waste production, water use, and carbon emissions throughout its supply chain. The business reduced its energy-related CO_2 emissions from manufacturing by 65%. IKEA has cut expenses and increased resource efficiency by emphasizing circular economy concepts like product take-back schemes and recycling materials. The company's brand image has been reinforced by its sustainability initiatives. By 2030, IKEA wants to be climate positive, emphasizing lowering carbon emissions and encouraging sustainable sourcing. The business has invested in renewable energy and established challenging goals for using sustainable materials.

SSCM significantly affects both environmental sustainability and business performance. Cost reductions, increased competitiveness, legal compliance, revenue growth, and enhanced brand reputation are all achieved through SSCM practices. In terms of the environment, SSCM greatly lowers pollution, waste, greenhouse gas emissions, resource conservation, and biodiversity protection. Businesses can attain zero-carbon targets, drive long-term economic success, and contribute to a more sustainable future by implementing SSCM strategies and technologies.

4 Benefits of Sustainable Supply Chain Management

The advantages of SSCM go beyond the interests of individual companies to include the environment and society at large. In addition to helping to meet zero-carbon goals, these advantages improve general sustainability and well-being. The main advantages of SSCM are highlighted in this section, which covers the three main spheres of business, society, and the environment. Operating costs can be significantly decreased by putting energy-efficient technologies into place and using available resources best. Cost reduction strategies include, for instance, implementing lean manufacturing techniques, enhancing the effectiveness of logistics, and utilizing renewable energy sources.

Businesses implementing SSCM can set themselves apart by showcasing their dedication to sustainability, drawing in eco-aware clients and investors. SSCM promotes ongoing innovation and adaptation, helping companies to stay one step ahead of rivals and adjust to shifting consumer needs and legal requirements. By abiding by environmental standards and regulations, businesses can avoid fines, legal repercussions, and reputational harm. Risks related to resource scarcity, supply chain disruptions, and climate change are decreased by SSCM (Kim and Min 2011).

Continuity and stability are ensured by implementing resilient practices and diversifying supply sources. Selling recycled materials, providing eco-friendly products, and breaking into green markets are just a few examples of how sustainable practices can generate new revenue streams. The long-term financial benefits of sustainability, such as lower operating costs and higher sales, outweigh the initial high costs and increase overall profitability. Businesses dedicated to sustainability frequently benefit from a favourable brand image, which improves their reputation and draws in clients who respect corporate responsibility.

5 Benefits for Society

Reducing emissions and pollutants from supply chain operations improves public health by lowering respiratory and cardiovascular disease incidence and improving air and water quality. By guaranteeing that employees in the supply chain receive fair compensation, work in safe environments, and are treated fairly, SSCM improves their general well-being and standard of living. By assisting small businesses and local suppliers, SSCM promotes inclusive growth within the community and economic development. Fair pay, moral work practices, and chances for financial growth are all part of this. Participating in and making investments in local communities fosters social cohesion and solid relationship building. Businesses can promote local infrastructure development, healthcare, and education, enhancing the general well-being of communities (GreenBiz 2018).

To prepare the workforce for upcoming challenges, the green economy's growth offers opportunities for education and skill development in new technologies and sustainable practices. SSCM can boost economic growth and stability by generating new jobs in several industries, such as green manufacturing, sustainable agriculture, and renewable energy. Businesses that support sustainability can increase consumer knowledge of social and environmental issues, enabling people to make wise decisions and support sustainable practices. To benefit society more broadly, SSCM encourages consumers to adopt more sustainable behaviors, such as cutting back on waste, recycling, and purchasing eco-friendly goods.

6 Benefits for the Environment

Using renewable energy, improved energy efficiency and logistics optimization are examples of SSCM practices that reduce carbon footprints and encourage the safeguarding of the environment. Employing blockchain and IoT technologies to track and report emissions in real time can assist businesses in mitigating their environmental impact. Ecological degradation and resource depletion are minimized through implementing a circular economy and sustainable sourcing strategies. For instance, materials are conserved when products are made to be long-lasting and recyclable (Matthews et al. 2016).

SSCM techniques include minimizing waste, cutting consumption, and optimizing energy and water use. Adoption of renewable energy sources and water-efficient procedures support conservation initiatives. Waste generation is reduced by applying lean manufacturing and waste reduction techniques. By ensuring that waste is recycled, reused, or repurposed, circular economy practices lessen the need to dispose of waste in landfills. Minimizing packaging waste and utilizing eco-friendly materials help alleviate the environmental impact. Obtaining raw materials from certified and sustainable sources facilitates protecting biodiversity and ecosystems.

SSCM has many advantages for companies, the community, and the environment. Businesses that implement SSCM experience cost savings, increased competitiveness, revenue growth, regulatory compliance, and improved brand reputation. Benefits to society include better public health, the advancement of social justice, more opportunities for work and education, and raised consumer awareness. In terms of the environment, SSCM helps with pollution control, waste reduction, biodiversity preservation, resource conservation, and greenhouse gas emission reduction. Businesses can accomplish zero-carbon goals, promote long-term economic success, and help create a more just and sustainable world by implementing SSCM practices.

By optimizing consumption and cutting waste, the integration of intelligent grid technologies enables improved energy management and distribution. Predicting energy demands and streamlining grid operations are becoming increasingly common uses for the IoT devices and AI algorithms. Biodegradable composites and recycled materials are two examples of environmentally friendly substitutes for conventional building materials produced by advances in materials science. Passive heating and cooling methods are incorporated into building designs to minimize the need for HVAC systems (Fransoo and Wouters 2000).

There is a tendency for businesses to prioritize short-term financial results over longer-term sustainability goals and investments. Adopting new technologies or practices may come with perceived risks, such as adverse effects on product quality or brand reputation or disruptions in supply chain operations. Adoption of SSCM practices can be hampered by organizational culture and change-averse employees, especially if sustainability is not incorporated into the company's guiding principles and business plans. The inability to precisely quantify and convey the social and environmental effects of SSCM practices can impede attempts to convince stakeholders of their benefits. To overcome these obstacles, a comprehensive strategy includes leadership commitment, cooperation throughout the supply chain, continuous education and awareness-raising, and the strategic alignment of sustainability goals with business objectives (Kim and Min 2011).

Invest in eco-friendly inventions and technologies that increase productivity while lowering environmental effects. Investigate emerging technologies via research and development, including advanced manufacturing techniques, IoT for supply chain transparency, and renewable energy. Evaluating viability and scalability can be aided by pilot programs and partnerships with technology suppliers. Encourage suppliers to implement SSCM practices by providing incentives and capacity building.

To establish trust and accountability, share developments and accomplishments regularly with internal and external stakeholders. Encourage a sustainable culture within the company. Integrate sustainable practices into the company's policies, values, and performance reviews of employees. Promote originality and ingenuity amongst departments and functions to identify long-term solutions (Christopher and Peck 2004).

Develop a relationship of trust and support with stakeholders for SSCM practices. Be open and honest when discussing sustainability projects, objectives, and advancements. Consult with clients, investors, NGOs, and local communities to improve responsiveness and credibility. Businesses can overcome obstacles to implementing SSCM practices and establish themselves as industry leaders by implementing these suggestions and solutions. This will promote social responsibility, environmental stewardship, and long-term economic viability.

7 Contributions to Academic Literature in Sustainable Supply Chain Management (SSCM)

The creation of theoretical frameworks, such as the triple bottom line (economic, environmental, and social), stakeholder theory, and systems thinking approach, incorporate sustainability principles into supply chain management theories. Empirical studies examine how SSCM practices are used and how they affect various industries and geographical areas. The research evaluates performance results, adoption factors, and optimal practices for incorporating sustainability into supply chain operations.

Comprehensive case studies focusing on industries or businesses that have effectively adopted SSCM techniques. These case studies shed light on the difficulties, approaches, and lessons discovered in accomplishing supply chain sustainability objectives. Creation of measurement frameworks and metrics to evaluate supply chains' sustainability performance. This comprises social and economic metrics (like labour practices, community involvement, and cost savings), environmental impact metrics (like carbon footprint and water usage), and ROI of sustainability initiatives.

They examine how international standards (such as ISO 14001 and ISO 26000) and government laws have shaped SSCM practices. Research looks at how regulatory frameworks impact supply chain choices and sustainability results. Studies on how SSCM practices improve supply chain resilience in the face of social risks, regulatory changes, and environmental disruptions.

Contributions to the scholarly literature on SSCM help to advance knowledge, inform practice, and direct policymakers toward supply chain strategies that are more resilient and sustainable. Effectively addressing the challenges of global sustainability requires ongoing research and collaboration among academia, industry, and government.

8 Policy Implications

Provide laws and policies that encourage SSCM practices. Provide rewards (such as tax breaks or subsidies) for businesses implementing sustainable practices. Enforce regulations to guarantee that social and environmental standards are followed throughout supply chains. Encourage cooperation between the government, industry, and academia. Establish channels for exchanging research results, technological advancements, and best practices in SSCM; encourage industry-led projects and collaborations for sustainable development. Provide funds to support studies on new developments and solutions in SSCM. Provide funding for scholarly investigations into sustainable technologies, supply chain resilience, and circular economy practices. Encourage pilot programs to test and expand innovations in the sustainable supply chain.

Mandate company sustainability reporting standards and enforce transparency in supply chain reporting and disclosures. Urge businesses to reveal their effects on society and the environment across their supply chains. Provide independent auditing and verification procedures. Offer SSCM training and capacity-building initiatives. Provide businesses with educational resources, workshops, and seminars to improve their comprehension and application of SSCM practices. Encourage projects that advance supply chain management and sustainable procurement practices. Engage in international cooperation to tackle sustainability issues in the global supply chain. To harmonize sustainability standards and practices, participate in multilateral forums and agreements. Through technology transfer and capacity building, assist developing nations in implementing sustainable practices and technologies. Globally reaching long-term environmental, social, and economic sustainability objectives depends on this cooperative approach.

Getting in line with international sustainability frameworks, like the UN Sustainable Development Goals (SDGs), supports more significant initiatives aimed at achieving social justice, environmental preservation, and economic prosperity on a global scale. By comprehending these consequences, interested parties can utilize the research's findings to promote advantageous effects on commercial activities, ecological responsibility, and community welfare.

9 Limitations of the Study

The study has a few limitations. The study is based on secondary data. The relationship between objectives and variables is causal. The present study is confined to only sustainable supply chains in the carbon era. The study mainly focuses on how SSCM contributes to achieving zero-carbon goals. A similar kind of study can be conducted on future operations and supply chain industries. A comparative study can be conducted with similar industries. Studies can be conducted in specific industries. Case studies can also be conducted in specific operations and supply chain organizations. The findings of the study are not confined to any specific industry.

10 Conclusion

SSCM is essential to address the modern environmental, social, and economic challenges businesses and societies face worldwide. By incorporating social responsibility, environmental stewardship, and economic viability into supply chain operations and SSCM fosters sustainable development. Despite its advantages, SSCM still faces many obstacles, including financial concerns, unknown regulations, problems with supplier preparedness, and cultural barriers within businesses. Companies, legislators, and other stakeholders have been advised to overcome these obstacles. These include the dedication of the leadership, cooperation amongst supply chains, financial investments in innovation and technology, and open measurement and reporting of sustainability metrics. Practical, policy, environmental, social, economic, and educational aspects are all included in these ramifications. Subsequent investigations in SSCM ought to tackle the pointed constraints, promote theoretical models, broaden the scope of empirical studies in various sectors and areas, and delve into the dynamics of global supply chains and developing technologies. Organizations can make a positive global impact on social justice, environmental preservation, and economic resilience by incorporating sustainability into supply chain strategies and operations. To fulfill these goals and produce value that will last for both the current and future generations, SSCM must continue to foster innovation and collaboration.

References

Acquaye A, Genovese A, Barrett J, Lenny Koh SC (2014) Benchmarking carbon emissions performance in supply chains. Supply Chain Manag Int J 19(3):306–321

Ageron B, Gunasekaran A, Spalanzani A (2012) Sustainable supply management: an empirical study. Int J Prod Econ 140(1):168–182

Bhattacharya A, Mohapatra P, Kumar V, Dey PK, Brady M, Tiwari MK, Nudurupati SS (2014) Green supply chain performance measurement using fuzzy ANP-based balanced scorecard: a collaborative decision-making approach. Prod Plan & Control 25(8):698–714

Cheng JH, Fu YC (2013) Inter-organizational relationships and knowledge sharing through the relationship and institutional orientations in supply chains. Int J Inf Manage 33(3):473–484

Christopher M, Peck H (2004) Building the resilient supply chain. Int J Logist Manag 15(2):1–14

Croxton KL, Garcia-Dastugue SJ, Lambert DS (2001) The supply chain management processes. Int J Logist Manag 12(2):13–36

Elgazzar SH, Tipi NS, Hubbard NJ, Leach DZ (2012) Linking supply chain processes' performance to a company's financial strategic objectives. Eur J Oper Res 223(1):276–289

Feng S, Li LX, Duan ZG, Zhang JL (2007) Assessing the impacts of south-to-north water transfer project with decision support system. Decis Support Syst 42(4):1989–2003

Fransoo JC, Wouters MJ (2000) Measuring the bullwhip effect in the supply chain. Supply Chain Manag Int J 5(2):78–89

Frohlich MT, Westbrook R (2001) Arcs of integration: an international study of supply chain strategies. J Oper Manag 19(2):185–200

GreenBiz (2018) Icebreaker's sustainability report sets the new standard to follow. www.greenbiz.com/article/icebreakers-sustainability-report-sets-new-standard-follow. Last Accessed 10 Feb 2019

Harris I, Naim M, Palmer A, Potter A, Mumford C (2011) Assessing the impact of cost optimization based on infrastructure modelling on CO_2 emissions. Int J Prod Econ 131(1):313–321

Iraldo F, Testa F, Melis M, Frey M (2011) A Literature review on the links between environmental regulation and competitiveness. Environ Policy Gov 21(3):210–222

Kim I, Min H (2011) Measuring supply chain efficiency from a green perspective. Manag Res Rev 34(11):1169–1189

Koh SL, Genovese A, Acquaye AA, Barratt P, Rana N, Kuylenstierna J, Gibbs D (2013) Decarbonising product supply chains: design and development of an integrated evidence-based decision support system-the supply chain environmental analysis tool (SCEnAT). Int J Prod Res 51(7):2092–2109

Laeequddin M, Sahay BS, Sahay V, Abdul Waheed K (2010) Measuring trust in supply chain partners' relationships. Meas Bus Excell 14(3):53–69

Lambert DM, Cooper MC, Pagh JD (1998) Supply chain management: implementation issues and research opportunities. Int J Logist Manag 9(2):1–20

Matthews L, Power D, Touboulic A, Marques L (2016) Building bridges: toward alternative theory of sustainable supply chain management. J Supply Chain Manag 52(1):82–94

Montoya-Torres JR, Ortiz-Vargas DA (2014) Collaboration and information sharing in dyadic supply chains: a literature review over the period 2000–2012. Estudios Gerenciales 30(133):343–354

Naslund D, Williamson S (2010) What is management in supply chain management?-A critical review of definitions frameworks and terminology. J Manag Policy Pract 11(4):11–28

Olga C (2012) Sustainable supply chain management: theoretical literature overview. Lund University, International Institute for Industrial Environmental Economics

Oxborrow L, Brindley C (2014) Disintermediation in the apparel supply chain. J Fash Mark Manag 18(3):252–268

Pullman ME, Maloni MJ, Carter CR (2009) Food for thought: social versus environmental sustainability practices and performance outcomes. J Supply Chain Manag 45(4):38–54

Schmidt CG, Foerstl K, Schaltenbrand B (2017) The supply chain position paradox: green practices and firm performance. J Supply Chain Manag 53(1):3–25
Walsh WE, Wellman MP, Ygge F (2000) Combinatorial auctions for supply chain formation. In: Proceedings of the 2nd ACM conference on electronic commerce, pp 260–269
Yang GF, Wang ZP, Li XQ (2009) The optimization of the closed-loop supply chain network. Transp Res Part E Logist Transp Rev 45(1):16–28
Zhu Q, Sarkis J (2004) Relationships between operational practices and performance among early adopters of green supply chain management practices in Chinese manufacturing enterprises. J Oper Manag 22(3):265–289

Sustainability Driving Eco-Innovation

Human Capital for Eco-Innovation Eco-System: Assessing the Impact of Sustainable HR Practices on Achieving Environmental Sustainability

Rubee Singh, Kamalesh Ravesangar, and Vikas Kumar

Abstract This part scrutinizes the vital role of human resource management practices in the ecosystem of company innovation. It provides a proposed framework for how strategic HR systems can enable workers and foster the creation of sustainable solutions. The framework examines how HR policies across the employee lifecycle contribute to environmental sustainability. This includes attracting and selecting environmentally conscious individuals, providing the knowledge and skills needed for sustainable practices, and driving sustainability considerations into business of research to promote a culture of environmental accountability. The discussion will focus on how employee engagement through networking, networking and recognition unleashes creativity, leading to innovative solutions to address environmental issues. Considering the impact of HR practices on the environment, this section provides a comprehensive framework for organizations aiming to shape a sustainable future through strong HR on. This chapter not only serves as pivotal reading material for scholars and students in human resource management, knowledge management, organizational innovation, and organizational behavior but also offers valuable insights and guidance for human resource practitioners, and professionals.

Keywords Eco-innovation · Human capital · Human Resource management · Environmental sustainability · Sustainable practices

R. Singh (✉)
Institute of Business Management, GLA University, Mathura, India
e-mail: dr.rubeerajput@gmail.com

Department of Management Studies, Kumaun University, Nainital, Uttarakhand, India

K. Ravesangar
Department of Accounting and Business, Tunku Abdul Rahman University of Management and Technology, Kuala Lumpur, Malaysia

V. Kumar
Faculty of Business Law and Social Sciences, Birmingham City University, Birmingham, UK

R. Singh et al. (eds.), *Zero Carbon Industry, Eco-Innovation and Environmental Sustainability*, World Sustainability Series,
https://doi.org/10.1007/978-3-031-80220-1_7

1 Introduction

Human-induced growth has pushed various ecosystems to their limits thus affecting them in a terrible way. Although the situation evolved, too little was done and no public response was generated. The economic growth of Southeast Asia is now more inclined toward agricultural development whose ecological impact is now environmentally friendly—such as low-carbon, ecological-circular, and green methods (Zaid and Jaaron 2020). This serious problem is forcing many national organizations to re-prioritize sustainability and the traditional notion of profit of a concern with the environment. Projects, environmental and community responsibility, increased social responsibility, and sustainable practices. The question that should be handled now is how working people from developing countries could secure their economic objectives while allowing for sustainable development in which they can also participate (Asadi et al. 2020). The environment, economy, and society are the substrates of sustainable development (Zaidi and Jamshed 2021) in the policy, different sector, action plans, and their execution. SDGs are web-like activities that the Earth has set for itself to tackle issues that are interconnected across sectors and bind everyone by their law. Basically, the goals are like connected nodes of the internet where all point to only the nodes they connect to. On the other side, the Sustainable Development Goals (SDGs) are objectives whose framework is as follows: "People, Planet, Prosperity, Peace, and Partnerships". In the human dimension, the SDGs deal with the issues of resource management, human capacities building, social inclusion, and labor conditions. One of the business targets is that companies should provide an environment that assures people's health and a well-educated workforce. At the same time, it enlightens the employees to the necessity of productivity which will in turn lead people to benefit from the country's development. The induction of SCPB is an essential aspect in the sustainable development. R&D professionals will decide the fate of the project. Gray areas show that some qualities are not yet gained. It has been observed that today's companies often spend too much money on marketing, administration, and customer acquisition but do not pay enough attention to their employees' performance. Many companies are not aware that there exists a link between the quality of leadership, specific knowledge, and expertise and their level of innovation. The development of community health services is one example of the healthcare workers driving organizational development. Along with the environmental issues, policymakers can focus on the development directions such as the Green Public Procurement (GPP) for sustainable consumption trends for households, and mobility. Climate, environment, and society are also the driving forces in this quest for sustainable development.

Over the past ten years, there has been a significant focus on eco-innovation in research, primarily due to the rise in the development of eco-innovative products and their integration into business operations. In essence, eco-innovation represents the technological progress aimed at achieving eco-efficiency, and it can be seen as a strategic move by companies to enhance their competitive edge (Zhang 2024). This pertains to the introduction of environmentally friendly products or the implementation of clean processes dependent on sustainable resources, focusing on waste reduction and potential environmental benefits. Companies engaged in eco-innovation must effectively manage resources and capabilities to ensure a strong linkage between innovation and environmental concerns, leading to favorable environmental and financial performance. Scholars concur that resources and capabilities are pivotal in enabling firms to realize returns on investments in eco-innovation (Ding et al. 2021). Despite the extensive and growing body of literature on eco-innovation, there is still a dearth of empirical studies directly measuring human capital specifically dedicated to eco-innovative processes within firms. Sustainable development entails meeting the integrated socio-economic needs of the present generation without compromising the ability of future generations to meet their own needs (Sharma 2020). It encompasses social, economic, environmental, and institutional dimensions. For instance, the innovation of new types of machinery, equipment, and other mechanical tools during the industrial revolution raised the graph of economic growth and living standard but also posed a threat to the environment and ecosystems. The other is the industrial and service uses of energy, driven by the quest for a more concentrated source of fuel due to the growing, big, and hungry needs of the industrial sector that demands energy sources on a huge scale, reliable, and flexible. There are various forms of technological advancements, such as new products and processes. Effects of technological product innovations include the introduction into the market of new or improved products and methods because of their implementation. For a product or method to be considered innovative, it needs to be able to show better benefits than those existing in an industry or the market.

In order to enlist the support of all employees for environmentally friendly initiatives within the organization, the Human Resources department will demonstrate its commitment to sustainability through Sustainable Human Resource Management (SHRM). This approach focuses on cultivating a workforce that embraces and implements eco-friendly practices, and upholds environmental objectives across all aspects of HRM, including recruitment, training, compensation, development, and advancement of human capital. SHRM serves as a guide for HR managers to integrate green practices and ensure that all employees contribute to environmental conservation. While SHRM shares similarities with traditional HR practices, it is centered on organizational sustainability and eco-friendly initiatives. The function of Human Resources is essential in advancing environmental conservation and promoting sustainable practices to position the organization as a champion of environmental protection. This entails embracing environmentally conscious recruitment, training, performance management, and other HR practices that align with environmentally friendly objectives. Management is responsible for taking proactive measures to raise awareness and drive action on climate change globally, while Human Resource Management (HRM) is entrusted with implementing and advocating for Sustainable HRM. SHRM represents a significant strategic intervention that can enable organizations to harmonize their HR strategies with environmental sustainability (Al-Ghazali and Afsar 2021). The establishment of eco-friendly policies and procedures would contribute to fostering an organizational culture that encourages and reminds employees to prioritize environmentally friendly practices. Consequently, this integration not only diminishes an organization's carbon footprint but also elevates its standing as a socially and environmentally conscientious institution. This chapter intends to provide a comprehensive examination of existing literature on HRM practices for achieving environmental sustainability, drawing upon foundational theories, conceptual frameworks, and practical implementations. Through the analysis of various studies and case examples, the chapter aims to underscore the opportunities and challenges associated with integrating environmental sustainability into the HRM agenda.

2 Literature Review

2.1 Human Capital for Eco-Innovation

A decade ago, the field of eco-innovation has experienced a huge leap among the researchers. There is a greater demand for eco-friendly products and services as well as environmental practice in many business sectors. It's a process of eco-innovation when a technology is used to produce or attain environmental efficiency. Companies are adopting eco-innovation to further strengthen their competitiveness by forming the creation of new environmentally friendly products or/and by applying environmentally friendly processes through the use of limited resources and waste avoidance. This move is based on different levels of environmental improvement (Antonioli et al. 2013). It is by mergers and acquisitions, project financing, and licensing which ideally bring the company to the market and at the same time make a return on investment. In addition to that, the introduction of new, innovative products on the market with better sustainability performances has also become one part of their strategy. Academia has identified the importance of a companyís resources and capabilities in determining the well-being of an eco-innovation investment. However, an assessment of all the human resources involved in the eco-innovation process of a firm is still not fully examined. Human capital is defined by the Organization for Economic Co-operation and Development (OECD) as the knowledge, skills, competencies, and personal characteristics that an individual has, which enhances personal, social, and economic life (Ardito et al. 2016). Human has specific qualities such as they can't be sold and there are no market platforms where HC assets can be exchanged. They can be visible in the lens of qualitative and quantitative as well as being general or specific and, all these considerations are rooted in external factors of the social and institutional environments. Researches have been conducted to explore the impact of human capital practices as mediating factors in the relationship between external knowledge breadth and innovation. In R&D, innovative teams have been the subject of multiple works due to the difference in their staffing size and geographic location (Triguero et al. 2016). Besides that, it is noticed that management of the human resources will also lead to their creativity in businesses (Lee et al. 2019). The main skills and personal attributes such as good communication, numeracy, and interpersonal skills like motivation and learning that are part of human resources have not been fully articulated.

There is a form of specification in regards to human capital in terms of health in education or mitigation, intellect, skills, training, employment features, values, and talent that comprises human capital quality to signify the quality of the workforce endowed with an economic value (Zhongwei and Liu 2022). However, it can be argued that health and education attributes could possibly be recognized as some of the most significant ones. In the last several years, many empirical and theoretical analyses have been estimated as to the relationship between human capital and ecological capital in the formation of economics and environmental protection. Some researchers have attempted to find correlation between economy and the environment, effects causing deterioration of the environment, plus methods of reversing the current status and enhancing the economy. As for those works that analyze the impact of resources on the development of nations that possess it, there is the one that demonstrates that resources stimulate development by fostering trade and production; on the other hand, there is works that claim that the large natural resource stock in a nation may actually cause protracted stagnation due to exploitation (Rouzi and Deng 2020). That is why different authors have written a number of research articles in English to produce such discourses; among the articles are Jena et al. (2022); Nathaniel et al. (2021). In the considered study on the topic of the resource curse, Langnel et al. (2021) discussed the topic and concluded the presence of the phenomenon in the analysed countries. Therefore the developing countries should guard the intellectual property and usher in supporting circumstances for innovations and growth of technologies. Governments may bring more efforts to the application and supervision of legal norms and regulations concerning Intellectual property right to stimulate enterprises' and individual's innovations. Moreover, it is necessary to increase financing for scientific and technological development, increase the index of purchasing power of the GERD and GDP, and provide higher status and positions for researchers in the labor market. The developing countries should enhance on technological advancement while at the same time creating organizational networks and affiliations with the rest of the world organizations for purposes of benchmarking on the best practices and innovations. For a nation, talent development and the encouragement of the relationships between institutions of scientific research and industries contribute to the quality improvement of inventions, as well as the acceleration of translation of technologies into applications to replenish the much-needed impetus for regional growth and development (Huan and Qamruzzaman 2019). Moreover, proper investments in good policies for future environmental issues are essential for developing countries and include such priorities as the decrease of CO_2 emissions, the increase of energy efficiency, and the usage of clean energy. The correlation between the two aspects of growth and the protection of the environment can only be realized if the developed countries embrace the principles of sustainable development which create the framework for future sustainable development.

Another relevant economic factor for promoting green technology innovation among educational human capital is a proximity of knowledge, technologies, skills, and experience in the sphere of green technologies. Hence, education has been found to have a particularly positive impact on the green economy itself, through three main mechanisms. First of all, a high level of educational human capital can increase awareness of the people in environmental protection and utilization of energy, thus fostering qualified people who can develop green technologies for business. Hence, improved education increases the tendency of organizations to adopt environmental standards, support environmentalism, and channel research and development towards enhanced energy efficiency, thus catalyzing green technology development. Secondly, high educational human capital means high income levels and better environmental conservation beliefs in which case, one can compare the environmental quality of local consumers with the international standards. Showcasing the populace to use less carbon, green industries, and sustainable lifestyles can impact the societal predisposition to the external environment aspect and improve the availability of green options among consumption products and raise the efficiency of the green economy. To summarize, learners with a considerable amount of educational human capital are more responsive to environmental pollution and ready to employ their time and funds to protest against such issues and establish organizations to influence governments and companies to reduce the rate of resource wastage (Chen and Hu 2019). Moreover, DH has spill over characteristics Educational human capital we used the term educational human capital because education has some characteristics of human capital and at the same time it has its own unique characteristics that distinguish it from the typical human capital. Places with HCA above average have a higher tendency of knowledge spillover which helps to transfer new knowledge and display an increasing effect gradient. The notion of cross-unit externalities does not generally present a concept where the impact lessens as the degree increases. Also, it can accelerate the fast development of superior foreign technology, which in turn can encourage local enterprises to invest more on green technology in compliance with strict environmental regulations they are bound to (Zhao et al. 2016).

The findings by the study show that a complete health status of the workforce is crucial in boosting the evolution of green technology advancement something that is in tandem with the findings on the importance of education standards among the workforce. The above state affects innovation in the following three aspects (Zhao and Li 2020). In the first place, increased psychic energy as well as general well-being of a person having healthy human capital, assists other employees to increase productivity and become an excellent frame work in the advancement of green technology. Secondly, through enhancing the positives of an employee's health status, healthy human capital may enhance cognitive ability and learning effectiveness and, in the process, impact educational achievements and the acquisition of educational human capital, thus giving a positive boost to the generation of green technology innovation. And, last but not least, a focus on wellbeing as a strategic goal is closely linked to higher investments in all parameters influencing peoples' healthy human capital, on providing a favourable environment for staff professional development and training, which in turn can strengthen enterprises' innovation activity in the sphere of green technology.

2.2 *Impact of Sustainable HR Practices on Achieving Environmental Sustainability*

There is a growing recognition of the impact of human resource management (HRM) processes on the environment, as HRM practices contribute to the establishment and maintenance of an environmental management system (EMS), which in turn helps organizations achieve improved economic performance. Green HRM, with its focus on environmental benefits, employee retention, and organizational attractiveness to workers, is expected to continue to play a crucial role in organizational management. Previous HRM literature has tended to examine individual practices as contributors to firm performance rather than as a combination of HRM practices. Siyambalapitiya et al. (2018) suggest that the proliferation of green practices in the workplace could enhance employee characteristics such as motivation and competence, potentially leading to improved financial performance. When a company integrates environmental responsibility into its recruitment policies or demonstrates a strong commitment to environmental conservation, it is likely to attract prospective applicants who value the eco-friendly nature of the company, as noted by Tariq et al. (2016) and Jamil et al. (2021a). Furthermore, promoting environmental interests and activities may result in increased skills, motivation, better retention rates, enhanced productivity, and an overall improvement in economic performance. Shah (2019) argues that as business organizations strive to incorporate sustainability into their operations and activities, they aim to foster a more sustainable corporate culture, leading to increased efficiency, cost reduction, and greater employee satisfaction. The relationship between embracing a greener culture and achieving higher sales with lower costs has been emphasized by Longoni et al. (2018) and Jamil et al. (2021b). Moreover, by increasing their focus on social performance activities, companies can guarantee that the social programs they implement bring advantages to both their internal community, including employees, and external communities, which consist of suppliers and customers. Several research studies have proven that companies that implement Green HRM practices, as evidenced by Mousa and Othman (2020), experience a beneficial effect on the health and overall well-being of their employees. This implementation ultimately leads to an overall improvement in the organization and the well-being of its employees.

Numerous organizations are presently adopting Sustainable Human Resource Management (SHRM) in their operations, while others have yet to embrace it due to various reasons. Some lack the knowledge of how to initiate SHRM within their organization. Additionally, some perceive it as an added burden for HR functions, while others are unaware that they are already practicing SHRM. The objective of this study is to provide guidance to organizations by exemplifying the successful implementation of SHRM and demonstrating its impact on environmental performance. SHRM entails the integration of environmentally friendly practices into HR functions such as recruitment, training, performance management, and benefits. If the public sector leads by example, it can inspire other organizations and private sectors to follow suit. Furthermore, SHRM is closely linked to environmental preservation and climate change mitigation, indicating an organization's commitment to protecting the planet. It is concerning that some individuals are apathetic to environmental concerns. Therefore, it is essential for all departments, including recruitment, training, and performance management, to incorporate eco-friendly practices. Environmental protection is a shared responsibility, and every employee needs to be mindful of its importance for future generations. The rapid pace of industrialization has led to increased environmental pollutants worldwide. Conversely, the rapid growth of air, water, and land pollution has reached such alarming levels that concerned authorities are now compelled to take necessary steps to protect humanity from disastrous ramifications to ensure a healthy and sustainable future (Rahman and Alam 2022). Recognizing the severity of the situation, the United Nations has listed environmental sustainability among its critical goals to be attained by 2030 (Arora and Mishra 2019).

Industry leaders bear the responsibility of reducing environmental degradation by implementing measures to decrease the release of harmful pollutants into the environment. Moreover, the adoption of green HR practices, including sustainable hiring, training, reward systems, and empowerment initiatives, can play a significant role in promoting environmental sustainability within organizations. Sustainable hiring involves expanding the workforce without the need for paper, thus conserving trees. Sustainable training focuses on educating employees on environmentally-friendly work practices. Sustainable rewards encompass both monetary and non-monetary incentives given to employees for their efforts in waste reduction and recycling. Sustainable empowerment entails providing employees with the necessary support for sustainable development. These practices fall under the realm of Strategic Human Resource Management (SHRM) and are anticipated to enhance environmental sustainability. The increasing prevalence of eco-friendly initiatives in business not only cuts costs but also boosts revenue, thereby offering a competitive edge. The subject of Sustainable Human Resource Management (SHRM) has become increasingly significant in the past two decades due to the escalation of natural disasters, pandemics, and human-induced harm to the environment. SHRM plays a crucial role in addressing environmental issues within different organizations through the implementation of environmentally friendly HR strategies, providing employees with eco-consciousness training, adopting eco-friendly recruitment processes, and incorporating eco-friendly performance management and incentives. These conclusions are in line with the studies conducted by Ojo et al. (2020) and Jehan et al. (2020), which highlight a positive correlation between eco-friendly HR practices and environmental sustainability (Fig. 1).

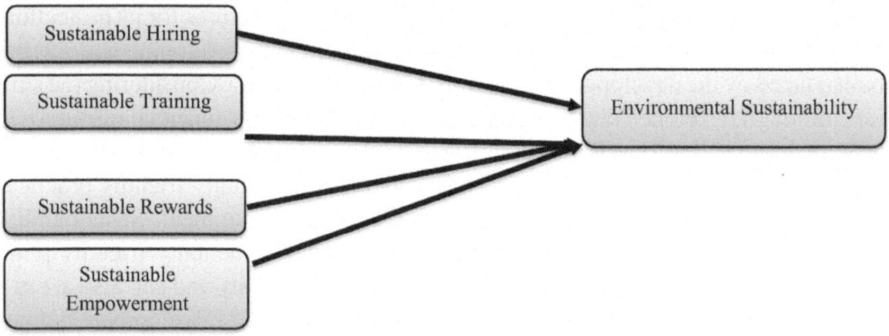

Fig. 1 Influence of eco-friendly HR strategies on environmental sustainability

2.3 Employees' Pro-Environmental Behavior and Environmental Sustainability

In supporting sustainability goals, sustainable behavior is crucial to advancing objectives of sustainability by employees that are prosocial and discretionary and are focused on environmental protection. Ahmad et al. (2021) have done a research that shows how behavior affects sustainability. HR professionals can use training, rewards, and empowerment to develop appropriate behaviour (Malik and Garg 2020; Zhang et al. 2019). Preventing the degradation of the environment necessitates the employees' actions towards its support as Kong et al. (2021) assert. This ties employee behavior with environmental sustainability through green HR practices (Cop et al. 2020). It acts as an arbitrator between green HR practices and environmental sustainability. For Carattini et al. (2020), cost plays a big part in influencing employee behavior or their willingness to embrace Pro-environmental Behavior (PEB). Employee involvement is essential for developing these green initiatives and policies necessary for effective environmental sustainability (Saeed et al. 2019). The need for employees to be involved in a variety of PEBs is imperative to support green companies (Lülfs and Hahn 2013; Saeed et al. 2019). The acknowledged role of employees has on business and sustainability strategies means that organizations must focus on promoting, encouraging and influencing employee behavior that fosters PEB as well as meeting corporate sustainability goals (Cheema et al. 2020; Paillé and Meija-Morelos 2019).

Environmental sustainability is all about the conservation of the natural environment functions for the future generation. In other words, it means conserving the environment from actions that are harmful to the environment such as emitting carbon dioxide into the atmosphere, emitting harmful chemicals into water bodies and contributing to the increased temperatures of the world (Silva and de Souza 2020). The rapid improvement in technology, urbanization and industrialization has undermined the potentials for human development to be sustainable. Environmental degradation resulted due to negligence of taking considerable efforts for saving the environment at the time of rapid industrialization (Ali et al. 2021). The situation has been getting worst due to the insufficient taking of proper measures for its protection. Environmental degradation is not only a problem linked with developing nations, but it is also faced by the developed ones as well (Cheng et al. 2021). Considering the situation, although developed countries have the financial and non-financial resources to deal with this problem, developing countries do not have the means to do so it is necessary to take them focus on a culture of fostering environmentally friendly practices among their employees rather than relying on imported equipment. Sustainability training that involves and empowers employees in environmentally friendly practices can have a positive impact on environmental sustainability and can also act as a force a it stimulates. The World Council on Environment and Development defines sustainable development as meeting the needs of the current generation without enabling future generations to meet their own needs is not destroyed. Environmental sustainability aims to improve the quality of the environment and prevent degradation and degradation (Louman et al. 2019). The process involves three steps: First, the organization complies with the environmental laws and regulations that provide the foundation. Second, the organization focuses on environmental protection through pollution reduction and related measures. Finally, environmental sustainability is ensured through active voluntary activities.

3 Theoretical Background

High performance work systems in the AMO model specify the nature of the strategic content of performance indicators more clearly by focusing on the conditions that motivate workers to go the extra mile. This model draws its basis from psychological theories that are associated with frameworks that define the systems attributes. Therefore, it highlights possession of ability, motivation to increase discretionary actions, and employees' enabling to support organizational objectives. Marin-Garcia and Tomas (2016) have also argued that the hardest one is the AMO model that enhances human resource management practices based on organizations' strategic plans. Thus, the transition to this model is not about following a specific number of practices but lays a lot on how these practices are interpreted by workers. Thus, there is a need to differentiate between the real and expected human resource management practices in this respect. Pointing to some scholars, it has been noted that more understanding comes from the AMO model applied from the employee perspective rather than the administrative one, as it helps to identify the mechanisms of how changes in employees' behavior are realized (Kundu and Gahlawat, 2018). From this point of view the perspective assumes that the right choice of hr practices and its application is necessary to implement also take into account the business leaders' capacity for executing it, encourages job performance, and requires organizational support which is needed to provide the necessary changes. The model is a three DOM that involves Ability, Motivation, and Opportunity in promoting the human resource management practices (Pak et al. 2019).

The preceding concepts of sustainable HRM have drawn strength from various theories in the domain of human relations and organization. A discussion, in detail, of these theories can be done to gain a deeper understanding of Green HRM (Arulrajah and Opatha 2016). The theory postulates that HRM can play a positive role through attracting, selecting, training, developing, and rewarding with adequate compensation an experienced able, and competent workforce so that their motivational factors and productivity are enhanced. This eventually makes them more effective and efficient, leading to more profits, maximum profitability, and enhanced quality. In this section, the fundamental elements of SHRM have been systematically evaluated in a chronological manner. One of the biggest challenges for an organization in today's competitive world, where every organization is trying to attract talented manpower to win over the competition, is the successful recruitment, training, and development of a capable and competent workforce. In this context, SHRM practices are being diffused at a rapid rate across various organizations in the world pertaining to employer branding and competitive advantage through creating awareness among young minds about environmental issues and sustainable business practices of the concerned organization (Chaudhary 2019). These current trends, leading to heightened pro-environmental employee behaviour, include using recycled paper, applying teleconferencing or video conferencing, carpooling, or a common van arranged for carrying the employees, so that firms may move towards greener practices. Public sector universities operating in the Gilgit and Baltistan region can apply the AMO theory effectively in making the universities foster sustainability and green practices through education and practical application.

4 Conclusion, Limitation and Recommendation

This emerging issue of Climate Change and Environmental degradation has promoted Environmental sustainability to be a foremost preoccupation to the whole world. Businesses which have long been recognized as entities with a strong negative impact on the environment, experience growing demands for them to take responsibility for the cost their actions impose on the natural environment. Thus, the trend towards incorporating sustainable aspects into their business processes can be observed. As a result of understanding the importance of employees to the success of an organization, there is an increasing trend towards the implementation of SHRM for advancing environmental sustainability. The consequences of four critical SHRM practices namely; e-coordination, e-selection, e-education, and e-motivation or green incentives, and e-empowerment in the realization of environmental sustainability in business organizations are explored in this study.

The key to a successful environmental strategy is the employees. Sustainable hiring practices focus on individuals with the right skills, experience, and a dedication to environmental values. This means hiring candidates who grasp the organization's sustainability objectives, care about environmental issues, and are willing to promote sustainable practices in their roles. Various methods can be used for green hiring.

Job descriptions can include environmental awareness, discussing the organization's sustainability efforts and the role's impact on environmental goals. During interviews, questions can center on the candidate's environmental beliefs, prior involvement in sustainable practices, and commitment to furthering the organization's environmental mission. Additionally, integrating environmental considerations into skills evaluations can identify candidates with expertise in green technologies and relevant best practices. By prioritizing green hiring, companies can establish a workforce that naturally aligns with their sustainability objectives. These employees are more inclined to identify chances for environmental enhancements, actively engage in green initiatives, and advocate for sustainable practices within the company.

It is imperative to target the moderating variables to achieve the green hiring objectives, but it is also important to equip the targeted people with the know-how that will enable them to practice what they believe in. Environmentally friendly training encourages the provision of knowledge and assistances needed by the employee in understanding the goals and responsibilities on the environment of the organization besides providing an understanding of the consequence of their operations to the environment and the environmental measures suited to their job description. Sustainable training also entails training on issues to do with environment, life cycle evaluation, waste minimization, energy conservation and use of environmental friendly technologies. Training could also involve various forms of training that can be in a form of classroom sessions, online, workshops and one on one coaching. Good green training plays a central role in developing organizational responsibility towards the environment. Employees become more knowledgeable about the environmental impact of their decisions and actions, as well as the tools that can be implemented for managing sustainable behaviour at the workplace, such as reducing paper consumption, energy saving, proper waste disposal etc.

In encouraging long-term behavioral changes, green incentive programs are crucial. These initiatives recognize and compensate employees for their efforts in advancing environmental sustainability. This motivates them to prioritize eco-friendly actions and participate in sustainable projects. These green incentives take different forms, such as integrating environmental standards into performance evaluations and providing financial rewards for achieving specific environmental targets. Incentives that are not monetary, such as acknowledgment programs, public recognition, or extra vacation time, also prove effective in promoting environmentally responsible behaviors. By linking positive reinforcement with eco-conscious behavior, sustainable incentive programs motivate employees to actively contribute to sustainability goals. Additionally, these programs foster a competitive and collaborative atmosphere among coworkers, promoting a supportive environment for reaching environmental targets. Encouraging sustainable initiatives is crucial for instilling a sense of environmental responsibility within an organization. By empowering staff to take the lead and share their ideas, a genuine dedication to sustainability can be established. Forming cross-departmental teams to generate and execute eco-friendly projects is one way to promote empowerment. This method ensures a wide range of insights and skills are considered. Additionally, facilitating open lines of communication for employees to propose environmental enhancements can further boost empowerment. Granting authority and resources for employees to act on their ideas can enhance their feeling of ownership and involvement, resulting in more effective and innovative approaches to meeting environmental sustainability targets.

Previous empirical evidence regarding SHRM is limited to investigating the primary form of human resource management, nevertheless, analyzing the effects of multiple forms of SHRM practices in an abstracter theories that comprise organizational culture, leadership type and the environmental constituent. The identified research articles may fail to capture all the possible research on the given area. Disadvantages include the fact that if it is done by journals or databases or keywords, there will be bias in the selection hence exclusion of other important works. To avoid this issue, it is recommended to use a more extensive search strategy. Despite presenting a qualitative synthesis of meta-analysis findings from various studies, a quantitative critique of the individual studies, with regards to strengths and weaknesses of each study's methodology, would propound a more accurate presentation of current research. Assembling data from different research works carried out employing different approaches and in different settings can be complicated. Therefore, this review may be lacking some of the subtle differences and specifics of the relationship between SHRM and environmental sustainability, as revealed in the literature.

References

Ahmad N, Ullah Z, Arshad MZ, waqas Kamran H, Scholz M, Han H (2021) Relationship between corporate social responsibility at the micro-level and performance: environmental: the mediating role of employee pro- environmental behavior and the moderating role of gender. Sustain Prod Consum 27:1138–1148

Al-Ghazali BM, Afsar B (2021) Retracted: green human resource management and employees' green creativity: the roles of green behavioral intention and individual green values. Corp Soc Responsib Environ Manag 28(1):536. https://doi.org/10.1002/csr.1987

Ali S, Dogan E, Chen F, Khan Z (2021) International trade and environmental performance in top ten-emitters countries: the role of eco-innovation and renewable energy consumption. Sustain Dev 29(2):378–387

Antonioli D, Mancinelli S, Mazzanti M (2013) Is environmental innovation embedded within high-performance organisational changes? The role of human resource management and complementarity in green business strategies. Res Policy 42:975–988

Ardito L, Petruzzelli AM, Albino V (2016) Investigating the antecedents of general purpose technologies: a patent perspective in the green energy field. J Eng Technol Manag 39:81–100

Arora NK, Mishra I (2019) United Nations sustainable development goals 2030 and environmental sustainability: race against time. Environ Sustain 2:339–342. (Springer)

Arulrajah AA, Opatha HHDNP (2016) Analytical and theoretical perspectives on green human resource management: a simplified underpinning. Int Bus Res 9(12):153–164

Asadi S, OmSalameh Pourhashemi S, Nilashi M, Abdullah R, Samad S, Yadegaridehkordi E et al (2020) Investigating influence of green innovation on sustainability performance: a case on Malaysian hotel industry. J Clean Prod 258:120860. https://doi.org/10.1016/j.jclepro.2020.120860

Carattini S, Gosnell G, Tavoni A (2020) How developed countries can learn from developing countries to tackle climate change. World Dev 127:1–3. https://doi.org/10.1016/j.worlddev.2019.104829

Chaudhary R (2019) Green human resource management in Indian automobile industry. J. Global Respons. 10(2):161–175

Cheema S, Afsar B, Al-Ghazali BM, Maqsoom A (2020) How employee's perceived corporate social responsibilityaffects employee's pro-environmental behaviour? The influence of organizational identification, corporate entrepreneurship, and environmental consciousness. Corp Soc Responsib Environ Manag 27(2):616–629. https://doi.org/10.1002/csr.1826

Chen GS, Hu JX (2019) The impact of human capital accumulation on environmental quality. Urban Probl 10:46–52

Cheng Y, Sinha A, Ghosh V, Sengupta T, Luo H (2021) Carbon tax and energy innovation at crossroads of carbon neutrality: designing a sustainable decarbonization policy. J Environ Manage 294:112957

Cop S, Alola UV, Alola AA (2020) Perceived behavioral control as a mediator of hotels' green training, environmental commitment, and organizational citizenship behavior: a sustainable environmental practice. Bus Strat Environ 29(8):3495–3508

Ding Q, Khattak SI, Ahmad M (2021) Towards sustainable production and consumption: assessing the impact of energy productivity and eco-innovation on consumption-based carbon dioxide emissions (CCO_2) in G-7 nations. Sustain Prod Consum 27:254–268

Huan Y, Qamruzzaman M (2022) Innovation-led FDI sustainability: clarifying the nexus between financial innovation, technological innovation, environmental innovation, and FDI in the BRIC nations. Sustainability 14:15732

Jamil K, Hussain Z, Gul RF, Shahzad MA, Zubair A (2021a) The effect of consumer self-confidence on information search and share intention. Inf Discov Deliv June. https://doi.org/10.1108/IDD-12-2020-0155

Jamil K, Liu D, Anwar A, Rana MW, Amjad F, Liu M (2021b) Nexus between relationship marketing and export performance of readymade garments exporting firms. Ind Textila 72(6):673–679. https://doi.org/10.35530/IT.072.06.202028

Jehan Y, Hussai D, Batool M, Imran M, Rasaizadi A, Askari M, Hayat N (2020) Effect of green human resource management practices on environmental sustainability. Int J Hum Cap Urban Manag 5(2):153–164

Jena PK, Mujtaba A, Joshi DPP, Satrovic E, Adeleye BN (2022) Exploring the nature of EKC hypothesis in Asia's top emitters: role of human capital, renewable and non-renewable energy consumption. Environ Sci Pollut Res 29:88557–88576

Kong L, Sial MS, Ahmad N, Sehleanu M, Li Z, Zia-Ud-Din M, Badulescu D (2021). CSR as a potential motivator to shape employees' view towards nature for a sustainable workplace environment. Sustainability 13(3):1499

Kundu SC, Gahlawat N (2018) Ability–motivation–opportunity enhancing human resource practices and firm performance: evidence from India. J Manag Organ 24(5):730–747

Lagnel Z, Amegavi GB, Donkor P, Mensah JK (2021) Income inequality, human capital, natural resource abundance, and ecological footprint in ECOWAS member countries. Resour Policy 74:102255

Lee J, Kim S, Lee J, Moon S (2019) Enhancing employee creativity for a sustainable competitive advantage through perceived human resource management practices and trust in management. Sustainability 11:2305

Longoni A, Luzzini D, Guerci M (2018) Deploying environmental management across functions: the relationship between green human resource management and green supply chain management. J Bus Ethics 151(4):1081–1095. https://doi.org/10.1007/s10551-016-3228-1

Louman B, Keenan RJ, Kleinschmit D, Atmadja S, Sitoe AA, Nhantumbo I, de Camino Velozo R, Morales JP (2019) SDG 13: climate action—impacts on forests and people. In: Katila P, Pierce Colfer CJ, de Jong WG, Galloway G, Pacheco P, Winkel G (eds) Sustainable development goals: their impacts on forests and people, pp 419–444. Cambridge University Press, Cambridge, UK

Lülfs R, Hahn R (2013) Corporate greening beyond formal programs, initiatives, and systems: a conceptual model for voluntary pro-environmental behavior of employees. Eur Manag Rev 10(2):83–98. https://doi.org/10.1111/emre.12008

Malik P, Garg P (2020) Learning organization and work engagement: the mediating role of employee resilience. Int J Hum Resour Manag 31(8):1071–1094. https://doi.org/10.1080/09585192.2017.1396549

Marin-Garcia JA, Tomas JM (2016) Deconstructing AMO framework: a systematic review. Intang Cap 12(4):1040–1087

Mousa SK, Othman M (2020) The impact of green human resource management practices on sustainable performance in healthcare organisations: a conceptual framework. J Clean Prod 243:118595. https://doi.org/10.1016/j.jclepro.2019.118595

Nathaniel SP, Yalciner K, Bekun FV (2021) Assessing the environmental sustainability corridor: linking natural resources, renewable energy, human capital, and ecological footprint in BRICS. Resour Policy 70:101924

Ojo AO, Tan CN-L, Alias M (2020) Linking green HRM practices to environmental performance through pro- behaviour environment in the information technology sector. Soc Responsib J

Paillé P, Meija-Morelos JH (2019) Organisational support is not always enough to encourage employee environmental performance. The moderating role of exchange ideology. J Clean Prod 220:1061–1070. https://doi.org/10.1016/j.jclepro.2019.02.192

Pak K, Kooij DT, De Lange AH, Van Veldhoven MJ (2019) Human resource management and the ability, motivation and opportunity to continue working: a review of quantitative studies. Hum Resour Manag Rev 29(3):336–352

Rahman MM, Alam K (2022) Impact of industrialization and non- renewable energy on environmental pollution in Australia: do renewable energy and financial development play a mitigating role. Renew Energy

Rouzi S, Deng F (2020) The nonlinear impact of human capital agglomeration on green economic efficiency. Soc Sci 8:96–102

Saeed BB, Afsar B, Hafeez S, Khan I, Tahir M, Afridi MA (2019) Promoting employee's proenvironmental behavior through green human resource management practices. Corp Soc Responsib Environ Manag 26(2):424–438. https://doi.org/10.1002/csr.1694

Shah M (2019) Green human resource management: development of a valid measurement scale. Bus. Strat. Env. 28(5):771–785. https://doi.org/10.1002/bse.2279

Sharma RR (2020) Green management and circular economy for sustainable development. Vision 24:7–8. https://doi.org/10.1177/0972262920912497

Silva LP, de Souza FT (2020) Urban management: learning from green infrastructure, socioeconomics and health indicators in the municipalities of the state of Paraná, Brazil, towards sustainable cities and communities. In: Filho WL, Tortato U, Frankenberger F (eds) Universities and sustainable communities: meeting the goals of the agenda 2030, pp 493–509. Springer, Heidelberg, Germany

Siyambalapitiya J, Zhang X, Liu X (2018) Green human resource management: a proposed model in the context of Sri Lanka's tourism industry. J Clean Prod 201:542–555

Tariq S, Jan FA, Ahmad MS (2016) Green employee empowerment: a systematic literature review on state-of-art in green human resource management. Qual Quant 50(1):237–269. https://doi.org/10.1007/s11135-014-0146-0

Triguero A, Moreno-Mondéjar L, Davia MA (2016) Leaders and laggards in environmental innovation: an empirical analysis of SMEs in Europe. Bus Strateg Environ 25:28–39

Zaid AA, Jaaron A (2020) GHRM bundle practices and sustainable manufacturing performance: understanding potential relationships. Int J Sci Technol Res 9:7125–7132

Zaidi SYA, Jamshed S (2021) The role of green human resource management practices towards attaining sustainable development goals through corporate sector. Sustain Bus Soc Emerg Econ 3:665–682

Zhang Y, Luo Y, Zhang X, Zhao J (2019) How green human resource management can promote green employee behavior in China: a technology acceptance model perspective. Sustainability 11(19):5408

Zhang Y (2024) Promoting sustainable growth through eco-innovation, energy transition, and human capital: a comparative analysis of G-7 economies. Econ Chang Restruct 57(2):1–29. (Springer)

Zhao X, Li X (2020) The influence of education and healthy human capital on labor productivity. Soc Sci Front 5:53–60

Zhao L-D, Zhang L, Xu L, Hu M-Z (2016) Mechanism of human capital, industrial structure adjustment and green development efficiency. China Popul Resour Environ 2016(26):106–114

Zhongwei H, Liu Y (2022) The role of eco-innovations, trade openness, and human capital in sustainable renewable energy consumption: evidence using CS-ARDL approach. Renew Energy 201:131–140. https://doi.org/10.1016/j.renene.2022.10.039

Assessing the Adverse Impacts of Innovation and Development Through Sustainability: A Global Perspective

Jacinta Dsilva, Godwin Francis, and Rekha Pillai

Abstract Sustainable transitions through innovations, although beneficial, bring significant challenges. Sustainability is the idea of meeting present needs without compromising future generations to achieve their needs. However, statistics and studies show that sustainable development has caused unintended social and environmental harm. This chapter discusses the harmful impacts of innovation in sustainable development, which include the disruption to social norms, unintended adverse environmental effects, and relative privileges a nation may hold because of their resource availability. Followed by a literature review highlighting the disparity between developed and developing nations in micromanaging sustainable actions. Next, the chapter analyses sustainable actions and the underlying consequences of those actions that specific developed countries have conducted. Then discusses the issues faced by developing nations to manage their development responsibilities with their available resources. Finally, the chapter discusses potential recommendation to combat these adverse consequences through a united front and international legal framework called Sustainability for the Future Organization (SFTFO) that provides countries worldwide with social, environmental, and economic needs to match personal goals in achieving the UN sustainable goals.

Keywords Darkside · Sustainability · Developed countries · Developing countries · Environment · Innovation · Policies

J. Dsilva (✉)
SEE Research Centre, SEE Institute, The Sustainable City, Dubai, UAE
e-mail: jacinta@seeinstitute.ae

G. Francis · R. Pillai
The British University in Dubai, Dubai, UAE

R. Singh et al. (eds.), *Zero Carbon Industry, Eco-Innovation and Environmental Sustainability*, World Sustainability Series,
https://doi.org/10.1007/978-3-031-80220-1_8

1 Introduction

The mammoth challenge of climate change, geo-political stress and armed conflicts with associated repercussions has paved way for accelerated interests in sustainability transitions thereby leading to an innovation spree (Grashof and Basilico 2023). Although these powerful innovations have the potential to address core inequalities and injustices in the current social systems, there is a dark side to it. Before we embark further into the former, it is imperative to know what sustainable development encompasses. Integrating the three sustainability dimensions, the World Commission on Environment and Development defined sustainable development as "meeting the needs of the present without compromising the ability of future generations to meet their own needs" (as cited in Brundtland 1987; p. 24). Additionally, Chaparro-Banegas et al. (2024) remark that for UNDP (United Nations Development Program), innovation, technology and entrepreneurship act as central pillars of the three dimensions of transition to sustainable development.

However, these transitions disrupt the established norms, rules and socio-technical systems by initiating and adopting revolutionary practices, disruptive technologies and adopting perspectives that challenge the status quo, inviting varying levels of displeasure and discontent from multiple stakeholders. Tomar (2023) elaborated on privacy erosions and rampant spread of misinformation due to information and communication technology (ICT) advancements leading to society polarizations and unfavorable socio-political-economic consequences. He added that artificial intelligence (AI) and internet of things (IOT) had enhanced organizational productivity at the expense of job loss and displaced labor in pursuit of new skills required for sustenance. The over reliance on AI systems to automate work process has led to faulty/discriminatory/biased credit scores (during company ratings) and personality scores (during hiring process). Autonomous weapons are now being utilized in unethical ways also snowballing to unintended harm that disrupt the moral paradigm. Moving on, sustainable ideas such as digital interconnectivity has led to cyber-attacks and espionage causing substantial loss to intellectual property and business intelligence, thus tarnishing company reputation, disrupting workflow and stock value slumps (Venkatachary et al. 2017). On another note, Munoz et al. (2018) report a negative business performance when excessive attention is given to sustainable social aspects, with Renko (2013) addressing pro-social entrepreneurial motivations posing as a hurdle in the business start-up process in comparison with their economically motivated counterparts. Furthermore, Chaparro-Banegas et al. (2024) warn about the social inequalities among countries as an after math of sustainable oriented innovation with Al Kez et al. (2022) cautioning massive carbon emissions, water shortage and land footprint as deleterious effects of digitalization. Focusing on carbon emissions, Ferreira-Filipe et al. (2021) mentioned the dark sides of sustainable packaging in the form of harmful emissions due to the nature of raw materials used, growing waste due to inadequate end of life infrastructure and uncompetitive prices compared to the other conventional products.

Similarly, rampant economic growth in the name of sustainability also aggravates loss of biodiversity, leading to unwarranted waste generation, all of which translate to adverse climatic conditions (Kallis et al. 2012). Adding on, Coad et al. (2020) refers to another dark side of innovation as 'end of product life considerations' which relate to nuclear waste disposals, mining waste remnants and hefty metal leakages from e-waste recycling. More surprisingly, Hansen et al. (2022) highlighted that waste (especially lead) from sustainable products such as solar panels are generally recycled in underdeveloped countries (especially Sub-Saharan African regions) who lack protective equipment or political support, thereby leading to unfavourable health issues. Nonetheless, the voluntary nature of sustainability adoption, without any standardized regulations or practices have also led to 'commodification' of these practices where customers are asked to pay a hefty price for products under dubious claims of sustainability (Ravalli 2021). In addition, biodegradable plastic is leading to negative social and environmental issues related to changes in land usage during feedstock and complicated waste management practices (Turkcu et al. 2022).

Interestingly, Pinheiro (2022), argued that the developed nations hold a privilege in resource availability (physical, financial and intangible), access to human capital and state of the art infrastructure which catalyze innovation and sustainable practices in comparison to developing nations who adopt simpler non-destructive technologies (Boschma et al. 2018). In this chapter we concur to Brandi and Weinhardt (2021) who employed World Bank's classification of 'developed countries' and 'developing countries' where the former enjoys a high-income status while the latter is home to low- and middle-income groups. They also refer to some of the emerging markets such as Brazil, India and China economically progressing towards the strata of developed nations. Prior research caveat of a void in sustainable literature which provides a comparative study on the varied effects of sustainable development and consequent inequalities in regions (Hartman et al. 2017). Additionally, Hansen et al. (2018) urge a critical reflection on sustainability transitions especially in developing countries. Therefore, to address the former gaps, the chapter entails two-fold objectives. First to identify the dark side of innovation and sustainable growth and secondly to demarcate their extent and intensity in developed and developing countries. By performing this exploratory study, the chapter can derive pertinent insights on the potential dark sides of sustainability practices and innovation in developed and developing countries, thereby enriching the sustainability transition literature. In addition to broadening our knowledge horizons on the shady areas sidelined under the name of sustainability, the research provides pragmatic value by identifying current and potential negative outcomes that entail policy implications to ensure a balanced development.

2 Literature Review

The developed countries are those characterised by a relatively high income per capita, advanced technological infrastructure and well-developed industrial bases that have both an international footprint and influence over global economic trends

(World Bank 2022). These are countries that have strong educational infrastructures, investment in research and development (R&D) and healthcare systems. Developed countries like the USA, and Germany are well known for their high GDP per capita and have a major share in global technological improvement (Mankiw 2018). Advanced economies have the competitive advantage of well-developed industries and high-quality institutional frameworks that are necessary for innovation, and productivity growth (Porter 1990).

On the other hand, less developed countries usually have limited per capita income and are emerging industrial bases where many of these developing economies face a technological world-front which goes along with not having total stability in its macroeconomics (Todaro and Smith 2020). These countries encounter problems unparalleled in terms of insufficient healthcare facilities, low standard education and infrastructural shortcomings. Many regions in Sub-Saharan Africa and parts of South Asia have elevated levels of poverty resulting from inadequate institutions limiting access to modern technologies (Sachs 2005). Social and economic inequalities further inhibit the development process in such countries, impeding overall growth (Sen 1999). On the one hand, developed countries have driven global trends in innovation and industrialisation by major economies with their economic strength to formulate global rules; on the other hand, developing countries often depend on technological transfer and foreign direct investment from some industries of advanced nations for spurring economic growth, meanwhile, such dependencies could cause policy deficits (Stiglitz 2002).

For both the developed and developing economies, it is an area filled with uncertainty at the corner point of technological innovation and environmental sustainability. Improvements mostly have achieved status as saviours of the environment, but if not built properly they can present new challenges. However, our understanding of the role innovation might play in achieving environmental sustainability is still gradual and varies from one regional or global context to another.

2.1 Developed Countries

In developed countries, technological progress is often praised to resolve environmental issues. But these interventions can also backfire if they aren't done holistically. To put it simply; Innovations take place within a socio-technical–economic-ecological system, and not fully realize this complex environment may result in simplifying and over-celebrating the potential of innovations or incorrect messages to advocate sustainability, tending towards dilemmas instead of answers (Quandary: Sustainability-Oriented Innovation 2022). The insatiable desire of capitalism for profit leads to overproduction, the squandering of natural resources and environmental destruction. This is a predicted measure, the unintended counter-final line of production treadmill theory in which economic development pursuits result in significant environmental damage without any cost-equivalent benefits for people.

Even with policies in place to foster eco-innovations, developed nations still face challenges of successfully innovating to meet societal, economic and environmental needs (Rodríguez-Sánchez et al. 2018). Although there have been some reductions in carbon emissions due to investments and technique improvements—such as energy efficiency, cleaner production or initiatives here discussed in the field—environmental sustainability is still inadequate. Degot et al. (2023) report that 10% of the companies surveyed previously in 2022 have showed no improvements as of 2023, while 14% of companies reporting a decrease in only 3pp as of 2023. This involves improvement in trade policies, political and legal institutions as well as the use of renewable energy sources etc. (Khan et al. 2021). For example, China's experience demonstrates the duality of technological advancement whereby ecological footprints are reduced in a relatively short term and long run, but continuous growth of economy/prosperity and population still worsen our environment which necessitates eco-friendly technologies/policies including carbon tax (Jun et al. 2024).

Technological innovation will need to play a key role as well, but it must connect economic growth and employment opportunities in countries throughout the world with environmental preservation (sustainability) taking into consideration societal needs which benefit stakeholders while minimizing undesirable impacts (Chovancová et al. 2023). One example is the food industry located in developing countries such as India, which exhibits a huge distance relative to attaining sustainable consumption and production patterns despite many innovations focusing on lower environmental pressures. This leads to environmental innovations directionally, and the technologies implemented create a positive impact on improving the state of nature at large; although it does still not mean that we can successfully transition towards an eco-economic paradigm with sufficient efficiency for countries where there is inefficient use experience or low energy recovery in battlegrounds without questions (Hussein et al. 2024).

For instance, using evidence from G7 countries Liu et al. (2022) suggest that certain innovations and policies can promote environmental sustainability through reducing carbon dioxide emissions but at the same time they show openness in trade may constrain this outcome by encouraging higher CO_2 emissions. Although innovations are furthering an environmentally sustainable future, their dark side–embedded within economic and social complexities and policy hurdles—must be considered and altered.

A complex and multifaceted relationship exists between technological innovation and environmental sustainability in developed countries. Previous technological advancements played major roles in significantly improving human well-being and economic growth but many unanticipated consequences, such as huge demands on natural resources through substantial amounts of waste generation and extensive pollution have led to significant environmental deterioration. Although earlier innovations were profit-gain oriented and hardly gave a second thought to any environmental effect of the innovation with time it gradually transformed to resource minimisation or equals sustainable development means combining economic gain through environment protection preservation/conversation together with social upliftment (Xiao and Su 2022).

Sustainable development has been defined as a process in which the needs of present generations are met without compromising future generation's ability to meet their own need and for this, there is required socio-technical change requires technological innovations integrated with changes in social (attitude and behavior) organization level towards sustainable growth (Seema 2018). In reality, however, the difficulty is that practices and experiments have to result in significant changes (Nicholas and Ashford 2002) frequently is often more work than almost any of their dominant organizations or agency partners are soon capable do concerning processes beyond merely incremental product but including transformative systemic steps. Although a range of national environmental policy instruments have been utilized to support eco-innovations in developed countries (for example standards, taxes, tradable permits and sometimes subsidies for producers; communication campaigns or covenants with stakeholders; (Kemp et al. 2012), these alone are often insufficient to deal with challenges associated with global consequences: international non-governmental organizations (NGOs) may therefore play an even more prominent role in raising the intrinsic motivation level of both producers and consumers.

Another point is that the economic growth was not decoupled from environmental pressure as much as we hoped for and improvements in resource productivity are far exceeded by an increase in overall resources and energy consumption (Baran 2021). This highlights the importance of a systemic approach to support environmental innovation deployment, due to often low awareness on their possibilities as well higher complexity in integrating them within operational structure (Liu et al. 2022). But the wider use of such potent industries by technology in others is proving tricky. Innovation and sustainable growth are key components in developing countries, which intend to provide well-being through value-added activities while preserving nature (Rodríguez-Sánchez et al. 2018).

Finally, sustainable development cannot be accomplished in developed countries unless an intensive innovation policy is implemented thus ensuring a significant reduction in Greenhouse Gas Emission without, at the same time, decreasing growth therefore new nations need to be incorporated into research and inside projects (Baran 2021). This systemic thinking is crucial for scaling innovation that tackles sustainability issues, as climate change and resource depletion (Liu et al. 2022).

These challenges are backed up by statistical evidence. For instance, the United States, one of its biggest contributors to E-Waste generated an estimated 6.92 million metric tons of e-waste in 2019 encapsulating only about 15% that was properly recycled (Forti et al. 2020). Again, in the European Union, where a harsh set of environmental guidelines exists (European Environment Agency 2019), emissions through transportation grew by nearly one percent during last year (largely driven by demands for goods and travel). These figures reflect poorly on the unintended spillover effects of technological progress and how innovation and environmental sustainability must be approached with a more holistic perspective.

2.2 Developing Countries

Despite the seeming benefits of technological advancements to less developed countries, other negatives often come with these innovations on their environmental sustainability side. On the other hand, while technological innovation is necessary for a sustainable development path it can all be done at a cost of the environment if proper management is not applied. This is true in the Belt and Road Initiative countries; while technology innovation will result in more renewable energy, enhanced institutional quality can help alleviate this indirect effect of carbon dioxide emissions (Gao et al. 2023). In China, too, technology has helped to shrink the ecological footprint in years gone by—but economic and demographic changes make it difficult for environmental destruction still (Jun et al. 2024).

Sustainable consumption and production patterns in India are only beginning to mature within the food sector of developing countries, with the combined focus on details like price over sustainability aspects such as nutrition rather than being considered at higher levels. Colombia and Mexico also have policy instruments that aim to foster sustainable practices in the manufacturing sectors, however, there are eco-innovations for feedstock materials which they lack either due to a knowledge deficit or economic barriers (Rodríguez-Sánchez et al. 2018). Challenges to adopting green innovations by small enterprises in Indonesia are faced both from within and outside firms, where the influence exerted by internal factors of customer pressure and management commitment outweigh that arising from environmental regulations (Endah et al. 2023). The making and maintenance of these solar PV systems create social benefits, but they are also entangled with a dark side and many environmental challenges are driven by the question of e-waste when this off-grid solar waste is disposed in the Global South (Hansen et al. 2022).

In Ukraine, inefficient use of natural resources and low energy conservation prevent a transition to eco-innovation and safe methods resulting in environmental protection problems which push the necessity for implementation of eco-innovations beside economic goals (Zinchenko and Filenko 2020). For instance, studies in the Global South (including India which really needs no introduction to this list of countries), Mexico and Nigeria demonstrate an increase in output around sustainability research cannot have meaningful impacts on practice across sectors as long as academic-corporate collaborations continue to be limited (Julian et al. 2021). The Mediterranean countries also demonstrate the importance of innovation in sustainable growth but may struggle to balance macroeconomic performance and environmental conservation (Ayfer and Derindağ 2021).

Innovative initiatives taking place in the developing world to grow their economies, though necessary and important for long-term economic sustainability, is also highlighting some dark sides that ironically undermine environmental sustainability. Innovations, sustainability-oriented ones in particular, must be embedded in the larger social-technical–economic-ecological background to avoid over-attribution of their impact and not accidentally romanticizing them as solutions to sustainability issues (Quandary of Sustainability-Oriented Innovations 2022). In developing countries, emphasis on financial inclusivity along with directing investments towards sustainable energy and technological advancements can intensify this paradoxical relationship in terms of environmental pollution if the importance of environmental consideration is not made a priority as seen from Environmental Kuznets Curve (Feng et al. 2022).

Furthermore, environmental sustainability depends strongly on innovation and institutions are not completely understandable; despite that trade openness, renewable energy consumption and foreign direct investment have a depressing relationship with carbon emissions while diminishing institutional quality indicators do any improvement of environmental quality (Khan et al. 2021). The example of Albania shows why even if there is an imperative for a standard calculation of sustainable growth, developing countries face extra hurdles and cannot be taken too fast regardless of how much greater bio-capacity can go in the future regarding new technologies now available, issues related to income inequality and lack in infrastructure jeopardize development (Developing Nations & Sustainability 2023).

In addition to the situation in terms of advances towards technological absorption and global competition acceleration prompted this mechanism is stimulated differently around several dark spots like corruption and top management's capability which may either be viewed as impeding or favoring innovation-led sustainability based on regional situations (Kuzman et al. 2018). This is, for example, demonstrated in the Bolivian renewable energy market where it has been shown that exploiting local resources and traditional knowledge can lead to more sustainable development as well as increased entrepreneurship at a regional level (Pansera 2013). Nevertheless, innovation in social and organizational systems is important when technologies alone are not enough because existing unsustainable competitive technologies might hamper (Seema 2018). In addition to addressing the needs of local customers, and business ecosystems in developing countries, sustainability-oriented innovations must be tailored to enable one-of-a-kind alleviation of poverty-related challenges.

Altogether, although innovations may be designed to drive sustainability in the global South; they likely come with positive and negative up and downstream effects which should not stand alone but instead need coping mechanisms such as border policies for instance if they can, given its potential based on worst case scenario it will deter a desirable transition towards Sustainability.

Statistical evidence provides significant challenges of environmental for developing countries. An example of that is India, where CO_2 emissions rose by 4.8% in 2019 that was driven mostly by industry activities and a strong reliance on coal for power generation (IEA 2020). As for Nigeria, rapid urbanization and weak governance have created many air quality problems: Fuelling conditions are responsible rant all 740 deaths in the country due to PM2. Higher than the World Health Organization's safe limits in major cities like Lagos and increasing health risks as well as environmental damage (World Bank 2019). The examples highlight the urgent need for well-managed integrated technological innovation to support equitable and sustainable development in these territories.

3 Case Studies on Dark Side of Innovation

Innovation is often indicated as the key to solving some of the world's most pressing problems, including those related to environmental sustainability. However, the unplanned consequences of technological and industrial advancements can sometimes aggravate environmental degradation. This part of the chapter will explore cases from around the world, highlighting the two-fold nature of innovation in both developed and developing countries.

Case 1: China's Air Pollution Challenges and Industrialization

The tradeoffs between economic growth and environmental stability poses a major challenge for developing countries since their economic development depends on heavy industry output, urbanization and infrastructure development leading to pressure on the environment and high pollution levels. China's rapid industrialization over the past few decades has become a major threat to the country and the world in terms air pollution and other nature-based challenges. China embraced rapid industrial innovation to drive competitive advantage and improved living standards, Cities like Beijing, Shanghai, and Guangzhou have become symbols of economic progress but also given rise to significant environmental cost, particularly in terms of air quality (Economy 2004).

The emissions from manufacturing industries and the reliance on coal for energy has led to extremely high levels of air pollution. As per a study conducted by the University of Chicago, there is social challenge that the people are facing in China, they found that air pollution reduces the life expectancy of its citizens by an average of three years. The scale of challenges has not changed despite efforts to adopt cleaner technologies and enforce environmental regulations. This highlights that complex trade-offs between economic growth and environmental sustainability will always be a battle for developing countries unless there are changes at multi levels of society and industry including the government.

Case 2: The Impact of Palm Oil Plantations in Indonesia

The reason for economic growth in Indonesia is the palm oil industry. The country realized the importance of Palm trees in the 1960's after the Palm oil boom. Due to the oil's high versatility and utility in everything from food to cosmetics to biofuels Indonesia surpassed its production compared to Malaysia. To meet global demand, Indonesia expanded palm oil plantations, often at the expense of its rainforests. As per Statista, approximately 289 thousand palm trees were produced in Indonesia in 2023 compared to 241 thousand trees in 2022 (Indonesia: Production of Palm Trees 2023). The deforestation associated with palm oil cultivation has catastrophic impact on the environment. It destroys habitats for endangered species like orangutans, disrupts local water cycles and contributes to greenhouse gas emissions. Moreover, the practice of unhealthy ways of clearing land for plantations often results in uncontrollable forest fires, adding to additional air pollution issues and health problems in the region.

While palm oil is an economically lucrative commodity, its production highlights the environmental costs of innovation in agricultural practices.

Case 3: The Future of India Jeopardized by Waste Management

India's pledge to achieving UN sustainable development goals has taken a significant jump over the past few years due to technological innovations; however, the damages being generated on India's ecosystem cannot be neglected. India has been one of the largest populations in the world for over 70 years as (Chavda, 2023) report, and with this came a huge toll on the environment to manage and sustain billions of citizens. For instance, a case study by Nelson (2023) reveals that India is currently the third largest greenhouse gas emitter in the world, and at this rate could become one of the most vulnerable countries to climate change. This can be attributed the geography and agricultural dependence of India. Additionally, Nelson (2023) reports that India's GDP could decrease by 35% and 6% if India does not decrease their carbon emissions and achieve the Paris agreement goals, respectively.

The increased demand of necessary technology and resources to supply the demand of the growing population creates an unstable situation as these resources need to be thrown out. This increased waste build up would cause the destruction of ecosystems and cause severe financial issues in the near future. In Delhi alone, more than 5000 tons of municipal solid waste is generated daily, according to Chattopadhyay (2008), and around 160 thousand metric tons of waste produced in 2021 alone, which requires landfills that often leak into the environment and create danger of ground water contamination that can cause sever health issues if not tackled promptly (Alves 2024). With this increased backlog of environmental problems India will face a sever situation which would require billions of rupees for cleaning air, water, landfills, and mountains of garbage and planting forests in areas that had previously been destroyed due to waste build up (Parikh and Lior 2009; Singh et al. 2024).

Interestingly, many parts of India have already recognized the issue and set up efficient methods to mitigate the harm. Mysore, a sustainable city in India, has adopted many ecofriendly initiatives to reduce carbon emissions and promote green living under the leadership of the government and private funded Mysore City Corporation (MCC). This includes efforts in installing solar panels and providing subsidies for citizens to install solar panels in their own homes (Earth5R 2023). Furthermore, Mysore has launched an initiative called the door-to-door garbage collection program meant to help segregate waste as biodegradable and non-biodegradable and transported by hand carts to compost and recycle solid waste (Indrajeet and Patil 2017). India has taken the necessary steps to help reduce environmental impacts but mostly in the small cities that make the bigger whole. Therefore, India would benefit greatly in improving sustainable projects taken by smaller cities to create a result representative of Indian culture and values towards the environment.

4 Recommendations

The adverse impact sustainable development innovations have had on many countries is clear through the disruption of societal norms and harm on the environment. These impacts could be attributed to the lack of a proper international protocol or organization to provide monetary or social aid in micromanaging the steps necessary to facilitate nations with the necessary tools to achieve their goals. The lack of an institutional body that oversees the implementation and efficiency of innovative technologies to tackle climate change would hinder the progress as it has till now. The United Nations (2024) SDG report reports the lack of progress made between 2015 to 2024 by UN Member States adding that a third of the Member states have stalled in their progression or are progressing backwards. This sluggish progress can be attributed to a lack of protocols and quotas to be met by nations in achieving and developing relevant innovative progress regardless of its developmental status.

Although conferences and meetings are held every few years, words would only hold power if actions were made as it is foundation, and societal disruptions are one of the first effects to fix. For both developed and developing countries, an international organization that oversees these innovations and implements initiatives to hold nations accountable for their role in supporting and meeting the United Nation's expected sustainable development goals is a necessity. Through cohesive rules and regulations, many problems could be solved while enabling an efficient method to manage social, economic, and environmental. The organization may be called Sustainability for the Future Organization (SFTFO). Like the UNDP, the SFTFO will exceed the expectations and extend beyond the responsibilities of the UNDP to physically, financially, and intellectually support nations in individually catered strategies to support developing and developed nations alike.

Through SFTFO, developing nations could get thorough checks on meeting environmental standards and provide increased educational opportunities for society to learn about being more environmentally conscious. Implementing more educational programs such as the Tbilisi Declaration (1977), which was regarded by Braus and Wood (1994) to be an ideal starting point to what an effective environmental education program should be is a key example of what the SFTFO should be able to carry out. This educational program would create an easier social environment to facilitate the implementation of innovative technologies as a part of social norms. The SFTFO could further establish local connections and consult with the local communities to facilitate easy implementation of sustainable development innovations. Despite the growth of sustainable development, a lack of a proper concept of sustainable development to help guide politicians presents an issue as they are necessary to guide the locals who present a key role in providing widespread collaboration to help sustainable development (Toniolo et al. 2023). With the help of proper guidelines and expectations that each nation must fulfil with the help of SFTFO, a system where innovative technologies that would not disrupt social norms and endanger future generations could be set up.

In addition to helping educational programs and guidelines a life cycle assessment (LCA) as part of the organization's main initiatives can be implemented to weed out the negative parts, including the manufacturing and extraction process, to minimize the effect these processes will have in adversely affecting the environment during their implementation. Although there exist many technologies to conduct assessments, one of the most used in the LCA because of its broad perspective beyond waste management (Ekvall et al. 2007). With the help of the SFTFO, the extraction and disposal could easily be managed, and strategies could be easily developed to help mitigate the harmful effects of the technology on the environment. The LCA could be a basis to what the SFTFO could build to create a more efficient method to tackle and resolve issues in the implementation of innovation in sustainable development.

Another strategy that the SFTFO will implement is the addition of corporate social responsibilities (CSR) in managing and adhering to environmental standards and sustainable implementations. With the responsibility of following SFTFO guidelines on CSR, which include prioritizing practices in environmental and ethical standards and providing complete transparency in social and environmental actions taken by the corporation. The purpose of a CSR is to provide a management that manages a corporate adhering to social and ethical standards while also being economically profitable (Carroll 1999, as cited in Kolk and Tudler 2010). With a more sustainable oriented future, it is important the companies start adopting a more sustainable approach that includes consideration for the environment and the future along with the CSR. Researchers claim that the importance of CSR extended beyond corporate responsibilities extending towards decreasing the environmental impact of economics activities because of the high profitability of more environmental conscious companies as compared to non-conscious ones (Ahmad et al. 2020). This responsibility will ensure that economic facilitations for developing nations to manage their environmental impact and net carbon emissions during trade and business are met by the SFTFO while the companies themselves experience profit.

5 Conclusion

In summary, the push for innovation in both developed and developing countries shares an intricate arena where technology can facilitate development but also bring considerable environmental challenges. Technological and geopolitical innovations are powerful enough to address the core issues. They have an unintended consequence in areas that require those innovations. Mainly because innovation is a required step to build a sustainable future; however, the underlying shadow of complex economic, social, environmental and political intricacies bring about unintended consequences that need to be controlled with utmost caution through organizations that specifically manage them.

Most importantly, the advancements of innovations help straightforward methods of research and connectivity which should in theory improve the environment and society; however, the overuse of technological innovations have made it harder for society and the economy to cope and even tougher for the environment to manage. The developments made in disruptive technologies and AI, in particular, have made it challenging for citizens to discern reliable and unreliable information which would make it harder to solve political discourse and may endanger social norms. Similarly, the over reliance on AI has been causing companies to focus more on maintaining company rating due to faulty and biased credit score and personality scores, making it harder for employees to find jobs. On the other hand, companies have, in an effort to be more sustainable, mostly did not recognize the expiry life of the innovation made to protect and preserve the environment causing adverse effects such as waste generation and biodiversity disruption. This waste generation and the lack of facilities in providing support for developing nations in recycling waste has caused many health detriments and significant issues to the environment and the biodiversity particularly in African and Asian nations, which bear the bulk of the issues. These innovations although meant to sustain and aid future generations, leads to economic and social discourse in developed nations, and a challenging front that most developing countries may struggle to manage, creating further barriers to the implementation of set innovations in sustainable development.

Altogether, an entity to oversee the application of innovative strategies and technologies seamlessly into society without causing disruptions to the environment while also supporting an overall economic profitability is an essential requirement. An entity such as the SFTFO which is meant to provide support in all manners on an individual scale to facilitate the efficient implementation of innovative technologies while considering and managing the underlying consequences. The SFTFO works on the foundation of many policies and strategies with the main ones being the education of students and citizens on the importance of innovative technologies and how to effectively cope with the changes they bring, the frequent life cycle assessments meant to weed out and mitigate potential risks to the environment through cohesive strategies and to prioritize environmental consciousness in corporate social responsibility as a key strategy to improve economic performance and environmental preservation.

While innovation play a key role in building a sustainable future, it is also important to consider the proper steps needed to efficiently implement these innovations without disrupting the peace and prosperity of the present. Developed and developing nations act as the pillars to push forward the importance of sustainable development, while organizations bear the responsibility to strategically apply necessary steps to maintain peace and citizens build the foundation to support and adapt to the changing needs of the world and change with it for a better future not only for the present generation, but also for future generations to come. As Margaret Mead emphasizes, "never doubt that a small group of thoughtful, committed citizens can change the world: indeed, it's the only thing that ever has".

Acknowledgements The authors would like to appreciate the efforts of our research support Ahammed Sebeeh for his valuable contribution for the timely completion of this chapter.

References

Ahmad R, Ahmad S, Islam T, Kaleem A (2020) The nexus of corporate social responsibility (CSR), affective commitment and organisational citizenship behaviour in academia: A model of trust. Empl Relat 42(1):232–247. https://doi.org/10.1108/ER-04-2018-0105

Al Kez D, Foley AM, Laverty D, Del Rio DF, Sovacool B (2022) Exploring the sustainability challenges facing digitalization and internet data centers. J Clean Prod 371:133633. https://doi.org/10.1016/j.jclepro.2022.133633

Alves B (2024) Topic: waste management in India. Statista. https://www.statista.com/topics/5586/waste-management-india/

Ayfer G, Derindağ MR (2021) Innovation, macroeconomic effects, and environmental degradation: the case of Mediterranean countries. In: Sustainable economic development, pp 300–320. IGI Global. https://doi.org/10.4018/978-1-7998-7391-4.CH009

Baran A (2021) Sustainable innovations – selected aspects. Econ Environ 76(1):12. https://www.ekonomiaisrodowisko.pl/journal/article/view/358

Boschma R, Coenen L, Frenken K, Truffer B (2018) Towards a theory of regional diversification: combining insights from evolutionary economic geography and transition studies. In: Transitions in regional economic development, pp 55–81. Routledge.

Brandi C, Weinhardt C (2021) The developed/developing country division in the global economy: losing traction in the COVID-19 era? Glob Perspect 2(1):Article 24318. https://doi.org/10.1525/gp.2021.24318

Braus J, Wood D (1993) Environmental education in the schools: creating a program that works! In: Google Books. North American Association for Environmental Education (NAAEE) in conjunction with the ERIC Clearinghouse for Science, Mathematics, and Environmental Education, The Ohio State University. https://files.peacecorps.gov/multimedia/pdf/library/M0044_Environmental_Education_in_the_Schools.pdf

Brundtland GH (1987) Report of the world commission on environment and development: our common future. World Commission on Environment and Development. https://docenti.unimc.it/elena.borin/teaching/2019/21812/files/lessons-and-recommended-readings-1/Our%20Common%20future%201987.pdf

Carroll AB (1999) Corporate Social Responsibility. Bus Soc 38(3):268–295. https://doi.org/10.1177/000765039903800303

Chaparro-Banegas N, Ibañez Escribano AM, Mas-Tur A, Roig-Tierno N (2024) Innovation facilitators and sustainable development: a country comparative approach. Environ Dev Sustain 26(4):8467–8495. https://doi.org/10.1007/s10668-023-03055-w

Chattopadhyay B (2008) Sustainable urban development in India: some issues. http://admin.indiaenvironmentportal.org.in/files/basudha_paper.pdf

Chavda J (2023) Key facts as India surpasses China as the world's most populous country. Pew Res Cent. https://www.pewresearch.org/short-reads/2023/02/09/key-facts-as-india-surpasses-china-as-the-worlds-mostpopulous-country/

Chovancová J, Majerník M, Drábik P, Štofková Z (2023) Environmental technological innovations and the sustainability of their development. Ecol Eng Environ Technol 24(4):245–252. https://doi.org/10.12912/27197050/162708

Coad A, Nightingale P, Stilgoe J, Vezzani A (2020) Editorial: the dark side of innovation. Ind Innov 28(1):102–112. https://doi.org/10.1080/13662716.2020.1818555

Degot C, Dimitrova D, Dong C, Meinecke H, Gudipudi R (2023) Why some companies are ahead in the race to net zero. BCG Global. https://www.bcg.com/publications/2023/why-some-companies-are-ahead-in-the-race-to-net-zero-and-reducing-emssions

Earth5R (2023) a case study on Mysore-a sustainable city. Earth5R. https://earth5r.org/mysore-a-sustainable-city-case-study/

Economy EC (2004) The river runs black: the environmental challenge to China's future. Cornell University Press

Ekvall T, Assefa G, Björklund A, Eriksson O, Finnveden G (2007) What life-cycle assessment does and does not do in assessments of waste management. Waste Manage 27(8):989–996. https://doi.org/10.1016/j.wasman.2007.02.015

Endah RL, Nimas MS, Sunyoto S (2023) Fostering green innovation in achieving sustainable performance. Nat Res Forum 47(1):120–135. https://doi.org/10.1111/1477-8947.12293

European Environment Agency (2019) Greenhouse gas emissions from transport. https://www.eea.europa.eu/data-and-maps/indicators/transport-emissions-of-greenhouse-gases/transport-emissions-of-greenhouse-gases-12

Feng Z, Zhang Y, Alharthi M, Zafar MW (2022) Environmental sustainability in developing countries: understanding the criticality of financial inclusion and globalization. Sustain Dev 30(4):1199–1210. https://doi.org/10.1002/sd.2350

Ferreira-Filipe DA, Paço A, Duarte AC, Rocha-Santos T, Patrício Silva AL (2021) Are biobased plastics green alternatives?—a critical review. Int J Environ Res Public Health 18(15):7729. https://doi.org/10.3390/ijerph18157729

Forti V, Baldé CP, Kuehr R, Bel G (2020) The global e-waste monitor 2020: Quantities, flows, and the circular economy potential. United Nations University/United Nations Institute for Training and Research (UNITAR), International Telecommunication Union (ITU), and International Solid Waste Association (ISWA)

Gao X, Liu J, Li Y, Zhang Y (2023) The role of quality institutions and technological innovations in environmental sustainability: panel data analysis of BRI countries. PLoS ONE 18(2):e0287543. https://doi.org/10.1371/journal.pone.0287543

Grashof N, Basilico S (2023) The dark side of green innovation? Green transition and regional inequality in Europe. Utrecht University, Human Geography and Planning (No. 23.14). http://econ.geo.uu.nl/peeg/peeg2314.pdf

Hansen UE, Nygaard I, Romijn H, Wieczorek A, Kamp LM, Klerkx L (2018) Sustainability transitions in developing countries: stocktaking, new contributions and a research agenda. Environ Sci Policy 84:198–203. https://doi.org/10.1016/j.envsci.2017.11.009

Hansen UE, Nygaard I, Dal Maso M (2022) The dark side of the sun: solar e-waste and environmental upgrading in the off-grid solar PV value chain. In: The dark side of innovation, pp 35–55. Routledge

Hartmann D, Guevara MR, Jara-Figueroa C, Aristaran M, Hildalgo CA (2017) Linking economic complexity, institutions, and income inequality. World Dev 93:75–93. https://doi.org/10.1016/j.worlddev.2016.12.020

Hussein TM, Ateeq RA, Shafi DI (2024) Eco-Economics: Navigating the financial effects of sustainability initiatives. Stud Syst Decis Control 353–360. https://doi.org/10.1007/978-3-031-71318-7_33

IEA (2020) Global energy review 2020. International Energy Agency

Indonesia: Production of Palm Trees (2023) Statista (2023), from Statista website https://www.statista.com/statistics/812862/production-of-palm-trees-in-indonesia/#:~:text=In%202023%2C%20approximately%20289%20thousand,241%20thousand%20trees%20were%20produced. Last Accessed 17 July 2024

Indrajeet CD, Patil DS (2017) Zero waste management system: case study-Kumbarakoppal, Mysore. Int J Eng Res Technol 6(5):768–770. https://doi.org/10.17577/ijertv6is050489

Julian DC, Mireia G, Katerina BC (2021) Innovation for sustainability in the global south: bibliometric findings from management & business and STEM (Science, technology, engineering and mathematics) fields in developing countries. Digital Libraries. https://doi.org/10.1016/J.HELIYON.2021.E07809

Jun X, Ai J, Zheng L, Lu M, Wang J (2024) Impact of information technology and industrial development on corporate ESG practices: evidence from a pilot program in China. Econ Model 139:106806. https://doi.org/10.1016/j.econmod.2024.106806

Kallis G, Kerschner C, Martinez-Alier J (2012) The economics of degrowth. Ecol Econ 84:172–180. https://doi.org/10.1016/j.ecolecon.2012.08.017

Kemp D, Owen JR, Van de Graaff S (2012) Corporate social responsibility, mining and "audit culture". J Cleaner Prod 24:1–10. https://doi.org/10.1016/j.jclepro.2011.11.002

Khan H, Weili L, Khan I, Khamphengxay S (2021) Renewable energy consumption, trade openness, and environmental degradation: a panel data analysis of developing and developed countries. Math Probl Eng 2021(1):1–13. https://doi.org/10.1155/2021/6691046

Kolk A, van Tulder R (2010) International business, corporate social responsibility and sustainable development. Int Bus Rev 19(2):119–125

Kuzman T, Bellos SK, Đulić K (2018) Determinants of innovation-based sustainability in transition and developing economies. Econ Themes 56(4):437–453. https://doi.org/10.2478/ethemes-2018-0024

Liu J, Ridwan L, Ibrahim AS, Mohammed M, Abdulaziz SA (2022) Exploring the heterogeneous effects of technological innovations on environmental sustainability: do structural change, environmental policy, and biofuel energy matter for G7 economies? Environ Plan C: Polit Space 40(1):185–204. https://doi.org/10.1177/0958305x221145922

Mankiw NG (2018) Principles of economics, 8th edn. Cengage Learning

Muñoz P, Cacciotti G, Cohen B (2018) The double-edged sword of purpose-driven behavior in sustainable venturing. J Bus Ventur 33(2):149–178. https://doi.org/10.1016/j.jbusvent.2017.12.005

Nelson M (2023) India: a case study in climate mitigation and adaptation. www.garp.org. https://www.garp.org/risk-intelligence/sustainability-climate/india-case-study-climate-mitigation-211018

Nicholas A, Ashford NA (2002) Government and environmental innovation in Europe and North America. Am Behav Sci 45(9):1412–1433. https://doi.org/10.1007/3-540-27298-4_9

Pansera M (2013) Innovation system for sustainability in developing countries: the renewable energy sector in Bolivia. Int J Innov Sustain Dev 7(1):27–45

Parikh J, Lior N (2009) Energy and its sustainable development for India, Editorial Introduction and commentary to the Special Issue of Energy – The International Journal. Energy 34(8):923–927

Pinheiro AB, Panza GB, Berhorst NL, Toaldo AM, Segatto AP (2023) Exploring the relationship among ESG, innovation, and economic and financial performance: evidence from the energy sector. Int J Energy Sect Manage 18(3):500–516. https://doi.org/10.1108/IJESM-02-2023-0008

Porter ME (1990) The competitive advantage of nations. Free Press

Quandary: Sustainability-Oriented Innovation (2022) Quemap Malunggay. https://doi.org/10.4324/9781003003588-15

Ravalli R (2021) The dark side of sustainability standards. https://euideas.eui.eu/2021/01/29/the-dark-side-of-sustainability-standards/

Renko M (2013) Early challenges of nascent social entrepreneurs. Entrep Theory Pract 37(5):1045–1069. https://doi.org/10.1111/j.1540-6520.2012.00522.x

Rodríguez-Sánchez P, González MH, Carmenza J, Arias JB (2018) Eco-innovation and sustainable production in developing countries. Cases Colombia and Mexico. Econ Bus J 12(1):228–238. https://www.researchgate.net/publication/327671202_ECOINNOVATION_AND_SUSTAINABLE_PRODUCTION_IN_DEVELOPING_COUNTRIES_CASES_COLOMBIA_AND_MEXICO

Sachs JD (2005) The end of poverty: economic possibilities for our time. Penguin Press

Seema A (2018) Technological innovations for sustainable environment

Sen A (1999) Development as freedom. Oxford University Press

Singh R, Kumar K, Khan S (2024) A comprehensive view of artificial intelligence (AI)–based technologies for sustainable development goals (SDGs). Artificial intelligence enabled management: an emerging economy perspective, p 183

Stiglitz JE (2002) Globalization and its discontents. W.W. Norton & Company

Tbilisi Declaration (1977) www.gdrc.org. https://www.gdrc.org/uem/ee/tbilisi.html

Tomar V (2023) The dark side of innovation: exploring the risks and challenges of emerging technologies. https://tomarvipul.medium.com/the-dark-side-of-innovation-exploring-the-risks-and-challenges-of-emerging-technologies-79c4ce949a2c

Toniolo S, Pieretto C, Camana D (2023) Improving sustainability in communities: linking the local scale to the concept of sustainable development. Environ Impact Assess Rev 101:107126–107126. https://doi.org/10.1016/j.eiar.2023.107126

Todaro MP, Smith SC (2020) Economic development. Pearson UK: London

Turkcu D, Tura N, Ojanen V (2022) A conceptual framework of the sustainability challenges experienced during the life cycle of biobased packaging products. Sustainability 14(17):10465. https://doi.org/10.3390/su141710465

United Nations (2023) The sustainable development goals report special edition 2023. https://unstats.un.org/sdgs/report/2023/The-Sustainable-Development-Goals-Report-2023.pdf

United Nations (2024) The sustainable development goals report 2024. https://unstats.un.org/sdgs/report/2024/The-Sustainable-Development-Goals-Report-2024.pdf

Venkatachary SK, Prasad J, Samikannu R (2017) Economic impacts of cyber security in energy sector: a review. Int J Energy Econ Policy 7(5):250–262. https://www.zbw.eu/econis-archiv/bitstream/11159/1315/1/1005320780.pdf

World Bank (2019) Nigeria air quality management. https://data.worldbank.org/indicator/NY.GNP.PCAP.CD

World Bank (2022) World development indicators: country classification. https://data.worldbank.org/indicator/NY.GNP.PCAP.CD

Xiao D, Su J (2022) Role of technological innovation in achieving social and environmental sustainability: mediating roles of organizational innovation and digital entrepreneurship. Front Public Health 10. https://doi.org/10.3389/fpubh.2022.850172

Supply Chain Decarbonisation and its Impact on the Environment and Society

Mananage Shanika Hansini Rathnasiri

Abstract This chapter presents an overview of supply chain decarbonisation, where topics such as the concept's meaning, importance, approaches, technologies, and environmental and social consequences are discussed. It seeks to afford a complete picture of how supply chain carbon emissions can be minimised for the general good, with a particular focus on the prospects and drawbacks. To undertake the study, a systematic literature review model that adhered to Preferred Reporting Items for Systematic Reviews and Meta-Analyses (PRISMA) was used to investigate the phenomenon. Accordingly, thirty-seven (37) seminal studies were taken as the sample for this study. The research discovered that the motivation that prompts organisations to move towards the decarbonisation of supply chains is regulatory, market and social. Numerous workable approaches towards emissions reductions include generating renewable energy power, energy-efficient technologies, green mobility and transport, waste management, the circular economy, and carbon footprint management through offsetting. Technological innovations such as the Internet of Things (IoT), artificial intelligence (AI), blockchain, and sustainable materials drive decarbonisation efforts. The environmental effects include a substantial cut in the emission of greenhouse gases, improvement of the quality of the atmosphere and water bodies, and ecological conservation. Benefits comprise the better quality of citizens' lives through an increase in overall health, the generation of new employment opportunities in environmentally friendly industries and the reduction in social disparities. Regarding climate change and sustainable development, the decarbonisation of supply chains holds commendable potential. At the same time, it is filled with economic, technological, policy and organisational barriers and enablers. Governments, organisations, and other stakeholders should craft favourable conditions, integrate cutting-edge solutions, and initiate sustainable development processes. Stakeholders interested in progressing supply chain decarbonisation will find this chapter insightful and packed with actionable advice. This research offers fresh input since

M. S. H. Rathnasiri (✉)
Department of Marketing Management, Faculty of Management Studies, Sabaragamuwa University of Sri Lanka, Belihuloya, Sri Lanka
e-mail: shanika@mgt.sab.ac.lk

R. Singh et al. (eds.), *Zero Carbon Industry, Eco-Innovation and Environmental Sustainability*, World Sustainability Series,
https://doi.org/10.1007/978-3-031-80220-1_9

it demonstrates the decarbonisation process's socially and environmentally complex nature, thus enhancing the existing literature.

Keywords Supply chain decarbonisation · Environmental sustainability · Carbon emissions reduction · Technological innovations · Circular economy · Society

JEL Codes Q54 · Q56 · O33 · R41

1 Introduction

Supply chain decarbonisation can also be described as cutting down carbon emissions in every link and stage involved in supply chain management (Koh et al. 2013). This pertains to all processes, starting from the extraction of raw materials, manufacturing, transport and distribution, consumption, usage, and disposal. The goal for supply chain management is to reduce by as much as possible the amount of carbon emissions relevant to specific operations within a supply chain network through the proper execution of sustainable activities, efficient technologies, and creative solutions. Supply chain decarbonisation reduces carbon emissions at every step to help combat climate change and enhance environmental sustainability (Zhang et al. 2022).

Supply chains have become elongated and highly integrated in the current world, characterised by globalisation and affiliation. They are also responsible for the emission of significant greenhouse gases, which have powered atmospheric change evolvements (Mathiyazhagan et al. 2014). The reasons for addressing the task of decarbonising these supply chains are based on the opportunities to decrease the negative impact of industrial and commercial processes substantially. Today's supply chain management processes are highly complex and are subject to a lot of pressure from regulatory agencies, customers, and shareholders to embrace sustainable strategies (Hsu et al. 2013). It assists in fulfilling firms' obligations based on environmental legislation and improves their positioning, health, and character. In addition, it creates awareness of business and market trends favourable for sustainability, leading to better control of the market environment and better business outcomes.

The effects of supply chain decarbonisation are far-reaching: Carbon emission reduction is associated with the fight against climate change, a decrease in air and water pollution and biodiversity protection (Habert et al. 2020). It also encourages the proper use of its resources in a manner that can enhance the health of systems to the benefit of those alive. In general, decarbonisation has an excellent effect on society as a whole. It can lead to the creation of green jobs since it forms part of the economic development of sustainable industries (Kuriyama and Abe 2021). It also improves the public's health standards by lowering the consequences of environmental pollution. It creates social equality by ensuring all parts of society equitably gain from the shift to a more sustainable economy. Supply chain decarbonisation is the need for a green economy and an opportunity to build a future that enables value for everyone. This chapter is organised as follows. First, the researcher discusses the methodology followed by the need for decarbonisation in the supply chain, strategies for decarbonisation in the supply chain, and the role of technological innovations. Then, the researcher discusses the impact on the environment and society. Furthermore, the researcher discussed the challenges and barriers followed by regulatory and policy frameworks for supply chain decarbonisation. Finally, the researcher examined the future directions, followed by the conclusion.

2 Methodology

This chapter used the systematic literature review approach according to the PRISMA to minimise research bias in acquiring and analysing information on supply chain decarbonisation. PRISMA model offers a coherent framework that can be quickly followed to make clear choices of the relevant studies and sources and critically appraise them, as emphasised by Karunarathna et al. (2024), Dewasiri et al. (2024), Rathnasiri et al. (2024a), Jayawardena and Dewasiri (2023), Rathnasiri and De Silva (2023), Dewasiri et al. (2023), Mishra et al. (2024), and Dewasiri et al. (2022). First, academic databases, including Google Scholar, were employed to find relevant literature; moreover, reports and white papers from industries and organisations such as Carbon Trust, the International Energy Agency (IEA), and the World Economic Forum (WEF) were considered. These criteria concerned the approaches to decarbonising supply chains, technologies, environmental and social effects, and policies of the studies, case reports/notes, and articles included. Furthermore, the selected items were analysed to eliminate the duplicated items, and the remaining abstracts were screened for relevance. Articles in the journals were then screened based on relevance using pre-determined topic areas and the quality of the research study. Information on the selected sources was collected, analysed and organised into themes corresponding to this chapter. Such a systematic approach would allow the reader to follow the author's rigorous approach to research that underpins the assertions made in this chapter, thus ensuring that the readers get a clear and comprehensive picture of the various dimensions of supply chain (SC) decarbonisation and their impact on the environment and society. Accordingly, thirty-seven (37) articles were selected as the sample of this study.

3 The Need for Decarbonisation in Supply Chains

3.1 Current Carbon Footprint of Global Supply Chains

Global supply chains have a powerful impact on cutting greenhouse emissions, fueling the climate change catastrophe. Such emissions involve essential material acquisition, processing, transport, and final disposal processes. For instance, the logistics sector, which consists of transportation and storage, is among the significant sources of global CO_2 emissions. Manufacturing procedures and the energy used also add to the lousy carbon imprint. Since supply chains become more extended and global, their impacts, the problems of their management, and emissions decrease have become more critical. Evaluating the current status of carbon emissions is the starting point for further efforts to decrease a company's carbon footprint.

3.2 Drivers of Decarbonization: Regulatory, Market, and Social Pressures

Supply chain decarbonisation is being informed by several dynamics as follows. The regulatory forces are growing as governments on the global level cherish intensified environmental laws and standards concerning greenhouse gas emissions (Steiner et al. 2023). Market pressures are also crucial because markets reward corporations that demonstrate concern for sustainable practices. It proves that the changes in market preferences force companies to become more environmentally friendly to remain relevant and appeal to ecologically sensitive clients. Social pressures also contribute to adopting ethical decisions since advocacy groups, non-governmental organisations, and society generally demand more responsible corporate actions. In combination, these drivers lead to formulating this argument as to why companies should decarbonise their supply chain.

Some segments are more dependent on carbon and can be deemed as clearly exhibiting the primary units that require decarbonisation. Road, air, and sea transportation make considerable emissions, relying heavily on fossil fuels. The manufacturing industry also contributes significant amounts of GHGs to the atmosphere, especially sectors that work under heavy production, such as steel, cement, and products. The agricultural industry otherwise contributes from practices that produce methane and nitrous oxide gases, potent greenhouse gases. Assessing the existing literature from these high-carbon impact industries helps to understand the potential and problems of carbon reduction on the supply chain. Those may show examples of successful approaches and new techniques that other sectors can employ to reduce their impacts on climate change.

4 Strategies for Decarbonizing Supply Chains

4.1 Renewable Energy Integration

One of the most efficient practices to eliminate carbon emissions in the supply chain is using renewable energy (Wang et al. 2020; Mishra et al. 2024). Replacing fossil fuels with green energy, such as solar energy, wind energy, hydroelectricity, etc., could help decrease the amount of carbon emission at various stages of the operation (Rathnasiri et al. 2024a). A firm can buy green energy credits, patronise renewable energy projects, or even produce green energy at its location. In doing so, they reduce greenhouse gas emissions and support the overall global transition to a sustainable energy economy. Also, adopting renewable energy enables firms to manage risks related to fossil fuel fluctuations and regulation changes.

4.2 Energy Efficiency Improvements

Energy efficiency improvement is another critical approach that needs to be employed for the decarbonisation process to be effective (Wesseling et al. 2017). This comprises utilising procedures that will reduce energy consumption while increasing efficiency. Some measures that can be adopted include LED lighting, control systems, and energy management software for the warehouse, manufacturing plants or offices. Maintaining the energy audit programs and following energy-efficient measures can highlight more energy-saving possibilities. In this way, highly energy-efficient approaches reduce environmental impact and increase company profitability by lessening operating costs.

4.3 Sustainable Transportation Solutions

Transportation occupies the paramount functions, offering a massive contribution to carbon emissions in the supply chain. This impact can be reduced by deploying sustainable transportation measures (Etukudoh et al. 2024). The measures are conversion to electric or hybrid vehicles, regenerative provisioning of routes with the minimum fuel consumption, and conversion to alternative fuels such as biodiesel, hydrogen, etc. Also, when deciding on logistics partners, choosing those that pay attention to environmental impacts and incorporating low-emission technologies into their practices is possible. Other means of transport, such as intermodal transport, including rail and road transport, also improve efficiency and reduce emissions. By implementing these steps, businesses could develop more efficient and sustainable means of transportation.

4.4 Circular Economy Practices

The principles of the circular economy must be integrated into supply chain decarbonisation. This involves employing strategies that minimise waste, reuse, and recycle products at the end of the lifecycle (Bocken et al. 2016; Rathnasiri et al. 2024b). Businesses can create products that are low in obsolescence and often reusable and recyclable, which minimises the emissions during manufacturing and at the end of the product's life. Take-back programs and the refurbishment or remanufacturing of products help to increase product lifespan and reduce environmental impact. In other words, adopting a circular economy offers capabilities for structuring a new sustainable supply chain network for the organisation with interdisciplinary collaboration (Murray et al. 2017).

4.5 Carbon Offsetting and Sequestration

Decreasing emissions and, at the same time, using carbon offsetting and sequestration as supportive tools can contribute to the process of global decarbonisation. Carbon offsetting is investing in projects that reduce or capture emissions at different places, like reforestation, renewable energy projects, or methane initiatives. It also means that these offsets can assist in eliminating emissions that the companies cannot offset completely (Nevzorova 2024). At the same time, carbon dioxide from the atmosphere is captured and stored through natural and technological processes called carbon sequestration. Some methods, such as biochar, afforestation, and carbon capture and storage (CCS), help to reduce the amount of CO_2 in the atmosphere. This way, the companies stated their willingness to go the extra mile, which will see them implementing carbon offsetting and sequestration, a significant undertaking in reducing emissions to meet international climate goals.

5 Technological Innovations

5.1 Role of Digitalization and IoT in Supply Chain Decarbonization

Digitalisation and IoT have presented new opportunities that may lead to the decarbonisation of the supply chain (Bastos et al. 2024). Organisations can also monitor energy intake, emission rates, and resources using IoT sensors and devices to track accurately the supply chain. It assists in monitoring carbon footprints and, as a result, can help identify areas of weakness. Predictive maintenance that IoT drives can help cut costs such as time loss and energy costs in production and transportation. Technological advancements through digital platforms make it easier to manage supply chain activities, resulting in minor transportation and storage emissions. In conclusion, implementing digitalisation and IoT improves the supply chain's transparency, the speed of its processes, and its endurance.

5.2 Advanced Analytics and AI for Carbon Footprint Tracking

Carbon footprints are also managed using advanced analytics and artificial intelligence (AI). With AI, big data can be collected and processed, and patterns can be analysed to predict the emissions and how to minimise or counteract them (Kaack et al. 2022). Humans can use machine learning models to improve carbon efficiency concerning supply chain plans and activities, including demand and supply prediction, inventory management, and transportation routing. Advanced analytics tools can determine the total effect on the environment across the different supply chain setups and assist organisations in selecting procedures that uphold sustainable practices.

5.3 Blockchain for Transparency and Accountability

Blockchain is a valuable tool for addressing problems in the supply chain, as it strengthens the security and audibility of the data (Khan et al. 2022). In particular, blockchain implementation enables companies to correctly register transaction records and emissions data throughout the supply chain to make the records unalterable. This transparency assists in checking organisations' fabricating claims concerning sustainability from their suppliers and partners. Blockchain also helps manage carbon credits and offset, ensuring carbon accounting with an immutable, auditable record. Thus, blockchain technology helps to develop trust and accountability in supply chains while making supply chains more environmentally friendly.

5.4 Innovations in Sustainable Materials and Packaging

Science and technology in the context of material processing and packaging design are critical in decreasing the carbon footprints of the supply chain (Meherishi et al. 2019). Some green products can be sourced from bio-based materials, recycled contents, and bio-degradable materials that can replace conventional materials with high environmental impact. Advancements like lightweight packaging and compact designs, returnable containers and packages, and even minimal packaging help save the materials and energy used to make and transport them. Also, innovative packaging incorporating IoT technology can enhance SC productivity by providing real-time product condition data and decreasing waste. Using sustainable materials and advanced packaging, businesses can create an efficient loop and reduce their CO_2 emissions drastically.

6 Environmental Impact

6.1 Reduction in Greenhouse Gas Emissions

Among all environmental consequences amenable to supply chain decarbonisation, the relative largest is the decreased emission of greenhouse gases (GHG). If more organisations embrace green energy production, reduce energy consumption through energy-efficient practices and adjust distribution, the supply chain's effect on the environment will be considerably reduced. Such an emission cut is necessary to reduce climate change and its impacts, including floods, drought, hurricanes, and global climate change. Lower GHG emissions support the struggle to achieve the goals of international treaties and conventions, including the Paris Agreement, to prevent climate change (Mishra et al. 2024).

6.2 Positive Effects on Air and Water Quality

Decarbonising supply chains have a positive impact on the quality of air and water (Zhang et al. 2022). Transitioning to renewable energy sources and minimising emissions from industrial processes diminish the release of various pollutants, including sulfur dioxide, nitrogen oxides and particulate matter, which lead to the rise of airborne contaminants. Reduced air pollution benefits the population since diseases associated with poor-quality air, such as respiratory and cardiovascular diseases, are less prevalent. Moreover, there are fewer harmful uses of chemicals and wastes in the supply chain; thus, there is a reduction in the pollution of water sources. Better water quality improves aquatic organisms' conditions and provides communities and farming with safe water.

6.3 Biodiversity Conservation

Decarbonising supply chains contributes to preserving the planet's wildlife as a direct result of negating the environmental pressures that give rise to habitat loss and, consequently, species loss. The sustainable sourcing of raw materials and measures taken to avoid harming the land are other factors that support the preservation of natural habitats. Thus, emissions and pollution decrease; natural habitats improve as various species of plants and animals benefit. Efforts such as afforestation and relevant activities, which fall under the category of carbon offsetting, positively impact the same. By implementing biodiversity conservation measures into the decarbonisation initiatives, companies support the conservation of ecosystems and the associated beneficial services (Fletcher et al. 2024).

6.4 Long-Term Environmental Sustainability

The movement toward decarbonised supply chains contributes to the sustained promotion of the efficient use of resources and the minimisation of ecological impacts. At the systems level, the nature of sustainable practices, including the circular economy system, makes it possible to cut on material use and waste, hence creating a more sustainable supply chain. Continued growth is also indicated by the funding for renewable energy sources and green energy technology to adequately support future generations with consistent and clean energy. In this way, sustainability becomes a criterion to guarantee that the supply chains of different manufacturing businesses will be prepared for sustainable change and will be able to sustain responsible performance for the long term.

The environmental effects are closely linked to the measures influencing the decarbonisation of supply chains (Liu et al. 2021). By reducing GHG emissions, improving air and water quality, protecting biological diversity, and ensuring sustainable resource use for the long term, decarbonising supply chains offers a downstream approach to most of the world's critical environmental issues.

7 Societal Impact

7.1 Improved Public Health and Quality of Life

Decarbonising supply chains directly affects people's quality of life (Penn et al. 2022). Through the reduced discharge of airborne hazardous substances like particulate matter, sulfur dioxide, and nitrogen oxides, there is always an improvement in air quality, hence the reduction in cases of respiratory and cardiovascular diseases. Clean energy sources and sustainable practices minimise pollution; hence, water and soil will likely be less contaminated, improving public health further. Healthier communities enjoy a better quality of life through lower healthcare costs and increased well-being.

7.2 Job Creation in Green Technologies and Industries

The shift to decarbonised supply chains results in employment in green technologies or industries (While and Eadson 2022). New jobs are created as organisations embrace innovations that focus on green energy, saving energy, energy conservation, cleaner transport, and sustainable usage of resources. Such jobs extend from research and development personnel, manufacturing, and installation of green technologies personnel to maintenance, monitoring, and management personnel of sustainable practices. Industry growths, specifically in environment-friendly areas, also unlock new talents and skills in the market, making the workforce more flexible and adaptable.

7.3 Social Equity and Community Development

Mitigating supply chain emissions can create opportunities for social equity and community development (Markkanen and Anger-Kraavi 2019). The decarbonisation process sometimes comes with community inclusion, which is necessary to ensure that the advantages of decarbonisation are widely shared. For instance, they can supply energy to remote regions at relatively cheaper rates, thus boosting the living standards of disadvantaged people. Approaches related to ensuring fair employment practices, sourcing materials from local suppliers, and improving efficiency can help enhance the community's relations and support economic growth. In addition, environmental equality should be established by reducing ecological risks so marginalised communities are not unduly exposed to pollution or climate change.

7.4 Case Studies of Societal Benefits from Decarbonized Supply Chains

Many examples prove that a shift to optimised and decarbonised supply chains is helpful for society. For example, firms with renewable energy deployed in the supply chain have realised some social benefits, especially for the local communities. In the areas where renewable energy in the form of solar and wind has been implemented, energy accessibility and energy prices have improved, thus increasing the provision for economic activities and living standards. Another example is adopting sustainable agriculture practices, which have enhanced soil quality, boosted food production, and improved farmers' earnings. Also, applying circular economy measures such as recycling and waste management has provided employment and helped dematerialise environmental impacts, making the community prosperous (Geissdoerfer et al. 2017).

All these societal impacts indicate the need to decarbonise the supply chain for environmental and people's welfare and economic prosperity. Decarbonised supply chains can drive positive societal change by prioritising public health, job creation, social equity, and community development, ultimately supporting a more sustainable future.

8 Challenges and Barriers

8.1 Economic and Financial Barriers

Decarbonising supply chains often face economic and financial barriers (Singh et al. 2023). The initial investment required for renewable energy adoption, infrastructure improvement, and the incorporation of new technologies is considerably high. Some firms and tiny and medium-sized enterprises (SMEs) need help accessing the required funds. Also, returns on investment in sustainable practices may take place over long periods, thus not effectively motivating business organisations with short-term financial goals. Unfortunately, there are still high costs linked to implementing research and development initiatives for green technologies, which are not affordable and disrupt existing business models and systems during the change-over phase.

8.2 Technological and Infrastructure Limitations

Lack of technology and infrastructure are the main obstacles to the decarbonisation of the supply chain (Li et al. 2015). Plenty of innovative technologies are seen today in the market; however, installing these technologies involves integration that can significantly change the existing systems. For instance, switching to electric vehicles requires the creation of widespread charging infrastructure. Likewise, applying renewable energy in the manufacturing process may entail extensive modification of structures. Especially in most regions, the absence of stable, cheap, and renewable energy sources adds another layer to the problem of reduction of carbon emissions. However, sometimes, the limited availability and flexibility of new technologies hinder their extensive adoption.

8.3 Policy and Regulatory Challenges

Another area that companies need help with in the process of supply chain decarbonisation is navigating the policies and regulations (Zhang et al. 2022). Legal systems in different countries vary significantly, meaning that conducting business on an international level is complex. Lack of clarity in the regulations and standards can present compliance challenges and may also result in high expenses. In addition, there needs to be more rigorous regulation on some of these practices in certain regions to motivate companies to incorporate sustainability in their operations. Another factor is the lack of coherent policies that effectively handle the changeover to low-carbon technology systems, including inadequate subsidies for renewable energy or the absence of any carbon pricing mechanisms, which hinder advancements. Effective policies providing guidance, promoting incentives, and supporting are essential for scaling specific initiatives and programmes to drive supply chain decarbonisation.

8.4 Organisational Resistance and Change Management

Business-level barriers include organisational resistance and change management as result-focused internal issues that arise when a firm seeks to introduce decarbonisation into its supply chain (Zhang et al. 2022). Some reasons for resistance are ingrained corporate cultures, lack of awareness of sustainability issues, and the distress of the unknown. Some people working in an organisation might need help to change their working culture and patterns. These challenges require robust change management strategies because they include stakeholder management, cultural change, and the creation of awareness of the benefits of decarbonisation. There are several essential factors requiring more attention from leaders. First, one can identify the commitment to sustainable values and ideas concerning changes and clear communication with all the employees as the major components to remove resistance and accomplish organisational changes.

Solutions to these challenges and barriers require financial commitment, technology application, sound fiscal and social policies, and organisational change processes. Addressing these challenges makes it possible to advance supply chain decarbonisation efforts and improve companies' contributions to a more sustainable world.

9 Policy and Regulatory Frameworks

9.1 International Agreements and National Policies

One can ascertain that the significant influence of global bargains and local regulation policies is the impetus for supply chain decarbonisation. Treaties like the Paris Agreement are international plans to reduce the emission of greenhouse gases and combat climate change (Lima et al. 2020). These agreements aim to present guidelines that different countries could use to formulate and implement national policies in recognition of international objectives. Policies such as carbon prices, emission trading schemes, and renewable energy mandates establish the legal frameworks that enable companies to commit to less carbon emission. States are equally enacting laws that compel firms to declare emission levels and formulate plans on how emissions would be lowered. These policies ensure that organisations comply with international standards and contribute to national and international decarbonisation.

9.2 *Role of Governments and NGOs in Promoting Decarbonisation*

To enable the supply chain decarbonisation process, governments and non-government organisations (NGOs) are valuable players to engage (Nalule et al. 2024). Some of the duties that governments can undertake include the development of policies that promote sustainability, funding of research on sustainable technologies, funding of sustainable manufacturing, and development of infrastructure that will support sustainable resources such as renewable energy and sustainable transport. It can also encourage the formation of public–private partnerships to create thought leadership and knowledge exchange. In this context, NGOs perform complementary functions by seeking to reinforce environmental standards, promoting awareness about the need for decarbonisation, and exercising pressure on businesses for their environmental responsibility. Such factors as certification programs, public campaigns, and other facilities provided by NGOs will assist in raising awareness among different companies and push towards more transparency and sustainability.

9.3 *Incentives and Penalties for Compliance*

Incentives and penalties are the most effective methods for enforcing compliance with decarbonisation policies (Alamoush et al. 2020). The following is a list of possibilities for governments to motivate companies to decrease their emissions: such incentives may include tax credits, subsidies for renewable energy initiatives, and financial support for energy efficiency advancements. Another factor influencing business decisions is the availability of low-interest credit for green initiatives and viable financing conditions for green projects. On the other hand, the penalties associated with noncompliance with environmental standards may discourage businesses from ignoring carbon reduction responsibilities (Sindhi and Kumar 2012). They may include fines, operations limitations, and licenses and permit suspension. In the same manner, policymakers thus encourage compliance through rewards and punishment of noncompliance so that the business is naturally motivated to work towards a decarbonisation policy.

Hence, the application and coordination of policies and regulatory frameworks play critical roles in providing stimulus to the supply chain decarbonisation. While international agreements prescribe ways and means for sustainable management of the earth's resources, national policies define its framework, and finally, there are other supporting roles by governments and NGOs. Rewards are a means to get people to do the right things, while punishments guarantee compliance with set standards. These constitute a holistic strategy for promoting low-carbon supply chain initiatives and attaining global sustainability goals.

10 Future Trends and Directions

10.1 Emerging Technologies and Innovations

New technologies and innovations are expected to be vital in the long-term supply chain decarbonisation process. Scientific developments in renewable energy, including better-quality solar panels, next-generation wind turbines, and improved efficiency in storing energy, will gradually decrease emissions in supply chains. Automobiles are expected to transition from the traditional internal combustion engine to electric and hydrogen vehicles as the battery has a longer life and increased fuel cell efficiency. Also, using increased sustainability in producing materials such as biodegradable plastics and superior composites will offer alternatives to high-carbon materials. AI and machine learning technology will be integrated to improve supply chain efficiency with predictive analysis and optimising logistics.

10.2 Evolution of Global Supply Chain Practices

There are several trends likely to shape global supply chain practices towards sustainability. There will be a more excellent implementation of the circular economy where there is a reduction of waste and reusing, as well as the recycling of products. More critical with this shift will be gains in remanufacturing and material recovery technologies. New approaches and technologies like blockchain and IoT will improve the workflow's overall transparency and traceability, making monitoring CO_2 emissions and guaranteeing sustainable compliance possible. There will be more collaboration among industries since many organisations already recognise the importance of joint sustainability practices and knowledge exchange. There will be increased localisation of the supply chains and diversification of transport to decrease the vulnerability of long-distance transport.

10.3 Predictions for the Next Decade

Over the coming decade, several trends are anticipated to define the future of supply chain decarbonisation. First of all, the use and production of renewable energy sources in all the value chain stages will be significantly increased, mainly due to the decrease in costs and the presence of policies. Second, decarbonisation will form a significant part of corporate agendas where corporations, irrespective of industry, declare ample targets for reducing emissions and incorporate sustainability as a fundamental selling proposition for organisations. Thirdly, the rules and regulations are expected to become further stringent with increasing carbon pricing mechanisms, emissions trading schemes and other standards. Fourthly, consumers' awareness of environmentally friendly products will increase, pushing companies to employ sustainable practices in supply chains. Finally, technology will further enhance the decarbonisation process as AI, blockchain, and the IoT will make the operation smoother and more environmentally friendly.

The projections of supply chain decarbonisation are relatively high, and many technological advancements and innovations will allow for fast progress in the identified area. Increased pressures from regulatory authorities and customers, alongside changes in global supply chain practices regarding sustainability and resilience, will also enhance the shift to a low-carbon supply chain. Depending on the companies' actions in the next decade, those that already adapt to these trends and actively implement decarbonisation plans will have the best shot at success in a more sustainability-driven global market.

11 Conclusion

11.1 Recap of Key Points

The decarbonisation of supply chains is one of the world's key priorities, and it is significant for overcoming climate change and achieving sustainable development. This chapter discusses different facets concerning decarbonisation, its meaning, and the need to cut carbon emissions in the contemporary supply chain. It also briefly discussed decarbonisation due to today's supply chain practices, carbon footprint, and the role of regulation, market, and social demands. Some discussed measures were decarbonisation through improved renewable energy, efficient energy utilisation, green transportation solutions, circular economy practices, and carbon offsetting. The researcher highlighted how IoT, AI, blockchain, and sustainable materials accelerate decarbonisation efforts. Relevant environmental and social consequences include decreasing greenhouse gas emissions, enhancing public health, new employment opportunities, and social justice were addressed. Economic, technological, policy, and organisational threats and the need for favourable policy and regulation were also discussed. Last of all, several scenarios concerning future trends, new technologies, and developmental changes in global supply chain management practices were discussed.

11.2 The Path Forward for Supply Chain Decarbonization

The future steps towards supply chain decarbonisation require a multi-faceted approach, including technologies, regulation, and cooperation. Businesses are obliged to go on implementing and investing in more sophisticated technologies that decrease the intensity of emissions and increase efficiency. Governments and other public authorities should develop solid policy frameworks encouraging sustainable practices and pushing firms to be responsible for their environmental impact. Cooperation among industries and sectors will be decisive in sharing experiences, generating innovation and provoking change. Developing sustainable and robust supply chains would prevent negative environmental consequences, provide economic opportunities, and lift social well-being. In the engagement of the whole world in delivering on the commitments of international conventions such as the Paris Accord, the process of supply chain decarbonisation will serve a very pivotal role in defining the future.

11.3 Call to Action for Stakeholders

All the stakeholders across the supply chain ecosystem, including businesses, governments, non-governmental organisations, customers, and investors, should act fast and in accord to accelerate decarbonisation efforts. Businesses should ensure they are sustainable in their operations, establish significant emission reduction goals for the company, and disclose progress towards them. Governments should implement regulations that require the process of decarbonisation, support through funding, and facilitate public–private partnerships. NGOs should raise awareness, pressure the entities, and promote community-based sustainability efforts. Consumers can demand more sustainable products and services through purchasing patterns. Investors should support organisations with sound environmental policies and promote sustainable investments. Last but not least, the decarbonisation of the supply chain is an essential process that needs the engagement of all stakeholders across society. The future also implies encouraging innovative solutions, nurturing collaboration, and adherence to sustainable practices that will help save the planet and reduce the carbon footprint. The time to act is now, and the path forward is clear; hence, everybody should work together to tackle decarbonisation, making this world safe and happy for all.

References

Alamoush AS, Ölçer AI, Ballini F (2022) Port greenhouse gas emission reduction: port and public authorities' implementation schemes. Res Transp Bus Manag 43:100708

Bastos T, Teixeira LC, Nunes LJ (2024) Forest 4.0: technologies and digitalization to create the residual biomass supply chain of the future. J Clean Prod 143041

Bocken NM, De Pauw I, Bakker C, Van Der Grinten B (2016) Product design and business model strategies for a circular economy. J Ind Prod Eng 33(5):308–320

Dewasiri NJ, Baker HK, Banda YKW, Rathnasiri MSH (2022) The dividend decision model: a possible solution for the dividend puzzle. In: Rana S, Sakshi, Singh J (eds) Exploring the latest trends in management literature (Review of management literature), vol 1, pp 249–267. Emerald Publishing Limited, Leeds. https://doi.org/10.1108/S2754-586520220000001013

Dewasiri NJ, Karunarathne KSSN, Menon S, Jayarathne PGSA, Rathnasiri MSH (2023) Fusion of artificial intelligence and blockchain in the banking industry: current application, adoption, and future challenges. In: Saini A, Garg V (eds) Transformation for sustainable business and management practices: exploring the spectrum of Industry 5.0, pp 293–307. Emerald Publishing Limited, Bingley. https://doi.org/10.1108/978-1-80262-277-520231021

Dewasiri NJ, Dharmarathna DG, Choudhary M (2024) Leveraging artificial intelligence for enhanced risk management in banking: a systematic literature review. In: Singh R, Khan S, Kumar A, Kumar V (eds) Artificial intelligence enabled management: an emerging economy perspective, pp 197–213. De Gruyter, Berlin. https://doi.org/10.1515/9783111172408

Etukudoh EA, Adefemi A, Ilojianya VI, Umoh AA, Ibekwe KI, Nwokediegwu ZQS (2024) A review of sustainable transportation solutions: innovations, challenges, and future directions. World J Adv Res Rev 21(1):1440–1452

Fletcher C, Ripple WJ, Newsome T, Barnard P, Beamer K, Behl A, Bowen J, Cooney M, Crist E, Field C, Hiser K (2024) Earth at risk: an urgent call to end the age of destruction and forge a just and sustainable future. PNAS Nexus 3(4):106–110

Geissdoerfer M, Savaget P, Bocken NM, Hultink EJ (2017) The circular economy–a new sustainability paradigm? J Clean Prod 143:757–768

Habert G, Miller SA, John VM, Provis JL, Favier A, Horvath A, Scrivener KL (2020) Environmental impacts and decarbonisation strategies in the cement and concrete industries. Nat Rev Earth & Environ 1(11):559–573

Hsu CC, Choon Tan K, Hanim Mohamad Zailani S, Jayaraman V (2013) Supply chain drivers that foster the development of green initiatives in an emerging economy. Int J Oper & Prod Manag 33(6):656–688

Jayawardena NS, Dewasiri NJ (2023) Food acquisition and consumption issues of South Asian countries: a systematic literature review and future research agenda. FIIB Bus Rev. https://doi.org/10.1177/23197145231194113

Kaack LH, Donti PL, Strubell E, Kamiya G, Creutzig F, Rolnick D (2022) Aligning artificial intelligence with climate change mitigation. Nat Clim Chang 12(6):518–527

Karunarathna KSSN, Dewasiri NJ, Singh R, Rathnasiri MSH (2024) What does artificial intelligence–powered ChatGPT bring to academia? A review. In: Singh R, Khan S, Kumar A, Kumar V (eds) Artificial intelligence enabled management: an emerging economy perspective, pp 3–13. De Gruyter, Berlin. https://doi.org/10.1515/9783111172408

Khan SA, Mubarik MS, Kusi-Sarpong S, Gupta H, Zaman SI, Mubarik M (2022) Blockchain technologies as enablers of supply chain mapping for sustainable supply chains. Bus Strateg Environ 31(8):3742–3756

Koh SL, Genovese A, Acquaye AA, Barratt P, Rana N, Kuylenstierna J, Gibbs D (2013) Decarbonising product supply chains: designing and developing an integrated, evidence-based decision support system–the supply chain environmental analysis tool (SCEnAT). Int J Prod Res 51(7):2092–2109

Kuriyama A, Abe N (2021) Decarbonising the power sector to engender a transition Japan: quantifying local employment impacts. Renew Sustain Energy Rev 137:110610

Li J, Pan SY, Kim H, Linn JH, Chiang PC (2015) Building green supply chains in eco-industrial parks towards a green economy: barriers and strategies. J Environ Manag 162:58–170

Lima MA, Mendes LFR, Mothé GA, Linhares FG, De Castro MPP, Da Silva MG, Sthel MS (2020) Renewable energy in reducing greenhouse gas emissions: reaching the goals of the Paris agreement in Brazil. Environ Dev 33:100504

Liu Y, Wang Y, Lyu P, Hu S, Yang L, Gao G (2021) Rethinking the carbon dioxide emissions of road sector: integrating advanced vehicle technologies and construction supply chains mitigation options under decarbonisation plans. J Clean Prod 321:128769

Markkanen S, Anger-Kraavi A (2019) Social impacts of climate change mitigation policies and their implications for inequality. Clim Policy 19(7):827–844

Mathiyazhagan K, Govindan K, Noorul Haq A (2014) Pressure analysis for green supply chain management implementation in Indian industries using analytic hierarchy process. Int J Prod Res 52(1):188–202

Meherishi L, Narayana SA, Ranjani KS (2019) Sustainable packaging for supply chain management in the circular economy: a review. J Clean Prod 237:117582

Mishra U, Rathnasiri MSH, Dewasiri NJ (2024) Net-zero economy and energy sustainability: a nepalese perspective. In: Singh R, Khan S, Kumar A, Luthra S, Chokshi H (eds) Net zero economy, corporate social responsibility and sustainable value creation, pp 145–161. CSR, sustainability, ethics & governance. Springer, Cham. https://doi.org/10.1007/978-3-031-55779-8_9

Murray A, Skene K, Haynes K (2017) The circular economy: an interdisciplinary exploration of the concept and application in a global context. J Bus Ethics 140:369–380

Nalule VR, Heffron RJ, Olawuyi DS (2024) Renegotiating contracts for the energy transition in the extractives industry. Springer Nature

Nevzorova T (2024) Carbon market for climate projects in Russia: an overview of nature-based and technological carbon offsets. Gases 4(3):153–173

Penn AS, Bartington SE, Moller SJ, Hamilton I, Levine JG, Hatcher K, Gilbert N (2022) Adopting a whole system approach to transport decarbonisation, air quality and health: an online participatory systems mapping case study in the UK. Atmosphere 13(3):492–499

Rathnasiri MSH, De Silva V (2023) The role of shopping values, cognitive and affective evaluations on the relationship between store environment and store patronage intention: a future research agenda, Sri Lanka. J Manage Stud 5(1):100–120

Rathnasiri MSH, Madushani KAC, Peiris TAK, Arman S (2024b) "Creative consumption of clothing": investigating the determinants of generation Z consumers' purchase intention for upcycled products. Int J Res Anal Rev 11(1):205–216

Rathnasiri MSH, Dewasiri NJ, Kumar AA (2024a) Policy framework and implementation strategies for Sri Lanka's transition to a net-zero economy. In: Singh R, Khan S, Kumar A, Luthra S, Chokshi H (eds) Net zero economy, corporate social responsibility and sustainable value creation, pp 43–60. CSR, sustainability, ethics & governance. Springer, Cham. https://doi.org/10.1007/978-3-031-55779-8_3

Sindhi S, Kumar N (2012) Corporate environmental responsibility–transitional and evolving. Manag Environ Qual Int J 23(6):640–657

Singh J, Pandey KK, Kumar A, Naz F, Luthra S (2023) Drivers, barriers and practices of net zero economies: an exploratory knowledge-based supply chain multi-stakeholder perspective framework. Oper Manag Res 16(3):1059–1090

Steiner B, Münch C, Küffner C, Hartmann E (2023) Mapping the intellectual foundation of low-carbon supply chains: a theory-based literature review on buyer-supplier relationships. J Clean Prod 138961

Wang M, Wang B, Abareshi A (2020) Blockchain technology and its role in enhancing supply chain integration capability and reducing carbon emission: a conceptual framework. Sustainability 12(24):10550

Wesseling JH, Lechtenböhmer S, Åhman M, Nilsson LJ, Worrell E, Coenen L (2017) The transition of energy-intensive processing industries towards deep decarbonisation: characteristics and implications for future research. Renew Sustain Energy Rev 79:1303–1313

While A, Eadson W (2022) Zero carbon as economic restructuring: spatial divisions of labour and just transition. New Polit Econ 27(3):385–402

Zhang A, Alvi MF, Gong Y, Wang JX (2022) Overcoming barriers to supply chain decarbonisation: case studies of first movers. Resour Conserv Recycl 186:106536

Integrating Sustainability into HRM: An Innovative Approach Towards Zero Carbon Industry

Rubee Singh and Akshay Singh

Abstract The concept of Net Zero has become increasingly significant for attaining sustainable development goals and it has given origin to an innovative approach for managing employees known as sustainable human resource management (SHRM). Although there has been a rise in research on SHRM and its implementation, this emerging research approach currently lacks cohesive principles and methodologies. Additional research is necessary for ascertaining the significance of inducing sustainable approach in HRM. The chapters aims to showcase the advancement of human potential within an organization and examines incorporation of sustainable aspects into different HRM practices in order to enhance comprehension of this innovative approach. We explore the rationale and need behind firms' adoption of SHRM practices and analyze the outcomes they have attained. Hence, our study enhances the theory base on SHRM by showcasing the results of the applied HRM practices, along with emphasizing the significance of contextual factors. The findings highlights that SHRM can yield long-term advantages for employers, society, and employees. The favorable consequences of SHRM encompass employee well-being concerns, such as prolonged careers, reduced absence rate, minimal turnover, and delayed retirements. Furthermore, there is a noticeable absence of employee engagement in the decision-making process about environment policies of the organization, which is another fundamental aspect of SHRM. The chapter has specific suggestions for HR managers. The chapter also provides suggestions about the under investigated scope that need more investigation.

Keywords Human Resources · HRM practices · Sustainability · SHRM · Zero carbon industry · Net zero economy

---

R. Singh (✉)
Institute of Business Management, GLA University, Mathura, Uttar Pradesh, India
e-mail: dr.rubeerajput@gmail.com

A. Singh
School of Management Studies, Indra Gandhi National Open University, New Delhi, India

R. Singh et al. (eds.), *Zero Carbon Industry, Eco-Innovation and Environmental Sustainability*, World Sustainability Series,
https://doi.org/10.1007/978-3-031-80220-1_10

1 Introduction

In the past decades, the globe has faced challenging periods of climate change. Consequently, the ideas of decarbonization and Net Zero for sustainability transition have been significant topics of interest for professionals and researchers. Hence, regulatory authorities have endeavored to compel companies to use precautionary steps in order to mitigate the adverse environmental effects they generate. This involves implementing decarbonization methods and enhancing their environmental performance in accordance with established standards (Gregurec et al. 2021). Pressure to deal with social issues and improve openness in the operational activities in order to maintain legitimacy has increased across organizations, as highlighted by Khan et al. (2021). For addressing the worldwide problem of environmental degradation, the United Nations (UN) Conference on Climate Change implemented the groundbreaking Paris Agreement in 2015, establishing ambitious objectives for all countries to achieve sustainability in the long run. Despite the fact that organizations are encountering numerous hurdles in their efforts to preserve competitiveness during the decarbonization process they are dedicated towards adopting sustainable measures into their functioning in response to the difficulties present in the way of sustainable development around the globe, such as environmental issues, increased population, and socioeconomic inequalities (Singh and Rao 2016). All these issues necessitate support of organizations and their approaches towards operation management. In business context, Sustainability refers to the adoption of a triple bottom line perspective, as proposed by Macke and Genari (2019), which involves a simultaneous focus on the performance of the enterprise in relation to economy, society, and environment.

The scholarly research about sustainability in companies have predominantly concentrated on the ecological facets of sustainability while HRM academics have given limited attention to the concept of sustainability (Ehnert 2009; Heikkinen et al. 2021). Primary aim of SHRM is to cultivate sustainable thinking in the ways the business organizations perform their functions including HRM. Stahl et al. (2020) argue that practices for managing human resources are referred as sustainable when they effectively promote long-term economic growth of the organization along with supporting the social wellbeing of employees and with least negative impact on the natural environment. Conversely, activities might be deemed unsustainable if they result in detrimental social, environmental, or economic consequences. These days, there focus on defining and advancing the notion of SHRM has increased (Lopez-Cabrales and Valle-Cabrera 2020). There are comprehensive studies on sustainable HRM which focused on the conceptualization of this emerging topic (Anlesinya and Susomrith 2020; Chams and García-Blandón 2019). Literature indicates a merging of strategic and sustainable HRM practices. Therefore, it appears that SHRM is based on certain concepts derived from strategic HRM based on the soft Harvard model (Beer et al. 1984). This model focuses on human element and their growth potential, thus pinpointing the role of the HRM within the firm and society as a whole. It emphasizes both short-term and long-term objectives. While the Harvard model is centered on the performance of organization coupled with wellbeing of employees, and societal welfare in HRM, it overlooks environmental objectives and primarily prioritizes shareholder value (Beer et al. 2015). In conclusion, both the Harvard model and SHRM are responsive to the context in which they are applied and emphasize the significance of many internal and external stakeholders, multiple objectives, and both immediate results as well as consequences of HRM in the future (Järlström et al. 2018). Therefore, it appears that the SHRM paradigm has elevated the former approach to a higher degree, because the conventional HRM techniques seems to be inadequate for maintaining the human capital of the organization for an extended time period (Ehnert 2009).

As a result of the insufficiency and ambiguity surrounding SHRM concepts (Aust et al. 2020), need arises for additional rigorous research for determining the significant role of SHRM. Creating a comprehensive list of SHRM practices appears to be an unattainable task due to their very contextual nature (Stankevičiūtė and Savanevičienė 2018). In addition, certain human resource management (HRM) strategies may possess sustainability characteristics, even if they are not explicitly designated as such (Guest 2017). Hence, there is a requirement for ascertaining the underlying goals and elements, including the concepts and values, for driving these behaviors (Ehnert 2009). This chapters seeks to examine the different methods where the adoption of sustainable practices is possible, as well as the benefits of implementing these initiatives. In addition, our study enhance the ongoing discussion on the topic of SHRM by incorporating the practices from employee recruitment to development of employees, from work-life balance to engagement and well-being, and managing the employees' performance to the decisions regarding their compensation. This indirectly aligns with the soft Harvard model proposed by Beer et al. (1984).

In this chapter, we proceed with a concise examination of the SHRM, its scope, and different practices for adopting SHRM, in the section of literature review, which is then followed by the methodology part. Subsequently, the implications are provided, and the chapter concludes with an examination of the constraints of study and potential avenues for further research.

2 Literature Review: Scope of Sustainable Human Resource Management

SHRM consists of deliberate and evolving practices and strategies for managing the human resources of any organization that are intended to facilitate the accomplishment of not only the economic objectives but also the environmental and social goals as well, while also maintaining the HR foundation in the long run. Ehnert et al. (2014) view strategic HRM as a means of promoting sustainable firm growth. They argue that sustainability encompasses more than simply environmental and economic aspects, but also includes considerations of fairness in appointing employees, their development, engagement and well-being as a part of the organization. The SHRM characteristics delineate the necessary attributes for HRM to achieve sustainability and elucidate how sustainability can be leveraged in HRM (Stankevičiūtė and Savanevičienė 2018) The three aspects are of utmost importance in SHRM that includes futurity, which expresses concern about the future and thus, highlights the the impact of present decisions on the future in light of perseverance (Bhushan and Singh 2024).

Regrettably, while making numerous significant decisions, the course of action that could be the most advantageous in the long term turns out to be the less suitable for the present scenerio (Singh et al. 2021). SHRM places importance on the well-being of employees, including their income, opportunities for balancing personal lives with work-life, and other relevant elements (Macke and Genari 2019), as well as ensuring the workforce remains healthy and productive in the long run (Ehnert 2009). Environmental concern is a crucial element of SHRM which belongs to the ecological outcome of strategic HRM (Kramar 2014). HRM should prioritize environmental responsibility by actively seeking to recruit skilled employees, especially from younger generations who are more conscious for the environmental issues (Jamil et al. 2023). As a part of SHRM, methods for ecologically conscious recruitment encompass the utilization of technology, conscientious branding, selecting applicants with environmental consciousness, and incorporating environmental considerations in recruitment communications (Anlesinya and Susomrith 2020). In a broader sense, engaging in environmentally responsible conduct in daily activities could involve practicing efficient electricity consumption, minimizing printing, transitioning to digital platforms from paper-based jobs, managing waste, and using those modes of transportation which have least adverse effect on environment (Lu et al. 2023).

However, it is not sufficient to effectively capitalize on strategic HRM and ensure long-term viability, a more extensive approach is necessary, in the form of SHRM for attaining sustainability goals (Järlström et al. 2018). Employee collaboration is fostered and enhanced through the knowledge sharing, the cultivation of mutual respect within the business, and the utilization of transparent and proactive communication. Additionally, it reduces absenteeism, enhances the level of productivity and overall job satisfaction and lowers the probability of labor disputes (Hirsig et al. 2014). In order to achieve strategic the objective of SHRM, it is imperative to actively guarantee that sustainability is evident in all elements of HRM, such as the hiring process, training and development of employees, performance evaluations, and employee wellbeing (Singh et al. 2024). Figure 1 presents incorporation of sustainability in different HRM practices.

2.1 Green Recruitment

Green Recruitment (GR) is a novel strategy that firms employ to prioritize environmental awareness as a key factor in their employment process. GR employs a digital method to reduce the potential environmental consequences (Ogbu Edeh and

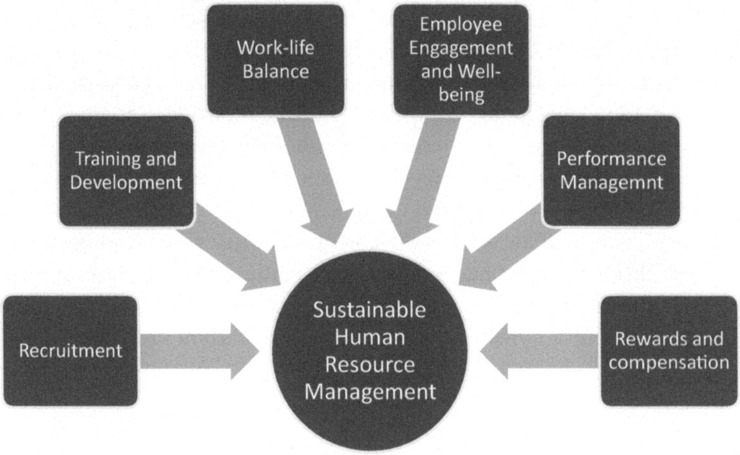

Fig. 1 Sustainable HRM practices

Okwurume 2019). This method employs platforms like websites, social networking sites, online job application forms, and email for targeting applicants worldwide. This method entails conducting interviews either through telephone or video conferencing, to minimize the adverse impact on environment linked to travel (Mwita and Kinemo 2018). Online recruitment allows recruiters to distribute a greater amount of information at the same time, which helps to minimize the negative environmental effects, unlike traditional media like brochures or newspaper advertising (Davidescu et al. 2020). It is argued that digital methods for recruiting promotes those individuals to apply who are concerned about sustainability and want to be a part of an organization that supports their ideology (Masri and Jaaron 2017). Similarly, the organization focusing on sustainability can hire employees who share the same values by creating jobs with profiles detailing the environmental goals.

The environment related initiatives of the organizations should be emphasized in the brochure available to the job seekers and also its website (Peerzadah et al. 2018). Organizations might employ diverse techniques to improve their SHRM practices throughout recruitment procedures. These efforts involve integrating health and safety responsibilities within job profiles and creating interviews for evaluating the compatibility of the potential employees with organization's goals. Recruiters have the ability of designing training programs for fresh employees for sharing details about the methods and processes of protecting environment (Masri and Jaaron 2017). The increasing recognition and apprehension about environment among job seekers and providers has emphasized the significance of integrating GR activities for promoting SHRM. Currently, numerous recruiting experts are keen and eager to be hired by a company that advocates for environmentally friendly recruitment techniques. This technique has resulted in advantages such as higher productivity, decreased costs, and improved employee retention (Lee and Ha-Brookshire 2017).

GR reflects firms' intiative for persuading and developing talented workers by implementing several SHRM methodologies (Jiang et al. 2012). Organizations in the several countries have altered their recruitment techniques due to ecological concerns. A recent survey found that highly competent graduates consider a company's ecological performance to be a crucial component in evaluating its reputation. All these practice aids the job seekers to examine an organization's sustainability before accepting the job offer for any position (Mwita and Kinemo 2018). Efficient execution of the GR process can facilitate staff involvement and offer green training and development possibilities in later phases (Pham and Paillé 2020). A study showed that nearly 75% of employees have a preference for working with firms that emphasize the implementation of a dynamic strategy to reduce carbon emissions (Masri and Jaaron 2017).

2.2 *Sustainable Training and Development*

Appropriate training is the primary tool for human resources development. Traditional approach to HRM assume that the competency based training should be prioritized by employers for fulfilling the objectives of business and maintaining competitiveness and agility (Schmidt 2007, p. 483). This entails that employees should develop only those skills which are necessary for meeting the demands of their current roles and fulfilling the job responsibilities assigned to them (Hinton 2018). Several authors examine the topic of sustainable HRM and specifically emphasize the HR functions, which encompasses HR training. The socially responsible HRM has extensively examined the topic of employee equality (Shen and Zhu 2011). It is essential that opportunities to participate in the training programs should be equally available to all the employees. Research has demonstrated that organizational justice significantly influences job satisfaction, and that contented employees are likely to be successful in carrying out various organizational endeavors related to sustainability (Singh and Singh 2019).

During crises, the competencies with high demand becomes more important. Despite being rendered redundant, an employee with valuable talents will readily secure alternative employment. The enhancement of human resources is intricately linked to a rise in employability (Blanco 2014). Thus, a company who cannot ensure stable employment fulfills a social duty by assisting employees in navigating a competitive job market. It is important to note that certain managers express concerns that employees who receive training paid for by their firm may ultimately depart to work for a competing company (McGrath 2015). Empirical studies have shown the significance of staff training in fostering employee retention (Khan 2018). Therefore, it is crucial to provide employee training to build the necessary skills for the future, especially considering that demographic shifts have led to employees staying in the employment market even beyond reaching retirement age (Fox and O'Connor 2015). Enhancing employees' knowledge about environment, considering the advancements in the expanding field of climate change is labelled as sustainable HR development involves (Yang et al. 2021). Hence, such development, through training, is seen as the primary step towards fostering an environment oriented mindset for establishing a green company culture.

According to Zibarras and Coan's (2015), HR training is viewed as a strategic method for influencing pro-environmental behavior among staff and green training has been proven to have a beneficial effect on improving one's understanding and abilities towards promoting environmental sustainability (Cabral and Lochan Dhar 2019). HR training benefits both individual and employees' group performance while inducing organizational citizenship behavior regarding environment protection along with team creativity (Pinzone et al. 2016). It may aid in addressing obstacles to the implementation of SHRM practices. There is a need to mention that the constant changes in technology impede the favorable correlation between training and creativity in addressing issues related to environment (Ogbeibu et al. 2020). External cooperation, refers to the act of collaborating with the educational system in the realm of HR development. For example the training courses for the employees may be advanced through further studies by collaborating with educational institutes.

2.3 *Work–Life Balance*

Work-life balance (WLB) is becoming a significant and crucial focus for a sustainable workforce maintenance. An organizational climate that fosters work-life balance and employee well-being is crucial for establishing and cultivating a sustainable work environment (Carlier et al. 2012; Kossek et al. 2014). From a theoretical standpoint, different ideas emphasize that the job role of an employee has impact on the family and vice versa (Mäkelä and Suutari 2011). Until recently, there has been a lack of research and practical connection between work-life balance, and sustainability. In recent years, governments in developed countries have emphasized the significance of creating a sustainable workplace (Lester 2013). They have urged organizations to prioritize strategies that promote a balance between employees' professional and personal lives, in combination with their overall well-being. This approach is seen as a crucial means of facilitating sustainable human resources (Tziner et al. 2015). Organizational work-life balance initiatives primarily prioritize time schedules, parental leave benefits, and amenities for childcare.

Before the time of COVID-19 pandemic, telecommuting was restricted to a select few job roles, primarily managers and white-collar with the purpose of lowering labor expenses, enhancing productivity, upholding work-life balance, and providing support for family members (Vyas 2022). Flexi-time arrangements empower employees to have control over determining the start or finish time for the working days, but the total number of hours are fixed for every day. Flexi-time is suggested as a highly favored method for enhancing workplace flexibility. It is most effective when used in conjunction with all WLB measures and can be included into various WLB alternatives provided by enterprises to simultaneously achieve their business objectives (Lazar et al. 2010). Remote work offers flexibility and enables employees to negotiate with their employers to work from a different location other than the traditional office (Bohlander and Snell 2010).

Due to the progress in information technology and the increasing trend of globalization, numerous firms view telecommuting as a viable alternative to minimize office space expenses, as well as transportation and mileage allowances (Parakandi and Behery 2016). Enabling employees to work remotely from their homes has significantly benefited firms by improving work-life balance and retaining talented employees. This has led to higher employee satisfaction coupled with positive impact on balancing personal life with work (Jose 2009). Part-time job is also an employment option where individuals work for less hours in a week when compared to full time working individuals who work for 40 h, allowing them to still fulfill their personal responsibilities (Asadullah and Fernández 2008). This arrangement provides an opportunity for individuals with impairments, health issues, and working mothers to continue their jobs and get valuable experience (Lazar et al. 2010). Part-time employment also contributes to the promotion of gender equality as it allows mothers to engage in work while still fulfilling their domestic responsibilities.

As per the annualized hour approach, employees cannot only select their desired working hours but working days as well within a specified time frame as per their convenience (Puigarnau and Ommeren 2010). An individual's working time is determined for the whole year and then distributed throughout the year in months, weeks, days, and hours. Under this approach, employees have complete autonomy for allocating annual hours based as per their own schedule, enabling them to effectively manage the balance between their professional and personal lives (Parakandi and Behery 2016). Employees can also adhere to a flexible schedule where each person commences work at their preferred hour. Within this particular work arrangement, employees have varying start times, breaks, and end times, allowing them flexibility (Holmes 2010). Flextime work arrangement defines the specific number of working hours numerically and distribute them in a non-traditional manner.

2.4 Employee Engagement and Well-Being

Employee engagement is crucial for the effective adoption of environmentally friendly activities. Encouraging employees to adhere to such practices and maintain a sustainable work-life balance is a difficult task for HR managers. Madero-Gómez et al. (2023) proposed that corporations should engage employees in efforts to diminish their carbon footprint, for fostering an environmentally friendly workplace and promoting waste reduction and recycling. Organizations can incentivize employees to engage in carpooling, video conferencing, and various forms of remote work. It has been highlighted that employee engagement fosters the development of environmentally conscious workforce and enhances the safety of the firm (Kossek et al. 2014). Employee involvement provides employees with the ability to contribute their ideas for the development of environmentally friendly practices within enterprises. It enhances the rapport between management and organizations, leading to the establishment of enduring trust between them. Guest (2017) advocated for employee involvement as a means to streamline the production process and enhance worker well-being and safety.

Research conducted on a global scale has shown that efforts to improve services can either be successful or unsuccessful based on how employers handle worker issues through factors such as motivation, leadership, and support (Larkins et al. 2014). Adopting appropriate adjustments to HRM strategies and practices allows organizations to achieve their long-term objectives while effectively handling unanticipated consequences such as negative feedback. SHRM entails prioritizing human resources by focusing on their growth, development, and well-being, while also ensuring the preservation of the environment for promoting the organizational sustainability. This approach can advance the growth of individuals and companies, focusing on preventing problems and ensuring long-term and genuine development of human resources (DiFabio and Peiró 2018). In addition sustainable HRM procedures promote a sustainable culture by establishing a connection between the competitiveness of the organization and excellence of its HR practices.

The impact of sustainable practices on the ability of firms to outcompete others is crucial for creating environmentally friendly workspaces. Workers require equitable remuneration, as mandated by SDG 4 (decent work); numerous nations have implemented minimum wage policies, aimed at guaranteeing that employees receive sufficient income to fulfill their essential requirements (Gülal and Ayaita 2020). Nevertheless, there is a contention that minimum salaries are inadequate in terms of enhancing employee happiness. Equitable remuneration encompasses factors beyond just adherence to legal requirements. Compensation in an employment relationship refers to the total amount of cash, physical services, and benefits that employees get (Milkovich et al. 2014). Financial returns encompass fundamental salary plus both immediate and future incentives. "Tangible services and benefits" including discounts for staff, medical benefits, paid leave, vacation time, and retirement are classified as perks. These incentives are not commonly provided in many occupations, and firms are not legally required to supply them. However, job satisfaction and wellbeing and receiving fair income are all positively related.

2.5 Sustainable Performance Management

The goal of performance management always remained a subject of debate among both professionals and academicians. Many employees frequently lack knowledge of the true aim of their performance management (Nduati and Wanyoike 2022). The aim of evaluating performance must be expressed and comprehended clearly by those whose performance will be evaluated as well as those who will appraise the performance at every level of organization, due to these reasons (Roberts 2003). If the aim of performance management is not properly comprehended, it is typically executed inadequately, and does not meet the employees' expectations. For instance, if an organization implements a focus on sustainability, the objective of the performance management system should align with this new emphasis. Although a strategic human resource management approach undoubtedly has advantages in integrating the strategy and HRM of an organization, it may also hinder performance management (Brown et al. 2010). It has been previously recognized that in a strategic HRM framework, organizations tend to prioritize short-term financial metrics. This negatively impacts the process of developing employee skills and abilities.

The implementation of SHRM, with its comprehensive and enduring approach to business, will greatly benefit the growth and progress of employees (Maley 2014). However, implementing this improvement would necessitate a significant change in mindset by senior executives across the corporations. Indeed, implementing such a drastic alteration could be perceived as an innovation and might be hindered by managers who are under immense pressure to generate immediate profits. This implies that the implementation of SHRM cannot be done gradually; instead, the shift to SHRM must be started from the corporation's main office and wholeheartedly supported by the top management of the corporation (Macke and Genari 2019). Specifically, the objective and the procedure of performance will necessitate focused consideration. Employees must comprehend the intended aim of their performance management as outlined by the corporation. In the presence of a robust HR atmosphere, corporations often guarantee that employees at all level has information about the aim of performance management for its effective implementation. However, in case of SHRM, managers needs to ensure that apart from clear communication of the aim of performance management its alignment with organization's sustainability commitment is equally important (Ravesangar et al. 2024).

It is essential to clearly define this aim within the SHRM scope for establishing harmony between expectations of the employee as well as the supervisor (Saratun 2016). However, this alone is insufficient. Ensuring uniformity in the expectations of supervisors and employees regarding performance evaluation necessitates more than just aligning their expectations (Enhert et al. 2014). Both parties should have a face-to-face interaction that is both reasonable and satisfactory. This entails frequent meetings for assessing the required level of performance, targets, actual results, and the growth prospects. Regular, brief in-person conversations found to be beneficial to maintain consistent expectations among stakeholders (Roberts 2003). When an organization embraces a sustainable human resource management (SHRM) strategy, it becomes evident that the goal of performance management should be flexible enough to align with this shift. When the organization prioritizes long-term sustainability as a key aspect of its competitive strategy, it becomes necessary for it to establish a performance management system that successfully communicates, represents, and executes this plan.

2.6 Green Rewards and Compensation

Rewards and compensation are the primary HRM practices that are used to rewards the employees for their contributions in the organization. These techniques are highly effective in aligning the personal interests of employees with the organizational objectives. Ramus (2012) suggests that awards and incentives are significant instruments for motivating the employees, leading them to exert their utmost effort in order to fulfill the goals of the firm. Gill (2012a, b) examined the role of employee involvement in environmental programs and concluded that paying employees for taking on responsibilities related to environmental responsibility boosted their commitment to environmental management programs. The most evident example of an effective green compensation strategy was found in 469 US companies across different industries with high levels of pollution (Berrone and Gomez-Mejia 2009). In addition to this, the companies with environmentally sustainable practices compensated their chief executive officers at a higher rate compared to companies that did not prioritize eco-friendliness (Ardiza et al. 2021).

The literature study demonstrates that incentives serve as a catalyst for CEOs to manipulate accounting and other practices in order to attain immediate outcomes and enhance their personal wealth (Frey 2007). While incentives and prizes can incentivize green actions in firms, it is impossible to eliminate all instances of malpractice. Designing efficient monetary incentives can be a complex task because of the challenge in correctly and impartially assessing environmental behaviors and performance. It has been argued that enterprises should develop energy-efficient products and create incentives to encourage their widespread use (Saka et al. 2021). Managers can promote environmentally-friendly behaviors by including sustainability related incentives in the pay program (Mandago 2018). In addition, managers have the ability to request that employees propose particular environmentally-friendly ideas that can be induced in the current roles. These ideas can be considered and incorporated into the targets of upcoming year and employees should be rewarded for achieving these goals.

Francoeur et al. (2017) suggested that employee compensation system can be adjusted to incorporate bonuses that are contingent upon the individual's appraisal ratings in behavioral terms towards safeguarding the environment and bonuses should be given to employees for their exceptional performance on their environmental efforts. Green rewards encompass a variety of advantages including carbon offsets, to involve individuals in the process of environment protection and acknowledge the tasks accomplished by them (Pillai and Sivathanu 2014). According to Jabbar and Abid (2015), receiving green awards is associated with the highest degree of job satisfaction and then to on environmental performance.

The implementation of green compensation schemes has a substantial influence on satisfaction related to environmental performance, as demonstrated by Renwick et al. (2013). Implementing such systems sustainability has a beneficial effect on employees' motivation to engage in green projects. Rewards serve as incentives to enhance motivation and encourage employees to demonstrate responsibility towards nature (Daily and Huang 2001). Compensation packages should be designed for incentivizing the learning and accomplishments of environmentally friendly behaviors (Das and Dash 2024). Employees can be rewarded for their green achievements via monetary-based, nonmonetary-based, and recognition-based incentives. Monetary rewards for efforts towards attaining sustainability may include pay raises, cash prizes, and bonuses. Recognition-based awards can showcase the environmentally friendly accomplishments of employees through extensive exposure and public commendation of their green initiatives by the CEO or other top executives (Francoeur et al. 2017).

3 Research Methodology

This chapter is completely based on the secondary data available on the topic in the form of book chapters, journal articles, and online published sources. The literature on incorporating sustainability into HRM practices have been explored that promotes the awareness among organizations regarding adoption of environment friendly practices for managing their human capital. The implications are also drawn on the basis of findings of earlier studies.

4 Implications

This chapter endeavored to ascertain whether sustainable human resource management (SHRM) is the solution to contemporary environmental issues. Affirmative, it is. Specifically, integrating sustainability into HRM will yield long-term advantages for the firm, its employees, and society at large. The HRM practices that solely concentrate on present requirements are not aligned with the principles of SHRM of adopting a long-term outlook and adaptability. SHRM is considered to be an exceptionally intricate undertaking in the field of HRM. The techniques and practices employed in SHRM aim to achieve financial, social, and ecological objectives. These techniques and practices have an impact both within the organization and outside it, and are implemented over an extended time period. Additionally, they are designed to prevent any unintended negative consequences and ensure effective feedback mechanisms (Ehnert et al. 2014).

It is possible for organizations to improve their SHRM practices throughout the recruitment process by employing a variety of different strategies. Among these initiatives include the incorporation of environmental reporting in the brochures of company. Additionally, the job descriptions should entail information related to the health and safety obligations. Further the interviews should be structured to assess the potential of interviewee's compatibility with the organization's goals about environment. The capacity to design an employee training program for newly appointed staff members that concentrates on providing insight into the organization's environment friendly approaches, objectives, and environmental goals is something that recruiters have the potential to do (Masri and Jaaron 2017).

HR development occurs through the generation of novel and innovative knowledge. Industrial firms have the opportunity to collaborate with universities in establishing research and development centers. This collaboration helps advance technological progress in the industry and fosters scientific development. Considering profitability, it is important to note that SHRM, aligns with the motive behind Triple Bottom Line. While financial is not an exclusive reason for measuring a company's success, they nonetheless hold significant importance. Profitability about HR development is measured in terms of skilled and motivated people to enhance company growth and provide innovative ideas, resulting in financial gains for the firm. When making decisions on training, such as the types of training, the selection of trainers, adoption of technology and digital learning platforms, both the costs and the long-term benefits should be taken care of.

Within the framework of SHRM, rewards and pay can be regarded as effective instruments for promoting and facilitating environmental initiatives within organizations. Contemporary firms should implement incentive programs to motivate their staff to undertake environmentally friendly efforts. This study suggests that such practices can effectively motivate employees to come up with eco-friendly initiatives. Moreover, it is essential to incorporate sustainability standards as a fundamental measure of success for top managers. The incorporation of performance management and its alignment with SHRM might serve as a significant aspect of the annual reports of the organization and perhaps be incorporated in its mission statement. As previously said, implementing sustainable human resource management (SHRM) strategies, has a beneficial impact on overall company's performance (Renwick et al. 2013).

The work of HR managers is vital in fostering a green culture inside the firm. HR managers must facilitate the integration of environmentally friendly practices while also encouraging and engaging employees to actively participate. Through the use of corporate reports and indicators, businesses are obligated to supply stakeholders with information that is topical, up-to-date, and easily understandable regarding their activities. The provision of this information enables businesses to fulfill their commitments in accordance with the ethics of responsibility and to develop policies that are sustainable and that allow for the flourishing of workers and their general well-being. The implementation of measures that promote economic growth, societal improvement, and environmental improvement is an essential component of sustainability. However, in order to guarantee the health, safety, and contentment of workers in their jobs, wellbeing must be evaluated using metrics other than the gross domestic product.

An important indicator of an organization's overall health and well-being is the degree to which employees are satisfied with their jobs. It varies depending on particular aspects and the nature of the work, such as the type and quantity of remuneration, advancements, colleagues, plus the environment of the organization. For instance, it can be different depending on the individual. Organizations that effectively utilize human resources in sustainable ways can enhance productivity, foster creativity and innovation, and promote the well-being of individuals. By aligning with SHRM, businesses can enhance their attractiveness to environmentally concerned prospective employees. Managers should utilize all forms of SHRM and determine the level of commitment they are willing to invest in its implementation. Due to the significant financial investments required for implementing SHRM practices, including employee and management training and organizational changes, there are practical constraints on the number of businesses that will choose to adopt these practices and the level of commitment they will be willing to dedicate to their implementation.

5 Conclusion

Global warming poses a significant concern to the world due to its adverse environmental effects. As a result, institutions and stakeholders are exerting pressure on companies to address carbon emissions and prioritize carbon accounting. Today's stakeholders prioritize environmental concerns and exert pressure on them to comply with regulatory guidelines in order to achieve sustainable growth. Stakeholders also emphasize the need of corporations maintaining their High-Density Centers (HDCs). Hence, companies must harness human resources to enhance the efficiency of their operations in the digital economy. Concurrently, companies formulate their strategies and develop their business models in alignment with environmental policies in order to mitigate their adverse environmental effects. The emergence of zero-based movements is contingent upon the integration of e-businesses, strategic management, the use of green technology, and the practice of responsible innovation. Existing literature emphasizes the importance of incorporating sustainability and innovative e approaches in their HRM practices. This requires implementing environmental management practices that focus on decarbonization processes and net zero to ensure environmental sustainability. This study has limited applications due to its conceptual nature. Through an analysis of the information presented in this paper, a comprehensive model can be developed by taking into account all necessary factors. Hence, this work proposes the further empirical investigation to validate the model and provide recommendations for enhancement of SHRM adoption.

References

Anlesinya A, Susomrith P (2020) Sustainable human resource management: a systemic review of a developing field. J Glob Responsib 11(3):295–324

Ardiza F, Nawangsari LC, Sutawidjaya AH (2021) The influence of green performance appraisal and green compensation to improve employee performance through OCBE. Int Rev Manag Mark 11(4):13

Asadullah M, Fernández R (2008) Work–life balance practices and the gender gap in job satisfaction in the UK: evidence from matched employer–employee data. IZA discussion paper no. 3582

Aust I, Matthews B, Muller-Camen M (2020) Common good HRM: a paradigm shift in sustainable HRM. Hum Resour Manag Rev 30(3):100705. https://doi.org/10.1016/j.hrmr.2019.100705

Beer M, Boselie P, Brewster C (2015) Back to the future: implications for the field of HRM of the multistakeholder perspective proposed 30 years ago. Hum Resour Manage 54(3):427–438
Beer M, Spector R, Lawrence P, Quinn Mills D, Walton R (1984) Human resource management: a general manager's perspective. Free Press
Berrone P, Gomez-Mejia LR (2009) Environmental performance and executive compensation: an integrated agency-institutional perspective. Acad Manag J 52(1):103–126
Bhushan M, Singh R (2024) Talent management facilitates net-zero transition through employee green behavior. In Singh R, Khan S, Kumar A, Luthra S, Chokshi H (eds) Net zero economy, corporate social responsibility and sustainable value creation. CSR, sustainability, ethics & governance. Springer, Cham. https://doi.org/10.1007/978-3-031-55779-8_7
Blanco TM (2014) The meaning of employability in the new labour relationships between company-employee: a model of training in companies. Procedia Soc Behav Sci 139:448–455. https://doi.org/10.1016/j.sbspro.2014.08.039
Bohlander GW, Snell S (2010) Managing Human Resources. Mason (OH): South Western Cengage Learning
Brown M, Hyatt D, Benson J (2010) Consequences of the performance appraisal experience. Pers Rev 39(3):375–396
Cabral C, Lochan Dhar R (2019) Green competencies: construct development and measurement validation. J Clean Prod 235:887–900. https://doi.org/10.1016/j.jclepro.2019.07.014
Carlier SI, Llorente CL, Grau MG (2012) Comparing work-life balance in Spanish and Latin-American countries. Eur J Train Dev 36:286–307
Chams N, García-Blandón J (2019) On the importance of sustainable human resource management for the adoption of sustainable development goals. Resour Conserv Recycl 141(6):109–122
Daily BF, Huang SC (2001) Achieving sustainability through attention to human resource factors in environmental management. Int J Oper Prod Manag 21(12):1539–1552
Das S, Dash M (2024) Green compensation and reward system: a novel approach towards the growth and sustainability of organization. J Adv Zool 45(2)
Davidescu AA, Apostu S-A, Paul A, Casuneanu I (2020) Work flexibility, job satisfaction, and job performance among Romanian employees—implications for sustainable human resource management. Sustainability 12:6086
DiFabio A, Peiró JM (2018) Human capital sustainability leadership to promote sustainable development and healthy organizations: a new scale. Sustainability 10:2413
Ehnert I, Harry W, Zink K (2014) Sustainability and HRM. In: Ehnert I, Harry W, Zink K (eds) Sustainability and human resource management: developing sustainable business organizations. Springer, pp 3–32
Ehnert I (2009) Sustainable human resource management: a conceptual and exploratory analysis from a paradox perspective. In: Contributions to management science. Physica/Springer
Fox K, O'Connor J (2015) Five ways work will change in the future. The Guardian
Francoeur C, Melis A, Gaia S, Aresu S (2017) Green or greed? An alternative look at CEO compensation and corporate environmental commitment. J Bus Ethics 140:439–453
Frey BS (2007) Awards as compensation. Eur Manag Rev 4(1):6–14
Gill M (2012) Res J Recent Sci 1:244–252
Gill M (2012) Green HRM: people management commitment to environmental sustainability. Int Sci Congr Assoc 244–252
Gregurec I, Tomicic Furjan M, Tomicic-Pupek K (2021) The impact of COVID-19 on sustainable business models in SMEs. Sustainability 13(3):1–24. https://doi.org/10.3390/su13031098
Guest DE (2017) Human resource management and employee well-being: towards a new analytic framework. Hum Resour Manag J 27(1):22–38
Gülal F, Ayaita A (2020) The impact of minimum wages on well-being: evidence from a quasi-experiment in Germany. J Happiness Stud 21:2669–2692
Heikkinen S, Lämsä A-M, Niemistö C (2021) Work–family practices and complexity of their usage: a discourse analysis towards socially responsible human resource management. J Bus Ethics 171(4):815–831. https://doi.org/10.1007/s10551-020-04458-9

Hinton S (2018) How the fourth industrial revolution is impacting the future of work. Forbes
Hirsig N, Rogovsky N, Elkin M (2014) Enterprise sustainability and HRM in small and medium-sized enterprises: a decent work perspective. In Ehnert I, Harry W, Zink K (eds) Sustainability and Human Resource management, pp 127–152. CSR, sustainability, ethics & governance. Springer, Heiderbeg. https://doi.org/10.1007/978-3-642-37524-8_6
Holmes B (2010) National flexible work—a pulse survey. CCH Australia in conjunction with Managing Work Life Balance International
Jabbar MH, Abid M (2015) A study of green HR practices and its impact on environmental performance: a review. Manag Res Rev 3(8):142–154
Jamil S, Zaman SI, Kayikci Y, Khan SA (2023) The role of green recruitment on organizational sustainability performance: a study within the context of green human resource management. Sustainability 15(21):15567. https://doi.org/10.3390/su152115567
Järlström M, Saru E, Vanhala S (2018) Sustainable human resource management with salience of stakeholders: a top management perspective. J Bus Ethics 152(3):703–724
Jiang K, Lepak DP, Han K, Hong Y, Kim A, Winkler A-L (2012) Clarifying the construct of human resource systems: relating human resource management to employee performance. Hum Resour Manag Rev 22:73–85
Jose S (2009) Cisco study finds telecommuting significantly increases employee productivity, work–life flexibility and job satisfaction. http://newsroom.cisco.com/dlls/2009/prod_062609.html
Khan N (2018) Does training and employee commitment predict employee retention? AIMS Int J Manag 12:153–160
Khan PA, Johl SK, Johl SK (2021) Does adoption of ISO 56002–2019 and green innovation reporting enhance the firm sustainable development goal performance? An emerging paradigm. Bus Strateg Environ 30(7):2922–2936. https://doi.org/10.1002/bse.2779
Kossek EE, Valcour M, Lirio P (2014) Organizational strategies for promoting work–life balance and wellbeing. In: Chen, PY, Cooper CL (eds) Work and wellbeing: a complete reference guide, vol 3, pp 295–319). Wiley & Sons, Ltd., West Sussex, UK
Kramar R (2014) Beyond strategic human resource management: is sustainable human resource management the next approach? Int J Hum Resour Manag 25(8):1069–1089
Larkins S, Panzera A, Beaton N, Murray R, Mills J, Coulter K, Stewart R, Hollins J, Matich P, Baird D (2014) Regional health workforce planning in North Queensland: starting with the end in mind. Health Workforce Australia, Adelaide, Australia
Lazar I, Osoian C, Ratiu P (2010) The role of work–life balance practices in order to improve organizational performance. Eur Res Stud J 13(1):201–212
Lee SH, Ha-Brookshire J (2017) Ethical climate and job attitude in fashion retail employees' turnover intention, and perceived organizational sustainability performance: a cross-sectional study. Sustainability 9:465
Lester J (2013) Work-life balance and cultural change: a narrative of eligibility. Rev High Educ 36:463–488
Lopez-Cabrales A, Valle-Cabrera R (2020) Sustainable HRM strategies and employment relationships as drivers of the triple bottom line. Hum Resour Manag Rev 30(3):100689. https://doi.org/10.1016/j.hrmr.2019.100689
Lu Y, Zhang MM, Yang MM, Wang Y (2023) Sustainable human resource management practices, employee resilience, and employee outcomes: toward common good values. Hum Resour Manage 62(3):331–353
Macke J, Genari D (2019) Systematic literature review on sustainable human resource management. J Clean Prod 208:806–815
Madero-Gómez SM, Rubio Leal YL, Olivas-Luján M, Yusliza MY (2023) Companies could benefit when they focus on employee wellbeing and the environment: a systematic review of sustainable Human Resource management. Sustainability 15(6):5435
Mäkelä L, Suutari V (2011) Coping with work-family conflicts in the global career context. Thunderbird Int Bus Rev 53:365–375

Maley J (2014) Sustainability: the missing element in performance management. Asia-Pac J Bus Adm 6(3):190–205
Mandago RJ (2018) Influence of green reward and compensation practice on environmental sustainability in selected service-based state corporations in Kenya. Eur J Bus Strat Manag 3(6):1–12
Masri HA, Jaaron AA (2017) Assessing green human resources management practices in Palestinian manufacturing context: an empirical study. J Clean Prod 143:474–489
McGrath M (2015) 5 Myths about employee learning. HR Magazine. https://www.shrm.org/hr-today/news/hr-magazine/pages/0515employee-learning.aspx
Milkovich GM, Newman JM, Gerhart B (2014) Compensation, 11th edn. New York, NY, USA, McGraw-Hill
Mwita KM, Kinemo SM (2018) The role of green recruitment and selection on performance of processing industries in Tanzania: a case of Tanzania Tobacco Processors Limited (TTPL). Int J Hum Resour Stud 8:35–46
Nduati MM, Wanyoike R (2022) Employee performance management practices and organizational effectiveness. Int Acad J Hum Resour Bus Adm 3:361–378
Ogbeibu S, Emelifeonwu J, Senadjki A, Gaskin J, Kaivo-oja J (2020) Technological turbulence and greening of team creativity, product innovation, and human resource management: implications for sustainability. J Clean Prod 244:118703. https://doi.org/10.1016/j.jclepro.2019.118703
Ogbu Edeh F, Okwurume CN (2019) Green human resource management and organisational sustainability of deposit money banks in Nigeria. J Manag Sci 3:224–235
Parakandi M, Behery M (2016) Sustainable human resources: examining the status of organizational work–life balance practices in the United Arab Emirates. Renew Sustain Energy Rev 55:1370–1379
Peerzadah SA, Mufti S, Nazir N (2018) Green human resource management: a review. Int J Enhanc Res Manag Comput Appl 7:790–795
Pham DDT, Paillé P (2020) Green recruitment and selection: an insight into green patterns. Int J Manpow 41:258–272
Pillai R, Sivathanu B (2014) Green human resource management. Zenith Int J Multidiscip Res 4(1):72–82
Pinzone M, Guerci M, Lettieri E, Redman T (2016) Progressing in the change journey towards sustainability in healthcare: the role of 'Green' HRM. J Clean Prod 122:201–211. https://doi.org/10.1016/j.jclepro.2016.02.031
Puigarnau E, Ommeren J (2010) Start time and worker compensation: implications for staggered-hours programs. Tinbergen Institute Discussion Paper TI 2010-073/3
Ramus CA (2012) Encouraging innovative environmental actions: what companies and managers must do? J World Bus 151–164
Ravesangar K, Ping LL, Pachar S (2024) A review on the sustainable HRM practices in building net-zero transformation: an emerging trend in the workplace. Net zero economy, corporate social responsibility and sustainable value creation: exploring strategies, drivers, and challenges, pp 131–144
Renwick DWS, Redman T, Maguire S (2013) Green Human Resource management: s review and research agenda. Int J Manag Rev 15(1):1–14
Roberts GE (2003) Employee performance appraisal system participation: a technique that works. Public Pers Manag 32(1):89–98
Saka N, Olanipekun AO, Omotayo T (2021) Reward and compensation incentives for enhancing green building construction. Environ Sustain Indic 11:100138
Saratun M (2016) Performance management to enhance employee engagement for corporate sustainability. Asia-Pac J Bus Adm 8(1):84–102
Schmidt SW (2007) The relationship between satisfaction with workplace training and overall job satisfaction. Hum Resour Dev Q 18:481–498. https://doi.org/10.1002/hrdq.1216

Shen J, Zhu CJ (2011) Effects of socially responsible human resource management on employee organizational commitment. Int J Hum Resour Manag 22:3020–3035. https://doi.org/10.1080/09585192.2011.599951

Singh B, Rao MK (2016) Effect of intellectual capital on dynamic capabilities. J Organ Chang Manag 29(2):129–149. https://doi.org/10.1108/jocm-12-2014-0225

Singh SK, Singh AP (2019) Interplay of organizational justice, psychological empowerment, organizational citizenship behavior, and job satisfaction in the context of circular economy. Manag Decis 57:937–952. https://doi.org/10.1108/MD-09-2018-0966

Singh R, Dsilva J, Centobelli P, Tripathi V (2021) A review on Sustainable HRM: a study evaluating sustainability for organisational development. Empir Econ Lett 20(2):1–10

Singh R, Ravesangar K, Narayanan S, Singh KKH (2024) HRM practices in creating net zero: supporting organizational sustainability. In: Singh R, Khan S, Kumar A, Luthra S, Chokshi H (eds) Net zero economy, corporate social responsibility and sustainable value creation. CSR, sustainability, ethics & governance. Springer, Cham. https://doi.org/10.1007/978-3-031-55779-8_1

Stahl GK, Brewster CJ, Collings DG, Hajro A (2020) Enhancing the role of human resource management in corporate sustainability and social responsibility: a multi-stakeholder, multidimensional approach to HRM. Hum Resour Manag Rev 30(3):100708. https://doi.org/10.1016/j.hrmr.2019.100708

Stankevičiūtė Ž, Savanevičienė A (2018) Designing sustainable HRM: the core characteristics of emerging field. Sustainability 10(12):4798. https://doi.org/10.3390/su10124798

Tziner A, Rabenu E, Radomski R, Belkin A (2015) Work stress and turnover intentions among hospital physicians: the mediating role of burnout and work satisfaction. Revista De Psicología Del Trabajo y De Las Organizaciones 31:207–213

Vyas L (2022) "New normal" at work in a post-COVID world: work–life balance and labor markets. Policy Soc 41:155–167

Yang R, Wong CWY, Miao X (2021) Analysis of the trend in the knowledge of environmental responsibility research. J Clean Prod 278:123402. https://doi.org/10.1016/j.jclepro.2020.123402

Zibarras LD, Coan P (2015) HRM practices used to promote pro-environmental behavior: a UK survey. Int J Hum Resour Manag 26:2121–2142

Sustainability Strategies: Positioning Business for Success in Zero Carbon World

Navdeep Kaur Dhillon

Abstract The urgency of addressing the climate change has prepared the Global ship to birds zero carbon industry which is given by the imperative of environmental sustainability and transformative potential of Eco innovation. With the increasing level of industrialisation and increasing pollution level leading to massive environmental deterioration the Global authorities are now trying their level best to shift towards and approach to industrialisation where they do not conduct any such activity which can lead to environmental damage. This chapter delves into the strategic issues surrounding the adoption of zero carbon industry practices exploring the interplay between innovative technology and sustainable environmental policies. It will start with the definition of zero carbon industry and Eco innovation highlighting the significance and current Global trends. The discussion will then shift towards the strategic challenges and opportunities in adopting these practices including policy framework economic consideration technological hurdles and social factors. Through detailed case studies and best practices, the chapter illustrate successful implementations and innovative approach across different segment and regions where improvement can be done to reduce the level of environmental damage which occurs as result of the common practices in those industries. The barriers to adoption such as economic technology social and regulatory of stacks are critically examine hair to provide a comprehensive understanding of the complexities involved. Along with these future directions and recommendations are presented which emphasize emerging trends policy measures and the strategies which can be adopted to overcome the identified barriers.

Keywords Zero carbon industry · Eco innovation · Sustainability · Strategies · Hurdles · Industry trends

N. K. Dhillon (✉)
Punjabi University, Patiala, India
e-mail: Kaurnavdeep2991@gmail.com

R. Singh et al. (eds.), *Zero Carbon Industry, Eco-Innovation and Environmental Sustainability*, World Sustainability Series,
https://doi.org/10.1007/978-3-031-80220-1_11

1 Introduction

The transition to a zero-carbon industry represents a fundamental shift toward reducing greenhouse gas emission and mitigating the climate change which happens because of the industry practices. This shift is driven by the pressing need to minimize the environmental impact of industrial activities and promote a sustainable future where we can protect the natural resources to the maximum extent possible which are now because of the bad industrial practices are at the verge of extinction. Apart from this Eco innovation which refers to the development of product and processes that contribute to sustainable development play the crucial role in the transition by fasting the creation and adoption of novel technologies and methodology that significantly reduce carbon footprints while maintaining economic viability and competitive advantage. The process which are going to be developed need to be efficient enough with the help of which maximum energy can be created with the help of least resources possible and at the same time the economy can maintain its efficiency and its competitive advantage across the globe otherwise the adoption will be a barrier.

Environmental sustainability and compassing the responsible management of natural resources reduction of pollution and conservation of biodiversity is increasingly recognised as essential for long term health for the planet and the society. It involves the integrated approach that integrates environmental social and economic dimension to ensure that current needs are met without comparing the ability of the future generation to meet their own requirements (Jiang et al. 2024). The focus of environmental sustainability is to ensure that resources are not over exploited that they do not exist for the future generation. The interconnectedness of zero carbon industry practices Eco Innovation and environmental sustainability highlights the necessity of strategic and comprehensive approach to address the multifaceted challenges of climate change. This chapter aims to explore the strategic issues associated with adopting zero carbon industry practices with a particular focus on the role of economic Innovation and the significance of environmental sustainability. Through the examination of policy framework economic consideration and technology challenges the chapter seek to provide a comprehensive understanding of the barriers and opportunities in the critical area. Along with this the chapter will present case studies and best practices and will offer insights into overcoming adoption barriers and propose you to direction and recommendation for stakeholders including policy maker industrial leaders and communities. Through the exploration the chapter will contribute to the broader disclosure on environmental sustainability, and it will offer actionable guidance for fasting a sustainable industrial future.

1.1 Zero Carbon Industry

Zero carbon industry also known as the net zero or Carbon neutral industry refers to the industrial sector which operates with net zero carbon dioxide emissions. This means the industry which either emits no carbon dioxide or offsets it emission through various means such as carbon capture and storage technique or the purchase of carbon credit refers to the zero carbon industry (Chatjhutamard et al. 2024). The goal of such industry is to achieve a balance between the amount of greenhouse gases produced and amount removed from the atmosphere effectively reducing the industries contribution to climate change. Zero carbon industry operates on some common concepts which include renewable energy such as solar wind hydro and geothermal energy with the help of which least amount of carbon dioxide is emitted through the industrial processes, energy efficiency which include implementation of such practices and technology is which reduce the energy consumption and improve the industrial operation meaning that the company will consume least amount of resources to produce the maximum output. Carbon capture and storage is one of the most leading trends in the zero-carbon industry which include capturing of carbon dioxide emissions at their source and storing them underground or using them in other industrial processes. With the help of these techniques and practices zero carbon industries contributes significantly to the reduction of carbon dioxide in the atmosphere (John et al. 2019).

Major trend of zero carbon industries can be seen in three major segments which include the energy segment manufacturing segment and the transportation segment. In the energy sector heavy investment in solar wind and hydroelectric power can be seen. The sector is exploring the use of green hydrogen in the attempt to develop a clean fuel which is required to run the industry. With regards to the energy sector, Orsted company in Denmark has transform from a fossil fuel-based energy company to a global leader in renewable energy (Mulatu et al. 2024). The company has digested significantly from fossil fuels and invested heavily in of shore wind farms with the aim to achieve carbon neutrality by the year 2025. This transformation shows a potential for traditional energy companies to progress towards sustainable energy solution.

Manufacturing segment also have started implementing the zero-carbon industry technology with the help of electric furnaces in steel making and alternative clinker material in semen production through which they are contributing to the reduction in their carbon emission. With regards to the steel industry and cement industry to of the major companies which include HYBRIT and Norcem Brevik, has taken great initiative to revolutionaries the common practices with the use of hydrogen produced from renewable energy instead of coal in the Steel Making process. Also, the company has integrated with CCS technology to capture and store carbon dioxide emissions to capture up to four hundred tonnes of carbon dioxide annually from the cement plant and store it in geological formations (Van Sluisveld et al. 2021). In Netherland, the Port of Rotterdam is one of the largest ports in the world and is leading in the effort to de carbonize Maritime transport. The initiative includes the development of green hydrogen production facilities the use of shore power for vessels to reduce emission while docked and the implementation of circular economy principle in Port operations. These efforts aim to make the port a zero-emission husband by the year 2050 (Etzel 2024).

1.2 Eco Innovation

Eco innovation refers to the development and application of new product process services or business model with the help of which the environmental benefits can be increased through the reduction of resource consumption waste minimization and decreasing level of emissions (Rawat et al. 2024). Type of Eco innovations include the product innovation such as biodegradable materials process innovation such as energy efficient manufacturing methods or organisational innovation which include green supply chain management and marketing innovation which include economic labelling. In all the organisational processes the company can improve with the help of adoption of renewable resources of energy, using such material with the help of which the company can reduce its carbon emission like less use of plastic materials in the production and in the labelling which usually creates land pollution when is disposed off. The focus is on improving manufacturing and operation processes to reduce environmental impact and involve adopting cleaner production techniques where the waste material can be recycled, and energy use can be optimised. In the current times majority of the organisational efforts goes into marketing. This type of innovation where marketing practices are modified to emphasize environmental benefits is the requirement of Eco innovation. A common example can be equal labelling green certification and environmental products declaration that inform consumer about the environmental attributes of products and encourage more sustainable purchasing decisions by the consumer (Yeap et al. 2024).

Example of successful Eco innovators include Tesla electric vehicles which have revolutionaries the auto motive industry by providing a sustainable alternative to fossil fuel powered cars and the development of biodegradable packaging by companies like Ecovatives. Eco innovation place a crucial role in promoting environmental sustainability by enabling industries to operate more efficiently and responsibly there by reducing their environmental footprint. It fosters a culture of sustainability that not only address immediate environmental challenge this but also pay the way for long term ecological balances and resilience. With the help of Eco innovation the company gets a way forward to think and progress in their usual day to day practices and how it can be modified so that it can be more environmental friendly and also profitable for the company in terms of their corporate image in the market and in the way of less costly resources to be used in the production process.

1.3 Environmental Sustainability

Environmental sustainability includes the use of natural resources in a reasonable and efficient way to meet the present needs without compromising the ability of the future generation to use those resources in their own way. It involves conserving the natural resources minimising pollution and waste and promoting biodiversity and ecosystem health. It is a crucial concept in the context of climate change as it addresses the route cause and impact of global warming. The unstructured and inefficient use of the natural resources is leading to ever increase in the environmental degradation rate. Some of the common strategies which can be used to achieve environmental sustainability include the transition to renewable energy improving energy efficiency in building transportation and industrial processes to reduce energy consumption associated with environmental impacts and sustainable land use and conservation where natural habitats can be protected through the implementation of sustainable agriculture practices and preserving biodiversity to maintain.

As illustrated in the above section that how various industries are shifting towards the renewable energy resources, the trend can be formalised with the help of proper regulations and policies through which the company have a set framework using which they can implement the framework in their own practice and become environment sustainable. For large organisation there is usually a regulation where the need to establish their own corporate social responsibility committee which should have a minimum investment in a financial period.

Across the globe especially in some of the industries and sectors there is a need of strict policy and regulation implementation so that the companies are aware of their own responsibility, and they can keep the carbon emission level below a certain level as per the regulation established by the government. Also, with the help of policy and regulations the company can promote sustainable lifestyle and faster the culture of environment is stewardship and responsibility. In the emerging trends there is a dire need of such measures with the help of which we can monitor the sustainability, identify the environmental performance, and track the progress towards sustainability goals (Oke et al. 2024). With the help of a proper measurement criteria, we can identify the areas where theme focus is required to bring the emission level below specific level. Some of the key indicators which are already in place to measure the environment degradation level include the greenhouse gas emission water and energy consumption waste generation and recycling rates and ecosystem health assessment. For these indicators, a common performance criterion can be developed which can be used to measure the level of improvement across all these areas which are crucial for environmental health. Some of the tools such as life cycle assessment environmental impact assessment commonly known as LCA and EIAs and sustainability framework reporting can be used to quantify environmental impact and measure the effectiveness of sustainability initiative (Bianco et al. 2024).

It is with the help of continuous improvement and tracking of the progress made towards environmental sustainability we can identify the strategies which can be implemented in place to reduce the environmental impact and drive meaningful progress towards achieving the sustainability goals.

2 Strategic Issues in Adoption

The adoption of sustainable practices particularly in the context of achieving environmental sustainability and combating climate change is faced with multiple challenges and consideration across policy economics technology and societal dimensions. There are numerous factors which prevent the complete adoption of the sustainable environmental practices across various sectors and establishment of zero carbon industry. Some of the common strategic issues revolve around policy and regulatory practices, economic consideration, financing and investment Strategies and technology challenges.

2.1 *Regulations and Framework*

For proper implementation of the sustainable practices across various industries and sector there is an important requirement which include the policy and regulatory framework which needs to be clear enough so that industries can adopt these practices as per the regulation established by the government. Various authorities at the national and international level continue to frame such policies and framework with drive adoption of the environment sustainable practice. At the national level government implement policies such as emission trading system carbon price in mechanism and renewable energy targets and environmental standards to drive adoption among various industries and companies. International agreements like Paris agreement provides for global emission target and encouraging corporation among Nations (Feng et al. 2018).

Such regulations significantly influence the adoption rate by shaving market incentive and compliance requirements. Emissions limits and environmental regulations compel industries to innovate and invest in cleaner technology is to avoid penalties and maintain competitiveness. The framework does not only support the implementation of sustainable practices but also set a guideline for the companies so that they can maintain their profitability and effectiveness in the market even after shift towards the sustainable processes followed in the industrial day today work. Regulatory uncertainty or inconsistence post barrier to adoption requiring government to balance stringent standards with flexible framework that accommodate technological advancement and economic realities (Mulatu et al. 2024). Inconsistent regulatory framework where the various regulations established by the government are changing on frequent basis in that case it creates a confusion in the mind of the business is for a clear framework which is to be established. Further the technological sector is enhancing at ever fast rate. It is because of this doing a significant investment in one area of the business operation can become outdated quite soon leading to more expense incurred by the business then the profit earned from the innovation.

2.2 Economic Consideration

Economic factor also plays a significant role in adoption decision influencing the feasibility and scalability of sustainable practices. Sustainable practices are also linked with technology with his required to be implemented in the business from the day-to-day practices to more strategic functions in the business through which the carbon emission can be reduce. However, this technology is required costly and not all the business operating in a sector can adopt this technology due to its heavy cost. Also the technology is quite subject with regards to the size of the business and its requirement and it increases the complexity of Decision made by the business with regards to the adoption because they won't be able to plan to what extent they will need to implement the technology in the business and what would be the approximate cost (Jiang et al. 2024). Cost benefit analysis is crucial for evaluating the economic viability of adopting technology such as renewable energy system or Carbon capture and storage facilities within the business.

2.3 Financing and Investment Strategies

Access to the financing an investment also plays a significant role in scaling of sustainable initiative. Government financial institution and private investors are increasing your directing the finances and capital toward green project through subsidies grants and low interest loan and venture capital investment. However, there is still some gap in the framework established for financing and investment in the sustainable technology. Proper market framework is yet to be established where the businesses can approach the financial institutions venture capitalist or government authorities to get the finance to implement the Strategies and them involve a complex process which creates hindrance in the process of business implementing the change move towards the sustainable practices. There is more requirement of creating awareness among the businesses with regards to the alternatives which are available with them through which they can get the access to the finance required to implement the technology within their business processes.

2.4 Technology Challenges

Technology Innovation and readiness are Central to overcoming barriers in the adoption of sustainable business practices and adoption of such technology through which the industry can become a zero-carbon industry. Significant advancement in renewable energy technology is energy storage solution and smart grid and sustainable manufacturing process are essential for reducing carbon footprints and enhancing the resource efficiency. This is a hurdle in the adoption of the technology in the industries because there is yet various technology which are required to be developed which can provide a unique and Specific service to the business. Each business has their own individual requirements when it comes to the technology implementation and adoption of the technology for minimising the carbon emission. However, the technology which are available up to date is quite standard and it cannot be made flexible with regards to the individual requirement of the businesses. It is because of this the adoption rate in some of the industry is quite affective like cement industry steel industry and transportation but other industries are lagging because no specific technology is yet developed which can provide cater to the individual requirement of those industries.

2.5 Integration of Modern Technologies

Another major strategic issue is the integration of the modern technology with the existing business processes. The adoption of the sustainable technology requires the business to significantly changed their day to day operation and strategic framework within which the operate which is a very long term process and require huge investment in workforce process development and the modification in the overall set up in the organisation so that the technology can be implemented effectively and the day to day operations are not impacted. The successful integration of modern technology hinges on collaborative effort among researcher industries stakeholders and policy makers. Due to lack of collaboration within these three major stakeholders the adaptation rate of sustainable technology has remained less over the period of time because researchers do not want to focus upon the individual and specific requirement of the business take hold and policy maker also provide or make regulations for the overall industry (Chatjhutamard et al. 2024). Cross sectorial collaboration and knowledge sharing platform is something which is required to facilitate innovation diffusion and accelerate the transition towards sustainable development pathway. Also, because the operational Strategies and framework is required to be modified after the implementation pilot projects and demonstration can be useful to identify and analysing the insights into technology performance operational challenges and the cost effectiveness which will help the business is to identify broader deployment strategies.

2.6 Social and Cultural Factors

Social and cultural adoption is another factor which is required for the macro scale adoption of any technology. The adoption of the technology by the society helps to implement the technology in the business. For example the business operates in a society and if there is a high demand within the society for the implementation of sustainable practices within the business so that the individual at their own level can also contribute to the reduction in the carbon emission then in that sector there is high chance that the business will implement the sustainable technology India day to day practices but in case the society or the cultural factor do not promote the use of the sustainable technology within the business to implement or invest in such technology through which they can reduce their carbon footprint. In the current time there can be seen a massive increase in the awareness regarding the environmental degradation by the industry practices and accordingly various public groups have been formed which operate for the main objective of helping the businesses to implement the sustainable technology. However, there is even more requirement of the increasing awareness among the consumers and the public so that the businesses can have the incentive to keep their customer base with them and implement on the technology as a mean to stay relevant in the market (Evro et al. 2024).

Increasing public awareness about the environmental issues climate change impacts and the benefit of sustainable practices is crucial for fasting support and demand for green technology. Education initiative media campaigns and community outreach programs play significant role in shaping attitude towards sustainability and encouraging individual and collective action. When more individuals are aware of the sustainable practices which are available within the industry and how with the help of implementation they can get the advantage of better practices less resource consumption and waste minimization it is likely that they will encourage the businesses to adopt is technology and it will shape the behaviour and perception of the company towards those initiative.

2.7 Community Engagement and Acceptance

Business cannot be developed in isolation. It operates in the society and their various communities which engage with the business in their day-to-day practices which includes stakeholders like government public customer suppliers and various other individual who stay associated with the company. It is very crucial for the business to engage these stake holders in the process of implementing the sustainable technology. Consulted approach participatory planning and transparent communication build trust and enhance the project acceptance and mitigate potential social conflict and resistance to the change (Heffernan et al. 2015).

Community engagement will also involve the competitors in the industry. Industrial unions can be a good example where healthy discussions can be done about the implementation of sustainable technology across all the industry so that a comprehensive approach can be developed to reduce the carbon emission at a smaller level which will then lead to a larger impact through collective approach. By addressing the concern of all the companies operating within the industry in a specific region a proper and well personalized strategy can be developed through which inclusiveness can be promoted.

All the above areas possess a significant challenge in the adoption of the sustainable technology by the businesses. To overcome all of these challenges the company needs to adopt a holistic approach which integrates policy coherence economic initiative technology Innovation and social engagement so that a more comprehensive approach can be developed through which the most efficient and effective implementation strategy can be developed by the business. By overcoming regulatory barriers and leveraging the economic opportunities the business can accelerate the transition toward a sustainable process through which the carbon emission can be reduced and a zero-carbon emission industry can be formulated.

3 Case Studies

The current state of the zero carbon industries varies widely across different regions and sectors. The pursuit of zero carbon industries and innovative eco friendly practices is a global effort driven by urgent need to address climate change and promote the environmental sustainability across different regions and sectors. There have been multiple successful examples and best practices across the globe which provide valuable lessons and insights into the pathways towards achieving zero carbon goals. Below are some of the successful case studies of zero carbon industries innovative eco friendly practices and lesson learnt from diverse reasons and sectors.

3.1 European Union

Some of the industries are making significant contribution towards achieving zero carbon status while other industries are still in the early stage of this transaction and researching about the process and techniques which can be used to make the industries practices more efficient. The European Union is a global leader in the push toward zero carbon industries. The European Union's green deal aims to make you of the first climate neutral continent by 2050 (Guo et al. 2024a, b). Some of the key initiative launched by European union include the emission trading system (ETS), stringent regulations on industrial emission where strict guidelines have been published for the industries regarding their emission levels and substantial investment in renewable energy and carbon capture and storage technology is has been done to reduce the overall impact.

3.2 HYBRIT Project

In the steel industry the hydrogen break through iron making technology project is a Pioneer initiative aimed at revolutionising the steel industry by replacing coal with hydrogen in the iron making process. The project was launched in Sweden in the year 2016 as a collaboration between SSB and LKAB, with the aim to create the world's first fossil free steel by replacing coal with hydrogen produced from renewable energy source (Bianco et al. 2024). The traditional steel production method emits a large amount of carbon dioxide which will be reduced significantly using hydrogen. The key achievement of the project was in the year 2021 when the project delivers the world's first fossil free steel to the customer marking a significant milestone in the process of industries decarbonisation journey.

3.3 Norcem—CCS Project

Norcem in the steel making industry in Norway launched initiative in cement industry focus done integrating carbon capture and storage technology to reduce emission. The company has initiated retrofitting in its Brevik cement plant with the aim to capture four hundred thousand terms of carbon dioxide annually. The key achievement of the project was that it is one of the first full scale carbon capture and storage application program in the cement industry demonstrating the potential of CCS mitigate the industrial emissions. Along with this the project also highlighted the successful integration of carbon capture and storage technology into the cement plant and how the integration can help the businesses to become a leader in the sustainable theme and production (Feng et al. 2018).

One of the key reasons for the success of the initiative was a regulatory support and funding which was deployed to the project during its development. Along with this it was highlighted that collaboration with different stakeholders in the industry is essential for achieving the overall success because with the help of collaboration the various knowledge and resources can be combined.

3.4 BASF's Chemical Industry Initiative in Climate Protection

BASF is one of the world's largest chemical companies and it has implemented a range of measures to reduce its carbon footprint and promote sustainability. With the large-scale operation in which the company was engaged in there was an elevated level of carbon footprint with the company was leaving due to its operation but with the help of its newly launched carbon management program the company is targeting to achieve Carbon neutral growth until 2030. The initial aim of the company is to reduce the carbon emission by 25% as compared to the 2018 level and for these the company is exclusively focusing on the energy efficiency and the use of renewable energy resources in the production processes (Yeap et al. 2024). Along with this continuous Innovation and investment in the modern technology is the key to achieve sustainability goals. This can be seen from the company's effort to reduce the carbon emission initially by 25% and then target more reduction (Aziz et al. 2024).

3.5 North American Efforts

Another country which is contributing a lot towards the reduction in emission level is not America where United States and Canada are also making considerable progress in the industries such a steel and cement industry. In contrast with Europe the country has adopted a technique of incentives where their providing the companies with rebate on certain taxes or providing them efficient working condition who are more involved in the uses of renewable resources. It is because of this various companies at their own level have implemented various technology with the help of which their carbon emission is reduced and due to this the countries making considerable progress towards the environmental sustainability.

4 Innovative Practices

There are several key factors which are required to be implemented to ensure consistent and collaborative growth in the field of sustainable development and in the promotion of zero carbon industry across different sectors and regions. Some of the key points which are required to be considered in this initiative includes policy support from the government, role of public and private partnership.

4.1 Policy Support

Policy support is critical enabler of sustainable practices across regions and sectors. Successful case study has highlighted the importance of clear policy frames which provide incentive to the industries also and regulate the mechanism using which the company promote the sustainable development (Xu et al. 2024). Policy refers to the set of guideline and principles which helps the industry to make decisions according to a set framework and it helps the industries to follow a consistent approach towards the development of sustainable technology to reduce the carbon emission. Incentive such a subsidies tax break and grants can accelerate the adoption of Eco-friendly practices by the company because by making the investment in the technology the company can save a certain amount of money which they can invest further for the development of business.

4.2 Public Private Partnership

Public private partnership has also provided to be effective in advance in sustainable initiative by leveraging the strength and resources of both the sectors. Public private partnership facilitates sharing of resources helps to mitigate rest and innovate to ensure success in the sustainable projects. One of the major barriers in the adoption of the industries of sustainable technology is the lack of finance and proper regulator requirements and these issues can be overcome with the help of successful collaboration between public sector and private sector technical and financial barriers can also be broken (Yeap et al. 2024). One of the major requirements of the successful public private partnership is effective communication and the alignment of goal between the public and private entities because public entities primarily operate with the aim to enhance public welfare whereas private entity operate with the aim to make profit. The better the goal alignment the more will be the chance of success in the project.

4.3 Community Engagement and Technology Innovation

Technological innovations and the successful integration of modern technology are fundamental to achieving sustainable goals. Certain projects which include BASF's climate protection initiative highlight the importance of continuous Innovation and the adoption of innovative technology. Investment in research and development is essential for advanced sustainable technology where the individuals are required to be trained about the new technology is which can be invented in order to solve the pressing problems of the current scenario and through pilot projects and demonstration facilities we can provide inside and pave the way for broader technology adoption by the industry helping them to become net carbon industries (Ohene et al. 2022).

5 Future Directions and Recommendations

Emerging trends and technology offer promising pathway to overcome existing barriers and accelerate the transition to zero carbon and sustainable practices. The advancement in the renewable energy resources and technology such as more efficient solar panels of shore wind turbines and next generation battery storage is making the clean energy more viable and cost effective in nature. Innovation is also leading to the promotion of circular economy where the resources can be utilised and then renewed and again utilise in the business which can be done through recycling technology and sustainable material reducing the amount of waste and promoting resource efficiency. With the emerging technologies the companies can now better implement the strategies using which they will be able to utilise their resources and generate more output through the consumption of less amount of input. Artificial intelligence block chain and internet of things are enhancing the process of monitoring optimising and management of sustainable practices across various sector. The traditional form of operation is becoming outdated at an amazingly fast pace giving an incentive to the businesses to improve upon their existing processes to remain relevant in the market and to take advantage of the emerging trends and technologies.

5.1 Policy Recommendations

As studied in the real life example of various industries becoming successful in adopting technologies through which they can achieve the net carbon emission rate through their operations, it can be understood that there is a significant requirement of proper policies and effective framework through which we can promote the adoption of the technology in the day today practices of the business and also provide a framework to the businesses to initiate their research and development practices through which we can attain improvement in the existing technology and take the way forward.

5.2 Overcoming Barriers

The various barrier which are studied including the financial technological educational and other factors which create the hindrance in the overall adoption of the sustainable technology to become zero industry can be overcome with the help of a multifaceted approach. By increasing the access to financing through green bonds grand and low interest loan can help address economic barriers where business is will be able to use the limited resources in the implementation of the green technology rather than being worried about the cost of the implementation. Investing in research and development and fasting innovation hub can overcome the technology where different businesses across industry can come together to put the resources in a single project and accordingly with the success of the project all the businesses can take the advantage of the outcome. Public education campaign and community engagement initiative can raise awareness and shift cultural attitude towards sustainability which will remove the friction in the adoption of the sustainable technology at the lower level.

5.3 *Role of Stakeholders (Government, Industry and Communities)*

The role of stakeholders is also crucial in driving the adoption of zero carbon and ecofriendly practices and three of the major stakeholders which are the centre of focus include the government industry and the community (Bianco et al. 2024). Government is responsible for framing effective policies and framework with the help of which it becomes easy for the businesses to adapt to the eco-friendly practices.

6 Conclusion

The transition to a zero-carbon industry driven by echo Innovation and commitment to environmental sustainability is a complex yet critical endeavour in the fight against climate change. There are numerous issues in the path of complete adaptation of the sustainable practices across various industries to become carbon neutral in the near or long term. The issues are widespread in terms of the economic technology social and regulatory dimensions, and it is through careful understanding and through addressing these challenges stakeholders can put away forward towards sustainable and resilient future. With regards to economic consideration, technological advancement, social and cultural engagement and the policy and regulatory framework it can be overcome with the help of cohesive and collective approach followed by the individuals and various stakeholders in the industry. Moreover, for certain barriers in the adoption of the sustainable technology in the industries to promote zero carbon emission public private partnership can be another way to achieve this milestone where the resources and the knowledge can be combined so that a more cohesive and efficient where can be found to achieve the desired goals. In conclusion addressing the strategic issues in the adoption of zero carbon industries Eco Innovation and environmental sustainability required and comprehensive approach and by overcoming economics technology social regulatory barriers and leveraging emerging trends and technology is we can accelerate the transaction to the sustainable future. The case studies of successful zero carbon industries across assorted reasons of Europe Sweden and Norway provides significant inside about what practices can be initiated at the ground level so that the desired objective of sustainable growth and development in the various sectors of the economy can be made a reality.

Initiative like HYBRIT and carbon capture and storage project in Norway demonstrate the potential of public private partnership and strategic investment in driving sustainable innovation. The successful implementation of circular economy practices by Philips sustainable agriculture practices by General Mills and sustainable urban development in Brazil further illustrate the drivers approaches to achieving sustainability goal. It is not only in the industrial sector we should try to achieve the sustainable growth and zero carbon emission, but various other sectors can also be identified in the economy in which work can be initiated so that an overall growth and development approach can be followed which is sustainable in nature. Stakeholders including Government industrial competitors and the public group have a crucial role in driving this growth towards the sustainable future and with a collaborative approach sharing of resources and knowledge and proper strategy followed towards a specific goal the practices can be improved and better operational plans can be prepared through which the carbon emission can be reduce significantly.

References

Aziz S, Ahmed I, Khan K, Khalid M (2024) Emerging trends and approaches for designing net-zero low-carbon integrated energy networks: a review of current practices. Arab J Sci Eng 49(5):6163–6185. https://link.springer.com/article/https://doi.org/10.1007/s13369-023-08336-0

Bianco V, Cascetta F, Nardini S (2024) Analysis of the carbon emissions trend in European Union. A decomposition and decoupling approach. Sci Total Environ 909:168528. https://doi.org/10.1016/j.scitotenv.2023.168528

Chatjuthamard P, Singh S, Jiraporn P, Lee SM (2024) Climate change exposure, shareholder wealth, and the adoption of the Paris agreement: a text-based approach. Int Rev Financ Anal 94:103327. https://doi.org/10.1016/j.irfa.2024.103327

Etzel RA, Weimann E, Homer C, Arora NK, Maimela G, Prats EV, Banerjee A (2024) Climate change impacts on health across the life course. J Glob Health 14. https://doi.org/10.7189/jogh.14.03018

Evro S, Oni BA, Tomomewo OS (2024) Global strategies for a low-carbon future: lessons from the US, China, and EU's pursuit of carbon neutrality. J Clean Prod 142635. https://doi.org/10.1016/j.jclepro.2024.142635

Feng JC, Yan J, Yu Z, Zeng X, Xu W (2018) Case study of an industrial park toward zero carbon emission. Appl Energy 209:65–78. https://doi.org/10.1016/j.apenergy.2017.10.069

Guo D, Li B, Yu W, Baroutian S, Young BR (2024a) A system engineering perspective for net zero carbon emission in wastewater and sludge treatment industry: a review. Sustain Prod Consum 46:369–381. https://doi.org/10.1016/j.spc.2024.02.033

Guo Y, Luo L, Liu T, Hao L, Li Y, Liu P, Zhu T (2024b) A review of low-carbon technologies and projects for the global cement industry. J Environ Sci 136:682–697. https://doi.org/10.1016/j.jes.2023.01.021

Heffernan E, Pan W, Liang X, De Wilde P (2015) Zero carbon homes: perceptions from the UK construction industry. Energy Policy 79:23–36. https://doi.org/10.1016/j.enpol.2015.01.005

Jiang Y, Guo Y, Bashir MF, Shahbaz M (2024) Do renewable energy, environmental regulations, and green innovation matter for China's zero carbon transition: evidence from green total factor productivity. J Environ Manage 352:120030. https://doi.org/10.1016/j.jenvman.2024.120030

John VM, Quattrone M, Abrao PC, Cardoso FA (2019) Rethinking cement standards: opportunities for a better future. Cem Concr Res 124:105832. https://doi.org/10.1016/j.cemconres.2019.105832

Mulatu KA, Nyawira SS, Herold M, Carter S, Verchot L (2024) Nationally determined contributions to the 2015 Paris Agreement goals: transparency in communications from developing country Parties. Clim Policy 24(2):211–227. https://www.tandfonline.com/doi/full/https://doi.org/10.1080/14693062.2023.2285519

Ohene E, Chan AP, Darko A, Nani G (2023) Navigating toward net zero by 2050: drivers, barriers, and strategies for net zero carbon buildings in an emerging market. Build Environ 242:110472. https://doi.org/10.1016/j.buildenv.2023.110472

Oke AE, Oyediran AO, Koriko G, Tang LM (2024) Carbon trading practices adoption for sustainable construction: a study of the barriers in a developing country. Sustain Dev 32(1):1120–1136. https://onlinelibrary.wiley.com/doi/abs/https://doi.org/10.1002/sd.2719

Rawat A, Garg CP, Sinha P (2024) Analysis of the key hydrogen fuel vehicles adoption barriers to reduce carbon emissions under net zero target in emerging market. Energy Policy 184:113847. https://doi.org/10.1016/j.enpol.2023.113847

van Sluisveld MA, de Boer HS, Daioglou V, Hof AF, van Vuuren DP (2021) A race to zero-assessing the position of heavy industry in a global net-zero CO_2 emissions context. Energy Clim Change 2:100051. https://doi.org/10.1016/j.egycc.2021.100051

Xu P, Adebayo TS, Khan KA, Özkan O, Shukurullaevich NK (2024) United States' 2050 carbon neutrality: myth or reality? Evaluating the impact of high-tech industries and green electricity. J Clean Prod 440:140855. https://doi.org/10.1016/j.jclepro.2024.140855

Yeap SP, Teng KH, Shaw A, Karunamoothei V, Sum JY, Omer ATA, Liu C (2024) Unleashing the contribution of nanoparticles in reforming low-carbon solutions: current status, trend, and prospects. Chem Eng J 149655. https://doi.org/10.1016/j.cej.2024.149655

Issues and Challenges in Zero Carbon Industry and Innovation

Exploring Barriers, Drivers, and Critical Success Factors of the Zero Carbon Industry: A Case Study from South Asia

Narayanage Jayantha Dewasiri

Abstract This study explores the barriers, drivers, and critical success factors of the zero-carbon industry, using Sri Lanka as a case study representing the South Asian region. This chapter begins with a literature review, and fourteen (14) in-depth interviews were conducted with industry experts, policymakers, renewable energy and sustainability experts, and representatives from critical non-governmental organisations in Sri Lanka. The significant barriers that have been pointed out include, firstly, economic barriers; secondly, there are a lot of technological barriers. Thirdly, there are policy and regulatory barriers. Last, there is much social and cultural resistance to change. Conversely, the drivers of the zero-carbon industry also fall into four key categories: economic, technological, policy and regulatory, and social and cultural. These drivers are interconnected with the barriers; when the respective challenges are effectively addressed, they can become powerful motivators for the progression. Critical success factors defining the transitional stage are strategic planning and management, stakeholder engagement and collaboration, innovation and technology advancement, financial resources and investment, policy support and regulatory compliance, market readiness and consumer acceptance. The chapter concludes with a reflection on policy-making processes, the interaction of various barriers and drivers, and research activities with a focus on the necessity of an integrated approach to reach the critical goal of having a zero-carbon economy. Perhaps it would be easier for stakeholders to upsurge the prospects for change toward a sustainable and resilient zero-carbon industry through economic and technological elements and the policy and social aspects.

Keywords Zero-carbon industry · Barriers and drivers · Critical success factors · Stakeholder engagement · Sustainable transition

JEL Codes Q1 · Q55 · Q58 · O33

N. J. Dewasiri (✉)
Sabaragamuwa University of Sri Lanka, Belihuloya, Sri Lanka
e-mail: jayantha@mgt.sab.ac.lk

Adjunct Faculty, University Centre for Research and Development (UCRD), Chandigarh University, Punjab, India

R. Singh et al. (eds.), *Zero Carbon Industry, Eco-Innovation and Environmental Sustainability*, World Sustainability Series,
https://doi.org/10.1007/978-3-031-80220-1_12

1 Introduction

The idea of zero carbon industry is becoming more important as countries and industries search for solutions to problems instigated by climate change. On the other hand, a zero-carbon industry means industries and businesses that emit no carbon dioxide and other greenhouse gases in their operations, which helps reduce global warming significantly. Repairing the world's carbons is among the most significant challenges and chances for the industry on this planet (Caineng et al. 2021). Therefore, it creates an environment that encourages the creation of new technologies, new markets, and new employment opportunities. The importance of this shift is that it can lead to the proper sorting of economies, set on the right path of future growth, which is economically and environmentally healthy.

This chapter aims to present the simulation and analysis of the various impediments, facilitators, and critical success factors toward achieving the intent of a zero-carbon industry; it is essential for stakeholders, such as policymakers, business leaders, and researchers, to comprehend these elements to adopt efficient strategies and policies to pave ways to increased zero-carbon operation. This chapter presents a conceptual framework to help strategise and facilitate the implementation of zero-carbon plans for various sectors, detailing the barriers, overcoming them by tapping into the drivers, and acknowledging the success factors.

Hence, the chapter examines the concepts of the zero-carbon industry's fundamental principles that help explain the dynamics of the industry. Justification is why zero-carbon practice does not often occur because of barriers, such as economic, technological, and legal constraints. On the other hand, drivers that compel industries to go for zero carbon include financial gains, technological development, and supportive policies. Essential concepts are the fundamental components that define critical success factors for zero-carbon measures, which encompasses planning, stakeholders, and innovation. These ideas will be discussed in this chapter and will give the reader more insight into what it entails to allow an entire industry to move to a low-carbon status.

2 Literature Review

This research is based on the epistemology of theories from environmental economics, sustainable development and technological innovation. Ecological economics is the branch that investigates the economic effects of the environment's policies and the ad from practical analysis of the carbon limit. Sustainable development theory underlines the strategy to satisfy the present's wants and needs while not limiting future generations' opportunities to meet theirs. Technological innovation theory is the theory that holds new technologies as key drivers to the realisation of sustainability objectives. These theoretical approaches give an insight into the zero-carbon industry and show how economic, environmental, and even technological factors are related.

2.1 *Sustainable Development Theory (SDT)*

Sustainable development practice is the foundation for the sustainable development theory (Lélé 1991; Steer and Wade-Gery 1992). Sustainable development theory holds that there is a cyclic way of attaining development that fulfils the present generation's needs and, at the same time, does not strain the next generation's abilities to achieve the same. Shi et al. (2019) divided the formation and development of SDT into three distinct stages, namely, the embryonic period, the molding period, and finally, the developing period, where the world is heading towards a net-zero economy. This theory relates to the zero-carbon industry, where economic growth and development are sought in share while the environmental impact is kept to the lowest.

2.2 *Triple Bottom Line Concept*

A critical component of sustainable development is the triple bottom line (TBL) concept, including the people, the planet, and the profit (Elkington 1997; Pereira and Martins 2021). In the case of a zero-carbon industry, this implies delivering value (profit) in the system where other components, referred to as people and the planet, are also well protected. These are the results of managing financial performance, social equity, and environmental sustainability in organisations adopting zero-carbon processes. For instance, investments in renewable energy sources and energy efficiency decrease operating expenses, identify polluting sources and boost public health (Mishra et al. 2024).

The other key idea within sustainability is intergenerational equity, which focuses on the effects of the current behaviour on the next generations. The future generations need to be granted equal opportunities to benefit from healthy surroundings and resources; thus, transitioning to a zero-carbon industry is vital. Through the application of zero-carbon technologies and practices, industries of the present society can work to reverse the effects of greenhouse emissions, fight climate change, and save on the resources needed for future use. This view creates awareness and commitment towards developing sustainable business strategies for the next generations and prosumers.

Also, TBL supports the integrated approach, which simultaneously solves economic, environmental, and social problems. Based on the described approach, the zero-carbon industry should establish policies and practices for economic growth, decarbonisation, and social well-being. This gradual change approach demands cooperation between governments, businesses, and civil society to develop proper policies and standards, fund changes, and public campaigns supporting sustainability.

2.3 Diffusion of Innovation Theory

The diffusion of innovation theory (DIT) posits how new ideas, practices, concepts, and technologies develop and spread in society and the industry landscape (Rogers 1995). DIT is highly relevant in achieving a carbon industry as it explains how zero-carbon technologies and practices can be introduced and adopted globally (Scheffran and Froese 2016). Hence, new technologies are significant in achieving an organisation's sustainability objectives. Thus, implementing and propagating innovative technologies are essential in breaking the barriers and utilising the drivers of the zero-carbon industry. Technological change is a critical element of the transition to zero carbon. Innovations in renewable energy sources, energy storage technologies, and energy-efficient technologies are crucial to cutting down on the use of fossils and, therefore, decreasing carbon emissions. For instance, designing innovative devices such as solar panels, wind farms, and battery systems that manage renewable energy provides a different kind of energy worthy of consideration.

DIT also explains how new technologies emerge in industries and societies. Knowledge of the diffusion process enables comprehension of measures to stimulate the implementation of zero-carbon solutions. For example, the early adopters and innovators ensure that those in the early majority and the laggards follow the same trend due to the proof that new technologies are viable and beneficial. The recognised strategies, as such, include communication, pilotage, and partnership that could promote the application of zero-carbon technologies.

In addition, technological innovation also benefits organisations by increasing their competitive advantages by improving efficiency, cutting costs, and expanding market opportunities. A firm can capture the market if it invests in zero-carbon technologies. For instance, companies that use efficient and green energy patterns can save considerable money, increase their ecological efficiency, and attract green-sensitive consumers and investors.

2.4 Integration with Other Theories

Combining sustainable development theory, triple bottom line concept, technological innovation theory with the resource-based view (Tcvetkov 2022), stakeholder theory (Demartini et al. 2023), and institutional theory (Dragomir et al. 2023) helps explain the shift towards the zero-carbon industry. Further, the theory of planned behaviour establishes the connection between social cognition and behaviours regarding sustainable practices. As a result, it comprehends how favourable social attitudes, perceived norms, and control regarding behavioural change contribute to adopting zero-carbon practices among people and businesses.

The diffusion of innovations theory is used to explain the processes that happen in the market regarding the adoption of innovative technologies, specifically the roles of the different categories of adopters, communication channels and social systems in the diffusion process. The strategic management perspective, known as the Resource-Based View, also emphasises the role of resources for competitive advantage, and out of this, it defines valuable, rare, inimitable, non-substitutable resources in the context of the zero-carbon transition.

Stakeholder theory emphasises different stakeholders' involvement in more extended sustainable activities; it accentuates the crucial role of stakeholder and employee, customer and supplier, investor, and community engagements for the effectiveness of zero-carbon activities. From the institutional theory perspective, this work examines the role played by rules and institutional environments governing organisations' actions, whereby the analysis of the regulations, industry standards and mimetic desires regulating climate-neural practices can be understood.

In this study, the zero-carbon industry has been discussed in detail in various themed articles in other research. Previous research in this area of the zero-carbon industry has closely focused on and analysed different features of this process. In the past, studies on cleaner energy sources, such as solar and wind power, have been pursued to lower carbon emissions (Mishra et al. 2024). Research has also been carried out on the effectiveness of energy conservation interventions and using lean technologies in industrial processes (Rathnasiri et al. 2024). Furthermore, examining policy initiatives and regulatory measures has brought the key to practising a zero-carbon economy into focus. For example, the following options: carbon pricing, subsidy for renewable power, and tightening emission standards have been recognised as feasible policy measures (Rathnasiri et al. 2024). Altogether, these works highlight that the zero-carbon industry is complex and that to approach its development, all the aspects need to be considered: technological, economic and policy-related.

To build up a comprehensive model of the zero-carbon industry, it is necessary to recognise the key barriers, drivers, and critical success factors from the existing literature (Pan and Dong 2023; Kannan et al. 2022). The evidenced challenges include relatively high costs, technology, and supportive infrastructure. Economic challenges are the significant costs and capital investment needed to shift to zero-carbon technologies and the suitability of such financial costs. The technological hurdles relate to the present state of the art in energy storage to support renewable energy and difficulties associated with the versatile implementation of such technologies into production lines. Policy and regulatory barriers refer to gaps and unfavourable business environments reflected through the lack of coherent and favourable policies and unclear regulations that lead to investors' hesitant decisions (Kannan et al. 2022).

On the other hand, the following drivers were found to have the potential to promote change towards achieving a zero-carbon sector. Other incentives are based on financial factors such as reduced costs connected with energy consumption and the opportunities to import new techniques that belong to eco-friendly innovations. Technological factors focus on the successive development in the renewal of energy technologies and the emergence of better energy storage systems. These relate to favourable government policies, legal frameworks, and fiscal incentives that encourage the firm to adopt green technologies in compliance with the internationally set carbon emission reduction targets. Social factors are also crucial, with more people becoming aware of the need for sustainable products and services and lobbying industries for the zero-carbon approach (Pan and Dong 2023).

Thus, critical success factors discussed in the literature outline the components that have to be present in a company for the effective execution of zero-carbon programs. Such factors include strategic planning and management, enabling entities to incorporate zero-carbon objectives in the strategic frameworks. Engagement of stakeholders is also essential as people and institutions from government, business, and non-governmental organisations can catalyse change towards the use of zero carbon practices. Innovation and technology cannot remediate new opportunities and the obliteration of current barriers. The common currency of the fight against climate change is the funding and money required for developing and deploying zero-carbon technologies. Lastly, the readiness and acceptance levels of markets and consumers are crucial determinants of the actualisation of successfully implemented zero-carbon practices.

3 Methodology

3.1 Research Design and Approach

This study's methodology involves using qualitative research as a qualitatively driven research question drives it (Dewasiri et al. 2018). An inductive approach is used to undertake in-depth interviews that help to capture the overall picture of the zero-carbon industry, including the barriers, the drivers, and the critical success factors. The qualitative approach enables the gathering of vast information about a particular issue and offers perceptions that other methods, such as quantitative, may fail to expose. The qualitative design is justifiable due to the exploratory study and the importance of gaining insights into the multiple stakeholders' views on zero-carbon transition.

3.2 Data Collection Methods

The sampling technique employed in the study is purposive qualitative interviewing, which forms the primary data collection instrument. These interviews were conducted to obtain information from professional representatives from three industry experts, four policymakers, four renewable energy and sustainability experts, and three representatives from critical non-governmental organisations (NGOs) in Sri Lanka. Their professional experience and activities within the zero-carbon industry were considered; therefore, the interviews provided valuable insights. The data saturation point recognised during the interviews determined the sample size.

It was helpful to introduce specific questions in some of the interviews while using some form of structure for all the interviews. The interview protocol produced questions on significant areas of interest such as economic, technological, policy, social enablers and barriers, and critical success factors in the zero-carbon industry. All interviews were conducted in a friendly manner and were online interactive sessions that lasted between thirty (30) to forty (40) minutes during April and May 2024, with the participants being allowed to express how they feel, what they know, and what they suggest.

3.3 Data Analysis Techniques

The researcher proceeded with the thematic analysis in conducting this study. The process will involve several steps, as outlined below.

Transcription: All fourteen (14) interviews were conducted with the participant's permission and recorded in audio form to enable the detailed transcribing of the findings.

Familiarisation: After that, the researcher read the transcripts several times to gain an overview of the data content and possible preliminary themes and impressions.

Coding: Therefore, the coding framework is developed based on the research objectives formulated for the literature review and the emerging themes. The transcripts were then analytically coded, with portions of the text labelled as barriers, drivers, and critical success factors.

Theme Development: The first step in data analysis was to work through the coded data to identify patterns and themes. The themes were formulated under areas corresponding to the research objectives to offer an organised approach to the significant observations.

Interpretation: The findings regarding the themes and patterns were then discussed from the theoretical frameworks and literature perspective. This interpretation focuses on the interconnections between various barriers, drivers, and success variables discussed and explains how they may affect the transition to a new zero-carbon sector.

Validation: An expert member check was carried out to increase the study's dependability and, thus, the validity of the findings. The selected interviewees were asked to review the findings and give their opinions on how correct and appropriate they were. This process assisted in narrowing down the analysis and ensuring that the findings were valid and accurate of the participants' perceptions.

The qualitative data gathered from the interviews with respondents offered a specific and detailed description of the factors influencing the zero-carbon industry; thus, the barriers, drivers, and critical success factors affecting the zero-carbon industry were well understood. The methodological approach proposed here helps accumulate knowledge that would, in turn, guide the formulation of proper strategies and policies for transitioning to a sustainable, zero-carbon industry.

4 Findings of the Study

4.1 Barriers to Zero Carbon Industry

A thematic analysis was adopted to analyse qualitative data from in-depth interviews to reveal the challenges of attaining a zero-carbon industry. Thematic analysis used in this research focuses on patterns (themes) within qualitative data to paint a picture of the main issues at the institution. Four barriers were derived from thematic patterns observed in the interview data, including economic, technical/Technological, policy/regulatory, and bearing societal/cultural forces. The thematic analysis offered a well-guided way of identifying these barriers and would help comprehend stakeholders' difficulties in the zero-carbon industry (Table 12.1).

4.1.1 Economic Barriers

The use of modern technologies and means of production requires quite a large amount of investment, which is one of the most critical hurdles to the zero-carbon industry. Most interviewees noted that a significant capital expenditure is needed for switching to the zero-carbon setup and is, as a rule, too high. A leader in the energy industry stated the following.

> The initial capital costs of incorporating renewables into the generation and improving facilities are very high. Sustaining and sometimes even creating the initial capability in renewable energy systems and technologies is expensive and demanding, which implies that sometimes, it will take a long time for the financial viability to catch up with the need. (Renewable Energy Expert 02)

Similarly, another industry respondent stated the following.

> There is always a danger that the technology invested in today may become a thing of the past tomorrow, implying a financial loss. (Industry Expert 02)

Table 12.1 Barriers identified through the thematic analysis

No.	Barriers
1	Economic barriers
2	Technological barriers
3	Policy and regulatory barriers
4	Social and cultural barriers

Source Author's own

4.1.2 Technological Barriers

Another factor that significantly affects the transition to a zero-carbon industry is technology. One of the remarks that were quite common during the interviews revolved around the absence of fancy technologies and the unknown efficacy of such methods. According to one of the professionals in the renewable power industry, there is excellent progress in developing green technologies; however, many sectors are kept from utilising the most modern solutions to minimise their carbon footprint. Moreover, there are problems with implementing new technologies in industrial practices. In the interviews, one sustainability expert stated the following.

> Redesigning current systems to accommodate renewable energy sources is difficult due to technological barriers. It entails drastic alterations and professional skills that are not always easy to obtain. (Sustainability Expert 02)

4.1.3 Policy and Regulatory Barriers

Another barrier that became rather apparent was the lack of direction in the policy and regulatory framework and the issues with their stability. Several participants stressed the need for policies that support better regulation and zero-carbon initiatives. Other policymakers stated the following.

> Companies do not know where they stand and get anxious about their exposure to regulatory risk; they are prolonging the period for investing in zero-carbon innovation. (Policymaker 01)

Other barriers noted were inadequate regulation and the absence of long-term policy frameworks. One of the non-governmental organisation's staff stated the following.

> We urgently need more profound, coherent policies to supply longer persisting positive incentives supporting the zero-carbon project. Without these, one will not be able to achieve this progress faster, and it will be in Custard. (NGO Representative 01)

4.1.4 Social and Cultural Barriers

As observed from the interviewees' responses, social and cultural factors, including resistance and ignorance, were cited much more often. Several respondents observed that the industry and the public do not understand the value and the necessity of moving to a zero-carbon economy. Another participant noted the following.

> The public does not properly understand carbon emissions and zero carbon, and there is a lack of acceptance. Therefore, the change is not implemented quickly enough due to ignorance. (Sustainability Expert 01)

Uncertainties of society and resistance within businesses remained another critical point of discussion. For instance, one industry leader stated the following.

> The greatest challenge with implementing a new policy is always the people's attitude; there is always resistance since people prefer to stick to what they know even if it is not best for the company. (Industry Expert 01)

Converting to a zero-carbon industry does not have general barriers but a collection of closely related barriers. The set-up is dictated by economic factors concerning available resources and capabilities, technology, policies and regulations, and social and cultural factors. Several actors, including governments, businesses, and civil societies, can only solve these barriers to provide the necessary environment to support zero-carbon initiatives.

4.2 *Drivers of Zero Carbon Industry*

Like the barriers mentioned above, the thematic analysis transformed these barriers into drivers of the zero-carbon industry. Therefore, there is a mutual interaction between the drivers and barriers on the path to a zero-carbon industry, whereas the drivers are the solution to the obstacles.

4.2.1 Economic Drivers

Among all the common reasons that work towards the change, economic drivers are some of the most influential factors that make the industry go for zero carbon. The two major economic issues in deploying zero-carbon technologies are high costs and financial risks. Accordingly, one industry expert stated below.

> The barriers can become drivers through economies of scale, cost reductions over time, and financial incentives. For example, through economic scale factors, the more firms invest in an area of technology, the more costs come down, thus making the investment more attractive to the rest of the firms. (Industry Expert 03)

Economic barriers like high initial costs elicit techniques and approaches that favour innovation in the long run. The initial economic enablers later became economic modulators as more organisations embraced the idea that decreased operational costs and increased efficiency have positive impacts. It was common for the interviewees to identify cost savings and market opportunities as significant motivations. In that regard, one industry executive remarked below.

> There is cost reduction in energy efficiency technologies over the long run. In light of the reduced operational cost, there is a good business prospect for utilising zero-carbon practices. The green technologies and sustainable products market also expands, offering opportunities. (Industry Expert 02)

Following the findings, a sustainability expert further posited below.

> There is growing interest in environmentally sustainable consumable goods and services. Organisations steering the development of zero-carbon solutions and technologies can

> increase their market share and thus improve the competitiveness of their business. (Sustainability Expert 01)

4.2.2 Technological Drivers

The absence of proven technology in processes can slow down the process. However, some technicalities can be surmounted as much as research and development develops. What used to be a hindrance because of the lack of innovations and technology becomes a booster. For instance, efforts to create efficient energy storage technologies can change constraints into opportunities that help advance renewed energy production. An energy expert stated the following.

> By nature, technologies develop as new answers to preceding technological problems. For instance, new and improved methods of renewable energy systems overcome the drawbacks observed in prior techniques in the technology field. This effectively transforms impediments that may ensue to motivational factors that fuel advances and continued improvement. (Renewable Energy Expert 02)

Therefore, there is a substantial dependence on technology as one of the primary foundations of the zero-carbon industry. A number of the interviewees underlined the importance of innovation in the field of renewable energy sources and energy storage. A renewable energy expert noted the following.

> Right now, with the advancement of technology in solar and wind, this has become a highly efficient energy source when complemented by the efficiency of battery storage. This makes these sources of energy highly accessible. (Renewable Energy Expert 01)

Technological advancement is not limited to the generation of energy. One energy expert noted the following.

> The novel technologies include smart grids, energy management systems, and carbon capture and storage, transforming how industries control and mitigate carbon. (Energy Expert 02)

4.2.3 Policy and Regulatory Drivers

Lack of proper regulation and turbulence in policies pose a hindrance to the journey towards zero emission. However, when governments and regulatory authorities understand sustainability's significance, they often adopt favourable regulations and measures. These policy changes start as constraints but become enablers with guidelines, incentives and policies that help an organisation adopt zero-carbon practices. This is what one policymaker was reported to have said.

> Policies and regulations get developed to reflect new challenges in the zero-carbon industry. It may start with some rather stiff policies that then transform into enabling policies that continue to promote further adoption through creating enabling environments. (Policymaker 04)

Consequently, favourable policies and friendly environments foster the zero-carbon transition. Some respondents specifically mentioned that government policies that are distinctive and beneficial to the development of zero-carbon solutions are crucial. Another policymaker expounded the following.

> Perhaps the most notable are the government policies like tax credits, production incentives, promotion of green energy initiatives through grant and subsidy schemes to help 'level the playing field' by reducing the entry cost of zero-emission technologies. (Policymaker 03)

The other are the intensified emission limits and global treaties. Another policymaker enlarges the below.

> Standards that include concrete and challenging targets of carbon emission push industries into functional zero-carbon models. International treaties such as the Paris Agreement also set an international guideline/standard and exert pressure on countries to adhere to it. (Policymaker 02)

4.2.4 Social and Cultural Drivers

In the social aspect, lack of awareness and resistance to change can be impactful. However, these can be turned into drivers when public awareness rises and society becomes more conscious of the environment. For instance, one of the interviewees highlighted the following.

> Increased environmental consciousness can result in increased consumer desire for introducing environmentally friendly products and, consequently, an increase in the number of companies that introduce zero-carbon practices. Perceived constraints include social and cultural factors like rural resistance to change, which translates into enhanced education and awareness. At the same time, people develop awareness of these issues, and their rejection weakens, thus encouraging society to expand the use of zero-carbon solutions. (NGO Representative 03)

Some respondents mentioned that the climate change issue is becoming more topical in the public domain, affecting consumers' choices. One industry expert noted the below.

> People do not want to spend their money with organisations that do not understand sustainable practices, and consumers do not want to spend their money with organisations that do not set zero carbon targets as their goal. (Industry Expert 02)

Social activism organisations are campaigning and making changes. Thus, an NGO representative argued the following.

> Grassroots movements and environmental organisations are essential for raising awareness among the population and campaigning for enhanced legislation in this sphere. Their actions are critical for developing social paradigms for the zero carbon transformation. (NGO Representative 02)

Hence, the drivers of the zero-carbon industry can be listed as economic, technological, policy and regulatory, and social and cultural drivers. When those are appropriately addressed, they are interconnected with barriers. Financial advantages, including cost reduction, market opportunities, technological advancement, and favourable policies, promote zero-carbon efforts. However, increasing awareness of the environment and customers' demands also enhances the movement towards a sustainable, zero-carbon world. Knowledge and application of such drivers help remove these barriers and help attain a zero-carbon industry.

4.3 Critical Success Factors for Zero Carbon Industry

The outcome of the thematic analysis identified six critical factors for the zero-carbon industry, as outlined in Table 12.2.

4.3.1 Strategic Planning and Management

Strategic planning and management strategies were critical success factors in executing zero-carbon activities. Some key interviewees stressed the centrality of zero-carbon targets to the company's business strategy. An industry expert expressed this below.

> To deliver on the net-zero promises, organisations require an overarching and well-integrated framework that supports companies' strategic plans for the long term. This entails establishing challenging but realistic targets, designing the plans in detail, and keeping track of the agenda. (Industry Expert 01)

The role of corporate governance in implementing these strategies cannot be overemphasised. This discrimination brings confidence among the staff or workers. An interviewee highlighted the below.

> You know that leadership commitment is crucial for any industry. In particular, self-generated information indicates that an absence of robust senior-level mannequins is problematic when assembling resources and sustaining zero-carbon endeavours over time. (Industry Expert 03)

Table 12.2 Critical success factors for zero carbon industry

No.	Critical success factors
1	Strategic planning management
2	Stakeholder engagement and collaboration
3	Innovation and technology advancement
4	Financial resources and investment
5	Policy support and regulatory compliance
6	Market readiness and consumer acceptance

Source Author's own

4.3.2 Stakeholder Engagement and Collaboration

Stakeholder involvement and participation are essential factors for a zero-carbon industry. For any organisation to achieve zero-carbon status, a conscious effort from all stakeholders, such as employees, customers, suppliers, and policymakers, is essential. A sustainability expert remarked the following.

> Successful zero-carbon projects often involve extensive stakeholder engagement. This means actively communicating with and involving all relevant parties to ensure their buy-in and support. Collaboration with suppliers and partners can also drive innovation and share best practices. (Sustainability Expert 01)

Another interviewee emphasised the role of public–private partnerships.

> Different authorities, such as government agencies, private companies, and non-profits, must collaborate. Coordination involves combining resources and sharing information, thus achieving impossible outcomes for each party in a synergistic effort towards a zero-carbon industry. (NGO Representative 01)

4.3.3 Innovation and Technological Advancement

Innovation and technological advancement are prerequisites for the zero-carbon industry. Several participants also underlined further work on the processes to address the existing barriers necessary for improving efficiency. An industry expert stated the following.

> It is obvious now that technological advancements remain at the centre of decarbonisation. Hence, companies must invest in proper research and innovation to identify and improve new technologies such as renewables, sustainable transportation, energy storage, and carbon capture technologies. (Industry Expert 02)

The role of technology in the business world and the ever-shifting technological developments make it essential for companies to be innovative and to look into what is ahead of them. Another interviewee stated the following.

> It is essential to be receptive and prepared to try out novel concepts because being a leader in states with full complicity means using only the latest solutions. (NGO Representative 01)

This proactive approach can help increase competitive advantage and lead to a change in the entire industry.

4.3.4 Financial Resources and Investment

Another critical key area of infrastructure is financial support for necessary zero-carbon measures. Due to the initial costs of installing zero-carbon technologies, it sometimes requires a significant investment. An industry expert highlighted the following.

> Funding is a significant issue. The project must obtain funding from internal budgets, government grants, and private investors. To attract investment, it is necessary to show zero-carbon projects' economic feasibility and benefits. (Industry Expert 03)

Moreover, support from heterogeneous financing models is vital. Supporting that argument, an interviewee stated the following.

> There is nothing wrong with diverse financial instruments such as green bonds, carbon credits, and financial instruments that can finance large-scale programs. (Industry Expert 02)

4.3.5 Policy Support and Regulatory Compliance

Policy support and compliance with the relevant regulations must be considered the key to the zero-carbon program's effectiveness. Adaptable and encouraging social policies enhance the confidence needed for investments and developments. One policymaker recalled the following.

> Government policies can create or destroy zero-carbon endeavours. Fundamental supportive policies include regulatory policies like carbon pricing and incentives for renewable power. These legal frameworks offer incentives like carbon credits and set expectations for firms. (Policymaker 04)

Companion regulatory conditions guarantee organisations' compliance in terms of environment, which may help boost the image and competitive edge of the firm. Indeed, another regulatory expert underscored the following.

> The country has to adhere to the environmental regulations; hence, doing it strategically is not a choice but a necessity. Sometimes, depending on the standards set by the authorities, investors and consumers tend to have a positive attitude towards those organisations that meet or even surpass these requirements. (Policymaker 03)

4.3.6 Market Readiness and Consumer Acceptance

There is also the market readiness and acceptance by the customers to help fuel the uptake of zero-carbon practices. The need for products and services that do not harm the environment might cause companies to shift to Zero Carbon Technologies. An industry expert observed the below.

> Demand for more environmentally friendly products is on the rise, and businesses that cater to this demand will be able to generate new sources of income and stand out from their counterparts. (Industry Expert 01)

At the same time, people should be explained the benefits of green products. Another interviewee stated the following.

> Increasing awareness of products' environmental friendliness and providing credible information on their impacts can help improve their acceptability level. As consumers grasp the positive implications of their choices, there are higher chances of embracing zero-carbon-related causes. (Sustainability Expert 02)

Strategic planning, stakeholder engagement, innovation, financial resources, policies, customers, and markets define the zero-carbon industry's critical success factors. All these factors are causally related and vital in accomplishing zero-carbon initiatives and making them sustainable. By tackling the above factors, the industry and the government can facilitate the shift to a zero-carbon industry.

5 Discussion

Therefore, the study implies a need to focus on eliminating economic barriers using incentives and introducing appropriate funding models. For instance, grants for cases to bring up renewable energies and exemptions for taxes to fund investments in efficient energies can significantly decrease the costs (Abolhosseini and Heshmati 2014). These results also imply the need to maintain the constant enhancement of technology. Significant changes to the generation technologies are needed to decarbonise electric power; better efficiency of photovoltaic panels and wind machines and progress in energy storage devices are crucial.

Moreover, the results also pointed out the great exigency of establishing sound policy backing. Pressing and predictable rules and lasting policies for decision-making minimise fluctuations and encourage business people to invest in zero-carbon technologies. The results also highlighted the opinions and experiences of stakeholders. The involvement and engagement of all stakeholders, such as the workforce, clients, vendors, and policymakers, to support the concept is indispensable in ensuring the success of zero-carbon projects (Falana et al. 2023).

Thus, the research results within the framework of the presented model correlate with the insights discussed in the current literature on the antecedents and enablers of the zero-carbon industry and its critical success factors. Previous research has observed that economic realities and financial risks can be sources of difficulty in moving to zero-carbon work activities (Kannan 2022). The literature also adequately covers the aspects of technological development and policies. For instance, it has been ascertained that industries in countries with comprehensive policy support and policies regarding deploying renewable energy technologies are inclined to report higher rates of zero-carbon practices (Rathansiri et al. 2024).

The findings consider stakeholder engagement to be another significant factor. Even though the literature affirms stakeholder engagement as a factor in transforming to zero carbon, our findings show it is paramount. On this basis, it is necessary to pay increased attention to developing cooperation and improving relationships within and outside the stakeholder management framework.

6 Conclusion

This chapter provides a detailed framework for change around the economic, technological, policy, and social transformations needed to move towards a zero-carbon industry. Accordingly, it has comprehensively analysed the barriers, drivers, and critical success factors in achieving a zero-carbon industry in Sri Lanka.

7 Summary of Key Findings

The potential economic loss was the most influential factor, as high fixed costs and general risks for the business were the main problems in adopting zero-carbon methods in the industry. However, these barriers can be addressed by providing good financial incentives, such as subsidies and tax credits. Developing renewable energy and energy storage systems technologies is a crucial factor, underlining the need to make periodic accruals of capital investments to boost research and innovation.

Policy support is another critical factor. The regulations have been defined well and are consistent, helping to provide stability for the target investments. A significant place belongs to the engagement of the stakeholders, as their participation is crucial for gaining their support and implementing the planned changes.

7.1 Implications for Policymakers, Industry Stakeholders, and Researchers

The following is the guideline of this exploration, and its findings have various implications for policymakers, related industry investors, and researchers. Consequently, the outcomes indicate policymakers' necessity to allocate impressive financial bonuses and initiate favourable legal bearings. It would also be beneficial to offer subsidies, tax relief and grants to minimise the costs that are always a turn-off when embracing zero-carbon solution technologies. Positive and stable policies also mean that there should be clear and understandable legislation for investment in sustainable practices (Jayarathne et al. 2021).

Thus, the study emphasises strategic planning and innovation practices among industries' stakeholders. Zero-carbon solutions should be part of a firm's strategic plan and invest in research to constantly adopt the right technology. Interactions with stakeholders involved and cooperation can lead to further innovations in the sphere and the distribution of best practices for the practical realisation of zero-carbon projects.

Subsequent studies of this issue can be based on the information on how economic motivation and policies can affect the utilisation of zero emissions (Pan and Dong 2023). However, additional research must also be conducted to assess the effect of stakeholders' engagement across various scenarios and buying sectors. Knowledge about relationships and effects between the different factors can be beneficial for developing better strategies and policies promoting the move to a zero-carbon economy.

Therefore, the results provide a clear and extensive picture of the specific barriers, fundamental drivers, and critical success factors in the global zero-carbon industry. Most interviewees also revealed that economic considerations should be eased, technological development encouraged, a firm policy framework offered, and stakeholders involved shifted to a low-carbon and sustainable economy in the future.

7.2 Path Forward

Getting to a state where industry-related carbon emissions are equal to zero involves several processes that are not only economic but also technological, policy and social aspects. Therefore, from this chapter, the stakeholders can create strategies that eliminate barriers, tap into drivers and foster the right critical success factors to support a sustainable future. Moving towards a low-carbon economy is necessary for the environment's survival and to enhance economic development, new technologies and social improvement. With the support of people and societies pioneered with professional collaboration, the fight for zero-carbon goals is possible—paving the way for a healthier planet.

References

Abolhosseini S, Heshmati A (2014) The main support mechanisms to finance renewable energy development. Renew Sustain Energy Rev 40:876–885

Caineng ZOU, Xiong B, Huaqing XUE, Zheng D, Zhixin GE, Ying WANG, Jiang L, Songqi PAN, Songtao WU (2021) The role of new energy in carbon neutral. Pet Explor Dev 48(2):480–491

Demartini M, Ferrari M, Govindan K, Tonelli F (2023) The transition to electric vehicles and a net zero economy: a model based on circular economy, stakeholder theory, and system thinking approach. J Clean Prod 410:137031

Dewasiri NJ, Weerakoon YKB, Azeez AA (2018) Mixed methods in finance research. Int J Qual Methods 17(1):1–13. https://doi.org/10.1177/1609406918801730

Dragomir VD, Dumitru M, Perevoznic FM (2023) Carbon reduction and energy transition targets of the largest European companies: an empirical study based on institutional theory. Clean Prod Lett 4:100039

Elkington J (1997) The triple bottom line. Environ Manage: Reads Cases 2:49–66

Falana J, Osei-Kyei R, Tam VW (2023) Towards achieving a net zero carbon building: a review of key stakeholders and their roles in net zero carbon building whole life cycle. J Build Eng 108223

Jayarathne PG, Sachithra V, Dewasiri NJ (2021) Sustainable practices through competitive capabilities in the apparel industry: Sri Lankan experience. FIIB Business Review. https://doi.org/10.1177/23197145211060398

Kannan D, Solanki R, Kaul A, Jha PC (2022) Barrier analysis for carbon regulatory environmental policies implementation in manufacturing supply chains to achieve zero carbon. J Clean Prod 358:131910

Lélé SM (1991) Sustainable development: a critical review. World Dev 19(6):607–621. https://doi.org/10.1016/0305-750X(91)90197-P

Mishra U, Rathnasiri MSH, Dewasiri NJ (2024) Net-zero economy and energy sustainability: a nepalese perspective. In: Singh R, Khan S, Kumar A, Luthra S, Chokshi H (eds) Net zero economy, corporate social responsibility and sustainable value creation, pp 145–161. CSR, sustainability, ethics & governance. Springer, Cham. https://doi.org/10.1007/978-3-031-55779-8_9

Pan Y, Dong, F (2023) Green finance policy coupling effect of fossil energy use rights trading and renewable energy certificates trading on low carbon economy: taking China as an example. Econ Anal Policy 77:658–679

Pereira THM, Martins HC (2021) People, planet, and profit: a bibliometric analysis of triple bottom line theory. J Manag Sustain 11(1):64–73

Rathnasiri MSH, Dewasiri NJ, Kumar AA (2024) Policy framework and implementation strategies for Sri Lanka's transition to a net-zero economy. In: Singh R, Khan S, Kumar A, Luthra S, Chokshi H (eds) Net zero economy, corporate social responsibility and sustainable value creation, pp 43–60. CSR, sustainability, ethics & governance. Springer, Cham. https://doi.org/10.1007/978-3-031-55779-8_3

Rogers EM (1995) Lessons for guidelines from the diffusion of innovations. Jt Comm J Qual Improv 21(7):324–328

Scheffran J, Froese R (2016) Enabling environments for sustainable energy transitions: the diffusion of technology, innovation and investment in low-carbon societies. Handb Sustain Transit Sustain Peace 721–756

Shi L, Han L, Yang F, Gao L (2019) The evolution of sustainable development theory: types, goals, and research prospects. Sustainability 11(24):7158

Steer A, Wade-Gery W (1992) Sustainable development: theory and practice for a sustainable future. Sustain Dev 1(3):23–35. https://doi.org/10.1002/sd.3460010306

Tcvetkov P (2022) Engagement of resource-based economies in the fight against rising carbon emissions. Energy Rep 8:874–883

Zero Carbon Leadership: Key Performance Indicators and Best Practices for Sustainable Industry

Vanshika Singh and Kamalpreet Kaur Paposa

Abstract Looking into the present worldwide concern for environmental preservation, business organizations must adopt measures to reduce their carbon footprints. This book chapter attempts to synthesize the currently available literature related to the zero-carbon industry and states the various indicators for performance measurement systems that are tailored specifically for the zero carbon industry. This chapter begins by examining the existing terrain of zero-carbon initiatives and the need for assessment. The chapter highlights the initiatives and the best practices organizations should adopt to contribute to the zero-carbon industry. Stemming from this groundwork, this chapter suggests an innovative approach to ensuring a balanced perspective in performance assessment in the context of the zero-carbon industry. It also deliberates the role Human Resource Management department can play for setting performance indicators for zero carbon industry. Moreover, the research work highlights the fact that keeping sustainability goals in mind is crucial for the organisations to reduce ecological damage alongside maintaining high economic output. It provides a structured and systematic means of measuring, managing, and enhancing environmental performance. The research work intends to contribute towards the aim of carbon neutrality by stating certain managerial implication contributing to sustainable industrial development in general, and supporting the global shift toward a low-carbon future thus optimizing the net zero industry.

Keywords Performance management system · Net zero · Zero carbon industry · Environment sustainability · Human Resource Management (HRM)

V. Singh · K. K. Paposa (✉)
University School of Business, Chandigarh University, Punjab, India
e-mail: kamalpreet.paposa@gmail.com

R. Singh et al. (eds.), *Zero Carbon Industry, Eco-Innovation and Environmental Sustainability*, World Sustainability Series,
https://doi.org/10.1007/978-3-031-80220-1_13

1 Introduction

The term "sustainability" is commonly used in policy research to describe the goals of public policies, which was derived from the Brundtland report in 1987 (Kuhlman and Farrington 2010). The United Nations' Agenda for Development defines sustainability as the development that aims to improve the quality of life for all individuals through various means. Sustainable development involves economic, social, and environmental development that are linked and mutually reinforcing (Agenda for Development; United Nations: New York, NY, USA, 1997).

Implementing net zero requires careful consideration of ethical, social, political, Economic, and technological considerations and potential pitfalls (Fankhauser et al. 2022). Sustainable Net Zero Economy (SNZEI) incorporates the notion of carbon neutrality into economic activity, ensuring that development does not come at the expense of environmental damage. Its implementation is important because it plays an important part in reversing the current state of climate change by aligning human efforts with the earth's natural boundaries (Vijay and Sathish 2024). Similarly, the literature addresses how sustainability has always been the top priority according to surveys, and how some companies are happier with environmental sustainability targets than economic benefit objectives (Rubio-Andrés and Abril 2024). Prioritizing equity, sustainable development, and poverty eradication has become an important agenda. It recognizes the importance of balancing responsibilities between countries at different levels of development, recognizes transitions tailored to "different national circumstances" and considers distributional impacts within each country (Fankhauser et al. 2022). The Sustainable Net Zero Economy Implementation (SNZEI) research across geographies and industries provides many options for reaching net-zero emissions (Vijay and Sathish 2024).

To turn net zero into a useful frame of reference for decision-makers, the global carbon constraint should be translated into individual decarbonization pathways for nation-states, sub-national entities, companies, and other organizations (Fankhauser et al. 2022). Studies have shown that developing countries try to maintain their economic growth by putting social and economic objectives ahead of environmental protection (Kuhlman and Farrington 2010). However, keeping sustainability goals in mind it is crucial to reduce ecological damage alongside maintaining high economic output. To achieve this stimulating goal, it's important to support net zero and implement a performance management system, which is one of the most important functions of the human resource management department of an organsiation. HR department plays a critical role in any organization, to increase employee performance. It is important to evaluate the performance regularly to examine the competencies and abilities and determine whether these competencies can contribute towards organisational vision and success. High-quality performance appraisals can enhance employees' attitudes, achievement, self-worth, and sense of contribution to the business (Paposa and Kumar 2015). Sustainable utilization is considered a major concern due to limitations and responsibilities, being top priorities. According to Din et al. (2024) implementing Green Human Resource Management (GHRM) practices can contribute towards Green Innovation thus leading to carbon neutrality for the environment in sustainability.

Previous literature established that carbon emissions should be minimized throughout the supply chain to support the environmental cause all over the world. Sustainable HRM or GHRM practices, followed by leadership and appraisal distribution in the industry would surely lead to sustainable organizations. GHRM practices monitored through performance assessment led to satisfying performance appraisals of the employees and also contributed towards environmental sustainability (Din et al. 2024). Thus, the main aim of this chapter is to synthesize the past literature on the zero-carbon industry, highlight the best practices of the zero-carbon industry, and identify the various performance measurement indicators of the same. In addition, research suggests the HR initiatives for performance assessment in a zero-carbon industry along with motivating factors and barriers towards implementing performance management systems in the zero-carbon industry (Din et al. 2024).

2 Research Methodology

To achieve the objectives of this study, a phase-wise literature review approach was adopted and review was conducted in three phases. It included planning the review process, conducting the review, reporting, and disseminating the findings. A comprehensive review of the existing literature was done by choosing appropriate data from existing literature. The initial stage of planning the SLR was done by finalizing the variables/keywords keeping in mind the scope of the study. The data were extracted from major databases like Scopus, Science Direct, and Google Scholar. The secondary stage of the data search was completed by finalizing the keywords, "performance assessment, performance management, net zero, zero carbon industry, and environment sustainability." As discussed above the keywords were used to extract the relevant articles from the databases. In the third stage of the review process, the literature from the 43 finalized articles was synthesised and presented in the chapter.

3 Zero Carbon Industry

During the 2021 G7 Summit, leaders from major industrial nations pledged to achieve a carbon–neutral world by the middle of the century. The agreement aligned with the Paris Agreement which limited global warming to 2 °C with an aspirational target of 1.5 °C above pre-industrial levels. At the COP 26 conference in Glasgow in 2021, major carbon-emitting nations including India, proposed timelines for reaching net zero carbon emissions (Wu et al. 2023). "Net Zero" or "Zero Carbon Industry" can be defined as the operational reduction of the emission of GHG from the activities in the industries. The activists aim to achieve the SDG (sustainable development goal) efforts (Wu et al. 2023).The reports clearly explain that the variations in definitions for "Zero Carbon Industry" arise from differences in indicators and their impacts. Globally, organizations are adopting green practices and striving towards net zero industry. This involves clean energy projects for production and quality standards (Reddy et al. 2023a).

Efforts to reduce carbon footprints in the industry include the calculation of carbon footprints, utilizing innovative carbon capture technology, and modifying industrial processes where subsidies by the government play a vital role in encouraging renewable energy production during the energy transition (Wu et al. 2023). Globally organizations are adopting green practices striving for net zero standards which participate in green habits initiating the clean energy domain for quality and production improvement (Reddy et al. 2023b).

Different industries are taking the initiative to contribute towards a Net Zero Economy. It has been studied that the airline industry contributes 3.5% of world carbon dioxide emissions hence to reach its carbon neutrality by 2050, the airline industry has shifted towards sustainability says Pimiä et al. (2024). Climate variations exist due to the airline and maritime industries. Therefore, various schemes like CORSIA (carbon offsetting and reduction scheme) for international aviation have been implemented. Some studies have highlighted that the automobile industry also explains the emissions and gives a global agenda of producing goods that provide sustainability towards the environment and reduction in carbon emissions. The negative effect of growth and development can be seen as environmental degradation (Yu et al. 2024). Similarly, the construction industry adds to the carbon emission target of a country. Scotland is one of the countries that have adopted higher energy-efficient buildings and implemented construction regulations (Moreno-Rangel et al. 2023). In the context of the Agricultural sector, the agricultural economy, architects have been looking forward to sustainable designs in rural areas that emit net zero emissions in rural areas, they planned to attain a feasible carbon goal. The innovative technology of space generators was introduced and described to the farmers so that it's easy for them to do chores such as crop selection, fertilizer application, and even irrigation and these practices have regulated carbon emissions (Yueyang and Yuan 2024). Focusing on the filmmaking industry, Studies show the transition to sustainable methods in the movie production business is an important step toward lowering the entertainment sector's environmental imprint. (Meilani 2021). The literature emphasizes the need to use environment-friendly methods such as green cinematography, and renewable energy sources throughout production. These programs not only serve to reduce carbon emissions but also foster a culture of sustainability inside the sector. The benefits of applying the practices are highlighted in the study by Meilani (2021). The shift to sustainable e-fuels is critical to reaching carbon neutrality in the various sectors of the economy. Green hydrogen, methanol, and ammonia have been discovered as feasible replacements for traditional fossil fuels and therefore, the literature delves into the production processes, energy efficiency, and environmental effect of these e-fuels. However, there are certain problems associated with its implementation, such as infrastructure needs, economic consequences, and regulatory assistance. (Reddy et al. 2023a, b). A Thematic analysis of selected articles on zero-carbon industry has led to the following outcomes.

- The DDP-LAC project demonstrates how LAC nations may achieve net-zero GHG emissions while experiencing economic development, improved air quality, cost-effective transportation, and other advantages. To implement these paths, policy packages must be economy-wide and sector-specific, adaptable to new knowledge, cover all emissions and development goals, and include regional requirements and conditions (Bataille et al. 2020).
- Zero-carbon energy reduces greenhouse gas emissions by 80% in the Australian economy. Renewable energy may directly reduce carbon emissions in electricity generation, as well as residential, commercial, manufacturing, mining, and land transport. As fossil fuels are phased out of the energy business, fugitive emissions will eventually disappear (Lu et al. 2021).
- The movie industry's large annual emissions contribute to the climate emergency, which is a pressing matter for everyone. The film industry has taken serious steps to address environmental concerns by adopting green production practices. The majority of manufacturing opts for low-emission vehicles, reducing CO_2 emissions by an average of 60% (Kannan et al. 2022).
- Focusing on the film industry, the researcher hopes that filmmakers may use these 5 main areas of sustainable filmmaking to respect the environment and improve their short, feature, documentary, and series productions. Materials can be reused, used for future manufacturing, or donated. Going paperless for production planning, reducing inefficient meetings and maintaining a production record for future investigations are some sustainable strategies (Kannan et al. 2022).

To deliver long-term benefits for every sector producing different carbon emissions, nature-based solutions require robust and resilient ecosystems that can withstand and recover from disturbances. Ecosystem connectivity and species diversity at several trophic levels have a significant role in determining ecological resilience (Fankhauser et al. 2022). Postindustrial release impact on human health and the economy. To stabilize the zero-surface temperature, global carbon emissions need to be controlled (Warzywoda et al. 2022). Economies are on the precipice of decreasing their environmental impact results in certain facets like complication, sociable well-being, and energy density others are still bent on using fossil energy to foster development and attain the proposed net zero emissions by 2050 and 2070. This requires much more than a mere reduction of the emissions that come with energy production. (Ardi et al. 2023). Clean energy attracts worldwide limelight as nations are initiating to eradicate their carbon footprints. In 2015 the term "digital India" was introduced to boost the IT sector and make it an integral part of the nation's economy (Reddy et al. 2023a).

Introducing a Net Zero emission initiative in India involves a multifaceted approach that leverage the country potential, technology framework, and technology advancements. Net-zero energy systems (NZES) is crucial to meeting the policy of climate change mitigation objectives. Nation is poised to play a significant participant in the discussion. This has been a hot topic for the world today. As the 1.5 °C study emphasizes, the world must begin lowering emissions now and reach NZE as soon as practicable (Chaturvedi and Malyan 2022).

There are just a few India-related studies on this topic, and only one presents the state of net-zero energy systems in India for 2050 and beyond across several scenarios. Several major features arise from the different peaking and net-zero-year scenarios. These include solar dominance in India's electricity systems, rapid electrification of the industrial sector where current electricity penetration is low, a much larger electricity system relative to business-as-usual, a drastic decline in fossil energy use if carbon capture and sequestration (CCS) is not commercially available, and a high share of bio-fuels in liquid fuels. While the quantitative aspects of India's net-zero energy systems have been discussed, the 'non-quantitative' aspects are just as important (Chaturvedi and Malyan 2022).

India will have much greater per capita income levels by 2050 and beyond, similar to other countries India is also stepping forward to enhance its indulgence in environmental activities reason being the media and internet promoting and awakening people about the cause. (Chaturvedi and Malyan 2022).

4 Best Practices of Zero Carbon Industries

The net-zero emission industry practices highlight the path towards implementing a zero-carbon industry (Yadav et al. 2023):

- The adoption of green technologies and sustainable practices which effectively utilize resources and align with the industry's SDG goals, include the implementation of the carbon emission trading scheme (CETS).
- To achieve radical sustainability goals through Net Zero, firms must adopt an advanced supply chain by merging modernized and digitalized energy, lean methods, and greener principles.
- No release of left-out fossil fuels in the environment promotes proper utilization of the resources (Yadav et al. 2023).
- Zero carbon practice can be implemented by SWOT analysis and implementation of value stream mapping.
- Promoting sustainable products that are less carbon-intensive elements can also contribute towards zero carbon industry.
- Encouraging fewer carbon-intensive trips, which means fewer trips, and fewer vehicles and also reassuring usage of electric vehicles.
- Promoting heat-resistant pumps, IGCCs, etc.
- Promotion of using machinery that consumes less energy and releases less toxic waste.
- Mechanism of carbon pricing which talks about the different pricing schemes, and implements recycling, reuse and recovery of raw materials in the industry (Din et al. 2024).

- The industry is at a peak range of carbon emission and low neutrality would bring transitions in the industry which has been the largest emitter of GHG (greenhouse gases). The advancement of zero carbon technology motivates carbon neutrality, innovation, and advanced biofuel usage (Din et al. 2024).
- Adapting electric vehicles is one of the best technological methods to attain carbon neutrality and generate low carbon emissions. Technological aspects leads to economic and industrial growth and also contributes towards environmental sustainability (Din et al. 2024).
- Using energy-efficient building materials and technology, such as high-performance windows, LED lighting, and HVAC systems, is vital for successful construction and implementation.
- Increasing the adoption of net-zero energy buildings in India, it's important to raise awareness of their benefits, especially as many buildings are privately owned and maintained. Encourage community engagement and stakeholder support for sustainable building approaches.
- Installation of On-site renewable energy sources, such as solar photovoltaic panels and wind turbines, to create clean energy and contribute to energy self-sufficiency.
- Implementing energy storage options, such as batteries, to store extra energy for use during low-generation periods might provide significant benefits in India's regions with longer daylight hours.
- Policies like net metering and feed-in tariffs can help integrate on-site renewable energy generation into NZEBs, making it financially feasible for building owners to invest in renewable technology.
- Collaboration among government agencies, business sector players, NGOs, and research institutes to exchange knowledge and resources. Collaborations with utilities for grid integration and demand response initiatives need to be promoted.
- Governmental regulations, such as tax or financial incentives, can help promote the use of NZEBs.
- Transitioning to renewables diminishes the motivation to invest in fossil technologies, leading to an equilibrium with poor fossil efficiency (Stern and Valero 2021). An important point triggered by particular policies may bring breakthrough technology, efficient and less-priced storage techniques and bring transition and advancement leading to sufficient productivity and zero carbon industry.

5 Performance Indicators for the Zero-Carbon Industry

Achieving net-zero capability in the supply chain performance is crucial for the zero-carbon industry. Performance indicators for this sector include financial and governance metrics that influence environmental sustainability. Effective governance, characterized by efficacy, rule of law, and political stability, plays an important role in mitigating environmental harm. Controlling corruption, utilizing renewable energy, and promoting sustainable economic growth are key governance strategies that contribute to a lower carbon footprint (Gyimah et al. 2023). Research indicates that while financial development alone may not reduce carbon emissions, it can lead to environmental damage if not managed properly. Therefore, financial indicators must also account for environmental impacts, promoting investments in green technologies and sustainable practices (Gyimah et al. 2023).

In the zero-carbon industry, specific performance indicators include adopting green practices like eco-friendly product design, green procurement, and emission reduction. These indicators help measure the industry's progress toward a sustainable future, ensuring that green practices lead to tangible reductions in carbon emissions and improvements in environmental sustainability (Gyimah et al. 2023; Wu et al. 2023).

CCTs (Carbon Capture Technologies) is a special focus on pre-combustion procedures, emphasising their potential as a promising strategy. The role of CCT in accomplishing the United Nations Sustainable Development Goals (SDGs) was thoroughly examined in previous studies, leading to the creation of several indicators to assess its impact. These indicators address the economic, environmental, and social elements of sustainable development. The findings show that CCTs contribute favourably to all 17 SDGs, particularly SDG-7 (Affordable and Clean Energy) and SDG-13 (Climate Action) (Olabi et al. 2022). The suggested metrics provide stakeholders with a thorough framework for ensuring that CCTs adhere to the goals of the SDGs. They have several benefits, including contributing to reaching the SDGs, giving vital information to decision-makers, boosting sustainability performance benchmarks, and improving risk management which can be a stepping stone towards achieving the target for zero-carbon industry as SDGs focus on environmental sustainability, optimizing data management, reporting processes, allocating resources more efficiently, lowering costs, increasing environment performance, decreasing social impact, and improving stakeholder communication Trivedi et al. (2023). Furthermore, the study tackles numerous impediments to CCT implementation, emphasizing the necessity for ongoing innovation and collaboration to overcome these problems. By linking CCT projects with SDG indicators, the research offers a strategic framework for combining environmental, economic, and social 'goals, building a more sustainable future (Olabi et al. 2022).

The Human Resource departments of every organization take various initiatives to put their efforts toward zero carbon industry (Paposa et al. 2024).

Human resource department initiatives play an important role in driving the performance assessment for net-zero industries. Practices for sustainability and green human resources are essential components of these initiatives. The HR department of an organization can significantly contribute to sustainable practices by advocating for environment-friendly policies and aligning workers' values with the organization's sustainability targets. This can enhance employees' understanding of ecological methods and foster a culture of sustainability within the organization. For instance, HR can assess the company's emission profile, set targets, and create a sustainable plan to work towards a net-zero state (Singh et al. 2024). Moreover, it is the responsibility of the human resource department to encourage staff members to adopt sustainable practices and reduce waste. This can be achieved by minimizing paper usage, adopting energy-efficient tools, and implementing flexible work arrangements such as remote or hybrid work (Ravesangar et al. 2024). Sustainable utilization is considered a major concern due to limitations and responsibilities, being top priorities. Green human resource management practices and green innovation strategies can lead to carbon neutrality thus contributing to environmental sustainability (Din et al. 2024). Sustainable HRM or GHRM practices, followed by leadership and appraisal distribution in the industry would surely lead to sustainable organisations. Sustainable policy-making is a significant aspect of this transition, as it involves mitigating environmental factors and embracing the latest trends to achieve environmental stability (Stern and Valero 2021). By following GHRM practices, industries can pave the way towards a net-zero or zero-carbon future, benefiting from technological innovation and socio-economic advantages. These practices help create a sustainable industry that balances productivity with environmental responsibility, ensuring long-term ecological and economic stability (Din et al. 2024).

6 Motivating Factors and Barriers for Implementing Zero Carbon Industry

The literature states that educational institutional awareness contributes to sustainable practices, it generates a sense of responsibility brings motivation in generations to keep the environment clean, and increases eco-consciousness in students. GMT (green method technologies) are promoted in educational institutions to make students aware of the zero-carbon industry and motivate them to contribute towards a net zero economy (Moghayedi et al. 2024). Increased financial savings, financial rewards, and cost reduction are the major aspect that motivates entrepreneurs to promote and implement further GMTs (Moghayedi et al. 2024). Moreover, employees should be trained and highly motivated to opt for GHRM to contribute to a sustainable industry with a zero-carbon target. The literature explains the motivating factors for the adoption of the carbon industry (Okereke 2007). These motivating factors includes Organizational profit, which has been the slogan for the business community, Competitiveness of businesses for creditability and subsequently financial obligations. Moreover, desire to prevent loss and damages at the business or industry level. Adopting zero carbon criteria is an ethical choice for every industry, significantly contributing to environmental sustainability. As mentioned above, the CCT and ZCI (zero carbon initiative) help in achieving the SDGs. Zero-carbon initiatives create economic stability and job opportunities, reducing poverty. It Supports sustainable agriculture by reducing emissions from farming processes, contributing to food security. By reducing air pollution from fossil fuels, CCT and zero carbon initiatives improve public health and reduce health-related costs. CCT initiatives create educational opportunities in environmental science and green technologies. Promoting gender-inclusive employment in CCT sectors supports gender equality. Green technology can provide equitable economic opportunities, reducing inequalities. It minimizes water pollution from industrial emissions, protecting water resources. These initiatives foster economic growth through new technologies and industries, creating green jobs and driving innovation in green technology and infrastructure, supporting sustainable industrialization. Lastly, it reduces urban air pollution, contributing to a healthier and more sustainable urban environment.

The adoption of Internet of Things in the renewable energy sector is hindered in many ways thus, limiting its uptake. The key is to prioritize the difficulty in ability so that one gets to recognize just how acute those problems are, and how to create methods of removing them (Reddy et al. 2023b).

Moghayedi et al. (2024) outlines different barriers in the application of zero carbon industries. Many respondents reported that the main challenge affecting GMT implementation was cost, which was associated with high cost is necessary and cost constraints arising from limited funds. The issue raised by a participant was to 'act on plans' in the university (Moghayedi et al. 2024). Hence, all departments of UCT buildings make them more ecologically friendly. This is the case because ideas and frameworks are available for its implementation.

Okereke (2007) explains that certain barriers can be considered while adapting net zero and assessing the net zero performance. A comprehensive motive, drivers, and barriers of companies' climate activities is essential for devising international climate policies. However, these elements should be contextualized about some key aspects for the continuing structural transformation of the climate system and the policy-making for the environment are mandated.

- The lack of a policy framework. Adapting and implementation have a difference.
- A large amount of cost is required to change the infrastructure and to adapt new technologies for the zero-carbon industry.
- Political and uncertainty of the policies of the government is also a barrier.
- Market instability is also a major concern as the products made with new technologies are not responded to well by the people. Reassurance faces a major barrier in this case, (Okereke 2007).
- Electrification involves transitioning end-use services like transport and heating to electricity, which is vital for decarbonization but faces challenges such as varying adoption rates and the need for technological advancements.
- Expanding the use of wind, solar, and other renewable energy sources is critical for decarbonizing the electricity sector but encounters challenges related to resource availability, infrastructure, and storage.
- Enhancing efficiency across all sectors to reduce overall energy demand significantly contributes to emission reductions but requires substantial investment and behavioral changes.
- Phasing out or significantly reducing the use of fossil fuels leads to a marked decline in their use, supplemented by CCS in some models, but necessitates infrastructure changes and impacts the fossil industries economically.
- Utilizing technologies like BECCS and DAC to remove CO_2 from the atmosphere is essential for offsetting hard-to-eliminate emissions but faces challenges in technological and economic feasibility and social acceptance.

- Differences in strategies and implementations across regions necessitate custom strategies based on local conditions, with challenges arising from diverse socio-economic contexts and policy differences.
- Ensuring comparability and consistency across different models is vital for clear policy guidance but faces challenges due to variations in assumptions, input data, and methodologies.
- Detailed analyses focusing on specific sectors and regions provide actionable insights for targeted policies but must balance granularity with broader applicability.
- Including behavioural and social factors in energy models enhances realism and predictive power but is complex due to the diversity of social variables (Bistline 2021).
- Further and extensive analysis is required to determine if this difference is the true one and if there are any interactions between these corners (Okereke 2007).

Achieving net-zero energy buildings is a tough but important aim for decreasing carbon emissions and addressing climate change. Implementing net-zero energy buildings in India requires several key success elements. Analysing crucial success variables for NZEBs in India is vital due to their urgent deployment in rural and urban regions.

7 Implications for Zero Carbon Industry Adoption

Integrating sustainability into business strategy is essential for managers as it not only enhances the company image but also appeals to environmentally conscious consumers and ensures legal compliance. Investing in sustainable technologies and renewable energy is crucial for achieving net-zero emissions. Managers must establish robust performance metrics to monitor carbon emissions and energy consumption, ensuring the effectiveness of sustainability initiatives. Engaging stakeholders, including employees, customers, and investors, is vital for gaining support for zero-carbon activities. Continuous improvement in sustainability practices fostered through an organizational culture of learning and innovation, is key to staying relevant with the latest advancements. Training and development programs are essential to boost employee commitment to sustainability goals. Risk management strategies are necessary to address potential disruptions, such as supply chain issues and regulatory changes. Encouraging innovation within the organization can lead to sustainable solutions to structural challenges. Collaboration with NGOs, research institutions, and government agencies can enhance sustainability efforts. Transparency in reporting sustainability performance builds stakeholder confidence. According to Okereke (2007), companies often employ multiple methods simultaneously, influenced by their specific location, industry, focus areas, and challenges. This means that significant steps for one organization might differ from another, underscoring the importance of tailored sustainability approaches. Implementation of these strategies can facilitate a smooth transition to a zero-carbon industry, ensuring long-term sustainability and success.

Lee et al. (2021) describes the implication of fair use of carbon, and various implications which can lead to a zero-carbon industry. The net-zero industry entails several complicated concerns, notably the equitable deployment of carbon dioxide removal (CDR) technology and its integration into national and international policies. Different normative assumptions influence net-zero objectives demanding a balanced strategy that can bring emissions reductions and CDR activities (Lee et al. 2021; Robiou du Pont et al. (2017). Furthermore, including CDR in international accords, such as Article 6 of the Paris Agreement, might encourage fair deployment through technology transfer and international funding, resulting in more ambitious climate initiatives (Lee et al. 2021).

8 Conclusion, Limitation, and Future Scope

While this research is faced with several limitations due to the lack of numerous existing research and the relative newness of the field, it provides a starting point toward identification of the components crucial for effective implementation. It points to guiding principles and key performance indicators that may be used in intentions toward the industry's net-zero emissions. One of the significant components in the work is the determination of the enabling technologies used in getting to zero carbon business processes. Numerous technologies are widely used to manage energy consumption, minimize waste, and track the environmental effects of streamlining. Each of the mentioned innovations not only optimizes business operations but also contributes to the concept of sustainability by offering relevant information for decision-making processes. Establishment of a performance management system for industries with zero carbon indicates the importance of responding to competent strategies in the attainment of sustainability goals. In addition, the chapter mentions the applied drivers and key success factors that encourage the usage of performance measurement systems (PMS) in the zero carbon industries. These are the regulatory factors, Corporate and Social Responsibility, and market considerations. Knowledge of these drivers helps to develop the framework that, in addition to being efficient, also attracts people's interest. It also looks at challenges to adopt zero-carbon industries including technological issues, financial issues, organizational issues and resistance to change. It is important to note that these are some of the barriers that can greatly help when designing interventions that might help in reducing these difficulties and improving the transfers (Kailay and Paposa (2023). The implications of this research for stakeholders are extensive. In this case, authorities in industries are required to take an active position in the adoption of zero-carbon initiatives, prioritizing the integration of sustainable principles into the organizational policy frameworks. The reviewed literature is concerned with a call to the management of educational institutions to assimilate knowledge of sustainable practices into their curriculum. It is also important to stress the visibility of female leaders and the number of NGOs with unique sustainability officers. Further research should focus on identifying new ideas and directions and finding ways to address the antecedent obstacles to the adoption of sustainability by various industries. Comprehensible review offers a scaffolding for making sense of the multiple approaches to the matter at hand, stress, and the systemic interconnection of several sectors toward a sustainable future. Zero-carbon companies have received attention in recent years, thus there is little literature and information on their study. As a result, future research should aim to overcome these gaps by conducting longitudinal studies and comparing frameworks across sectors. This is a significant limitation because there are not many published research papers that discuss the topic of zero carbon industry performance assessment. Thus, comprehensiveness and depth may be affected by this limited pool of literature. There is little knowledge about the concept of zero carbon industry as it is still new to many people; hence, theories, models, and principles have not been formed about it. This novelty makes it difficult to find previous methodologies and validated frameworks thus affecting how reliable the findings will be. Some research findings can become quickly obsolete due to rapid technological development and advancements in sustainability practices. This situation calls for continuous updating of the study to keep pace with changes to make relevance an issue; hence precision entails a difficult task for maintaining accuracy in the study. However, there are some obvious limits. Net-zero commitments are not a substitute for immediate and comprehensive emissions reductions. Net zero requires prioritizing the elimination of challenging emissions sources. Net zero requires social and environmental integrity, which limits the scope, timing, and governance of carbon dioxide removal and offsets. Not all of these elements are currently well understood. The socio-political interpretation of net zero is a rich study goal requiring input from several disciplines, including climate science, biology, geology, anthropology, law, and economics. Thus, this chapter presents a theoretical background of the major factors essential in enhancing the formulation and integration of performance assessment frameworks for zero-carbon industries. Due to the coverage of the key areas which include; Technology integration motivational drivers, barriers, and implications for all the stakeholders, the chapter provides a rich resource for the policymakers and leaders in the industry. The conclusion reveals that the client's strategic goals cannot be met exclusively through innovative, sustainable planning and management strategies but require the consideration of several interrelated interests. Therefore, it is expected that as the field progresses, additional studies will be needed to improve the frameworks presented here and make them even more relevant for addressing the constantly emerging needs of sustainable development.

References

Ardi A, Cahyadi H, Sarwono R, Limaho H, Meilani YFCP, Pramono R (2023) The importance of a chief sustainability officer (CSO) in multinational and state-owned enterprises. J Hum, Earth, Future 4(3). www.HEFJournal.org

Bataille C, Waisman H, Briand Y, Svensson J, Vogt-Schilb A, Jaramillo M, Delgado R, Arguello R, Clarke L, Wild T, Lallana F, Bravo G, Nadal G, Le Treut G, Godinez G, Quiros-Tortos J, Pereira E, Howells M, Buira D, Tovilla J, Farbes J, Ryan J, De La Torre Ugarte D, Collado M, Requejo F, Gomez X, Soria R, Villamar D, Rochedo P, Imperio M (2020) Net-zero deep decarbonization pathways in Latin America: challenges and opportunities. Energy Strat Rev 30. https://doi.org/10.1016/j.esr.2020.100510

Bistline JET (2021) Roadmaps to net-zero emissions systems: emerging insights and modeling challenges. Joule 5(10):2551–2563. https://doi.org/10.1016/j.joule.2021.09.012. (Cell Press)

Chaturvedi V, Malyan A (2022) Implications of a net-zero target for India's sectoral energy transitions and climate policy. Oxf Open Clim Change 2(1). https://doi.org/10.1093/oxfclm/kgac001

Din AU, Khan MIM, Yang Y, Khuram W (2024) Innovative technological solutions for environmental sustainability in Chinese engineering practices. Eng Technol Appl Sci Res 14(2):13648–13657. https://doi.org/10.48084/etasr.6935

Fankhauser S, Smith SM, Allen M, Axelsson K, Hale T, Hepburn C, Kendall JM, Khosla R, Lezaun J, Mitchell-Larson E, Obersteiner M, Rajamani L, Rickaby R, Seddon N, Wetzer T (2022) The meaning of net zero and how to get it right. Nat Clim Change 12(1):15–21. https://doi.org/10.1038/s41558-021-01245-w. (Nature Research)

Gyimah J, Hayford IS, Nyantakyi G, Ofori EK (2023) Battling for net zero carbon: the position of governance and financial indicators. Environ Sci Pollut Res Int 30(57):120620–120637. https://doi.org/10.1007/s11356-023-30358-2

Kailay M, Paposa KK (2023) Embedding sustainability towards the growth of the developing economy: a study on sustainable management practices of the leading corporates. Int J Indian Cult Bus Manag 30(4):515–536. https://doi.org/10.1504/ijicbm.2023.136202

Kannan D, Solanki R, Kaul A, Jha PC (2022) Barrier analysis for carbon regulatory environmental policies implementation in manufacturing supply chains to achieve zero carbon. J Clean Prod 358. https://doi.org/10.1016/j.jclepro.2022.131910

Kuhlman T, Farrington J (2010) What is sustainability? Sustainability 2(11):3436–3448. https://doi.org/10.3390/su2113436. (MDPI)

Lee K, Fyson C, Schleussner CF (2021) Fair distributions of carbon dioxide removal obligations and implications for effective national net-zero targets. Environ Res Lett 16(9). https://doi.org/10.1088/1748-9326/ac1970

Lu B, Blakers A, Stocks M, Cheng C, Nadolny A (2021) A zero-carbon, reliable and affordable energy future in Australia. Energy 220. https://doi.org/10.1016/j.energy.2020.119678

Meilani M (2021) Sustainability and eco-friendly movement in movie production. IOP Conf Ser Earth Environ Sci 794(1). https://doi.org/10.1088/1755-1315/794/1/012075

Moghayedi A, Michell K, Hübner D, Le Jeune K, Massyn M (2024) Examine the impact of green methods and technologies on the environmental sustainability of supportive education buildings, perspectives of circular economy and net-zero carbon operation. Facilities 42(3–4):201–222. https://doi.org/10.1108/F-12-2022-0161

Okereke C (2007) An exploration of motivations, drivers and barriers to carbon management: the UK FTSE 100. Eur Manag J 25(6):475–486. https://doi.org/10.1016/j.emj.2007.08.002

Olabi A, Obaideen K, Elsaid K, Wilberforce T, Sayed ET, Maghrabie HM, Abdelkareem MA (2022) Assessment of the pre-combustion carbon capture contribution into sustainable development goals SDGs using novel indicators. Renew Sustain Energy Rev 153:111710. https://doi.org/10.1016/j.rser.2021.111710

Paposa KK, Kumar YM (2015) Impact of performance management system on job satisfaction of faculty members: a study on Technical Education Institutes of Nagpur. Manag Labour Stud 40(1–2):159–175

Paposa KK, Paposa SS, Chibber P (2024) Human resource management practices: facilitating sustainable value creation in organisations for a net-zero economy. In: Singh R, Khan S, Kumar A, Luthra S, Chokshi H (eds) Net zero economy, corporate social responsibility and sustainable value creation. CSR, sustainability, ethics & governance. Springer, Cham. https://doi.org/10.1007/978-3-031-55779-8_10

Pimiä A, Urponen A, Kolehmainen A (2024) Green horizons ahead: navigating carbon neutrality in European Transportation by sky and sea. Econ Cult 21(1):149–158

Ravesangar K, Ping LL, Pachar S (2024) A review on the sustainable HRM practices in building net-zero transformation: an emerging trend in the workplace. In: Net zero economy, corporate social responsibility and sustainable value creation: exploring strategies, drivers, and challenges, pp 131–144

Reddy NSS, Krishankumar R, Priya SS, Cavallaro F, Mardani A, Ravichandran KS (2023a) Fermatean fuzzy-based personalized prioritization of barriers to IoT adoption within the clean energy context. Inf (Switz) 14(6). https://doi.org/10.3390/info14060309

Reddy VJ, Hariram NP, Maity R, Ghazali MF, Kumarasamy S (2023b) Sustainable E-fuels: green hydrogen, methanol and ammonia for carbon-neutral transportation. World Electr Veh J 14(12). https://doi.org/10.3390/wevj14120349. (Multidisciplinary Digital Publishing Institute (MDPI))

Robiou du Pont Y, Jeffery ML, Gütschow J, Rogelj J, Christoff P, Meinshausen M (2017) Equitable mitigation to achieve the Paris Agreement goals. Nat Clim Change 7(1):38–43

Rubio-Andrés M, Abril C (2024) Sustainability oriented innovation and organizational values: a cluster analysis. J Technol Transf 49(1):1–18

Singh R, Ravesangar K, Narayanan S, Singh KKH (2024) HRM practices in creating net zero: supporting organizational sustainability. In: Singh R, Khan S, Kumar A, Luthra S, Chokshi H (eds) Net zero economy, corporate social responsibility and sustainable value creation. CSR, sustainability, ethics & governance. Springer, Cham. https://doi.org/10.1007/978-3-031-55779-8_1

Stern N, Valero A (2021) Research policy, Chris Freeman special issue innovation, growth and the transition to net-zero emissions. Res Policy 50(9). https://doi.org/10.1016/j.respol.2021.104293

Trivedi A, Trivedi V, Chaurasia S (2023) Modelling the critical success factors of net-zero energy buildings in India. E3S Web Conf 455. https://doi.org/10.1051/e3sconf/202345501012

Vijay V, Sathish AS (2024) Advancing towards a sustainable future: a systematic review of sustainable net zero economy implementation studies. Qubahan Acad J 4(1):137–149. https://doi.org/10.58429/qaj.v4n1a279. (Qubahan Organization for Development)

Warzywoda N, Dargusch P, Hill G (2022) How meaningful are modest carbon emissions reductions targets? The case of Sumitomo electrical group's short-term targets towards longer-term net zero. Sustainability 14(7):4287

Wu H-H, Liu W-Y, Huang MC (eds) (2023). Moving toward net-zero carbon society. Springer International Publishing. https://doi.org/10.1007/978-3-031-24545-9

Yadav S, Samadhiya A, Kumar A, Majumdar A, Garza-Reyes JA, Luthra S (2023) Achieving the sustainable development goals through net zero emissions: innovation-driven strategies for transitioning from incremental to radical lean, green and digital technologies. Resour Conserv Recycl 197. https://doi.org/10.1016/j.resconrec.2023.107094

Yueyang W, Yuan PF (2024) A parametric approach towards carbon net zero in agricultural planning. Comput Des Robot Fabr Part F2072:305–314. https://doi.org/10.1007/978-981-99-8405-3_26. (Springer)

Drivers of Innovation for Environmental Sustainability: Experience from the South Asian Apparel Industry

P. G. S. Amila Jayarathne, Narayanage Jayantha Dewasiri, Amit Joshi, and Rubee Singh

Abstract The apparel industry is one of the vital industries in the global and South Asian context due to its contribution to the economy. However, this industry is recognised as one of the industries that adversely affect the environment due to the nature of the operation. Thus, initiatives are made to achieve environmental sustainability mainly through innovative techniques and processes. Addressing the need for context, the current study aims to explore the drivers for innovation for environmental sustainability with the experience from the South Asian context. Employing a qualitative approach, the study interviewed 37 respondents from Bangladesh, India and Sri Lanka, representing government and industrial experts relevant to the apparel industry. Regulatory drivers, market and consumer-related drivers, technological support, economic incentives and corporate social responsibility motives are identified as critical drivers for ensuring environmental sustainability through innovation. The main contribution of the current study is to derive vital drivers contextually relevant to the South Asian apparel industry using empirical evidence. The implications for both corporate and government sectors are discussed, and a few suggestions are highlighted for scholars.

P. G. S. A. Jayarathne
De Montfort University, Leicester, UK
e-mail: amila.jayarathne@dmu.ac.uk

N. J. Dewasiri (✉)
Department of Accountancy and Finance, Faculty of Management Studies, Sabaragamuwa University of Sri Lanka, Belihuloya, Sri Lanka
e-mail: jayantha@mgt.sab.ac.lk

Adjunct Faculty, University Center for Research and Development (UCRD), Chandigarh University, Punjab, India

A. Joshi · R. Singh
Department of Management Studies, Kumaun University, Nainital, Uttrakhand, India

A. Joshi
Institute of Business Management, GLA University, Mathura 281406, India

R. Singh et al. (eds.), *Zero Carbon Industry, Eco-Innovation and Environmental Sustainability*, World Sustainability Series,
https://doi.org/10.1007/978-3-031-80220-1_14

Keywords Apparel Industry · Environmental sustainability · Innovation · South Asia

1 Introduction

The apparel industry is one of the most crucial industries of today's economy in terms of different aspects such as global economic contribution in manufacturing and retailing and employment opportunities created around the world (Jia et al. 2020; Koszewska 2018; MacCarthy and Jayarathne 2010; MacCarthy and Jayarathne2013; Jayarathne et al. 2021; Susitha et al. 2024). However, it has also been identified as one of the principal offenders in environmental exploitation, owing to the nature of the industry (MacCarthy and Jayarathne 2012; Jayarathne et al. 2021; Shaw et al. 2006). In particular, due to the extended supply chain starting from the extraction of raw materials used in the manufacturing process that involves many chemicals with highly toxic, dying, washing, in turn, harmful disposal in addition to the standard factory operational related environmental harmful disposal (MacCarthy and Jayarathne 2012). These result in high water consumption, extensive use of chemicals, and the emission of greenhouse gases, among others (Bailey et al. 2022). The manufacturing activities require numerous resources where. The fashion industry contributes 10% of the global carbon production and 19% of the world's water pollution from textile production (Geeneva Environment Network 2024; European Parliament 2019; Global Fashion Industry Statistics 2021; Hassan and Carr 2018). We can discuss the critical need for change as the apparel industry's environmental impact remains colossal.

In this context, innovation is central to attaining environmental sustainability within the industry (Harsanto et al. 2023; Rudrajeet and Jayarathne 2021). Thus, it is possible to state that the industry can minimise adverse environmental effects through the further integration of new technologies, the search for environmentally friendly materials, and optimising production. Technological solutions like water-less dyeing processes, ecologically friendly fabrics and fabrics that are made out of recycled products, efficient ways of using resources, and recycling form an excellent platform for the reuse of resources (Luo et al. 2021; Abbate et al. 2024). The outcomes of these sustainable practices, thus, reduce the ecological impact and improve the robustness and profitability of the industry to satisfy the increasing number of customers and compliance with environmental standards. The industry has already made initiatives by being in the broader category of digitalisation of the entire supply chain operations to streamline the operations (Susitha et al. 2024; Rudrajeet and Jayarathne 2021) and implementing specific environmentally sustainable practices (e.g., Abbate et al. 2024).

The South Asian perspective is pertinent regarding the demands that may facilitate innovation on the sustainability axes within the apparel business. South Asia, home to many apparel-producing countries like Bangladesh, India, Pakistan, and Sri Lanka, is an essential supplier in the global apparel supply chain (McKinsey 2024). The role is apparent and even more significant now because South Asia, on its own, constitutes a large part of global apparel exports (International Finance Corporation, World Bank Group 2023; McKinsey 2024). Nonetheless, environmental problems experienced in these countries due to high industrialisation call for a specific effort in green innovation. Signing the sustainability agreement thus makes economic sense in this southern Asian country, given its socioeconomic structure of a large labour force that relies on the apparel industry.

Therefore, addressing such requirements of the existing knowledge, this research study aims to identify drivers that influence innovation for environmental sustainability in the South Asian Apparel Industry. The chapter is organised in five parts. A brief discussion of the literature follows the introduction. Then, the methodology is presented. After that, the main findings are presented, followed by a discussion of them and making the concluding remarks.

2 Literature Review

2.1 Global Apparel Industry

The international clothing sector is a vast and constantly developing segment of the world economy in which different countries actively participate (MacCarthy and Jayarathne 2010; MacCarthy and Jayarathne 2012; Jayarathne et al. 2021). Textiles comprise clothes and accessories and cover fashion designing, manufacturing, and marketing of these products; this industry plays a significant role in employment and income generation in many nations. Basic structure of the global apparel industry from a supply chain perspective includes all the players in the global apparel industry with a holistic approach. The textile and accessory producers supply the clothing plants, which feed into distribution and logistics systems to enable garments produced in dispersed global networks to meet anticipated demand in specific retailers at their stores.

The world market for apparel has grown tremendously over two or three decades due to globalisation, innovations, and trends. Statista (2021) shows that the global apparel market worth is estimated at USD 1. 5 trillion at the beginning of 2020 and is expected to reach USD 2. 25 trillion by 2025. This growth signifies that the industry is expanding despite the current challenges, such as the COVID-19 pandemic. Further, employment generated by the apparel industry is manifold around the world. Thus, the garment industry obliges 60 million individuals globally, where women constitute most of the working force (ILO 2021). It is more critical in developing nations because it improves the economy and reduces absolute poverty rates.

Similarly, the apparel industry contributes significantly to international trade. This is because the industry is extensively involved in the international exchange of goods through outsourcing. Countries produce various stages of apparel, ranging from crop cultivation such as cotton, fabric, and accessory production to apparel manufacturing, logistics, warehousing, retailing, recycling, and reversed logistics. Major exporting nations are China, Bangladesh, Vietnam, and India; major importing nations are the US, EU, and Japan (WTO 2023).

There has been growing attention regarding the effects of the apparel industry on the social and physical environment. Concerns like the violation of labour rights, health hazards at the workplace, and pollution have given rise to demands for better and ethical practices. Some labour conditions in the apparel industry include low wages, long working hours, and deplorable working conditions. The latest tragedy that captured people's attention and demonstrated how severe conditions can be in some spheres of the industry was the collapse of the Rana Plaza factory in Bangladesh, which claimed the lives of over 1,100 workers in 2013 (Clean Clothes Campaign 2013). Regarding environmental concerns, the apparel industry is considered one of the largest polluting industries globally (Earth.org 2023). Textile and apparel are other significant products produced in large quantities and contribute to waste production. For instance, as of now, it contributes to roughly 10% of emissions of carbon in the environment, and it is forecasted to increase up to 50% in 2030; it uses much water in addition to the contribution to 20% of global water pollution; and of the 100 billion garments produced each year, 92 million tons end up in landfills (Earth.org 2023).

Based on these challenges, numerous activities have been developed to improve the sustainability and ethical aspects of the apparel industry. These include the Sustainable Apparel Coalition (SAC), which was established in 2010 and is a group of multiple stakeholders primarily interested in decreasing the environmental and social footprint of apparel and footwear products (SAC 2021). The Fashion Revolution is another milestone. It is an international campaign that promotes transparency, fairness, and ecological responsibility in the apparel and fashion sector, supporting consumers to demand from the clothing companies: Who made my clothes? (Fashion Revolution 2021). Circular fashion is another major initiation that grabs global attention. The idea related to circular fashion is using and recycling old clothes, textiles, and accessories to reduce wastage and enhance product durability. Companies like Patagonia and H&M have introduced take-back programmes and recycling schemes (H&M Group 2021).

In the current technology-driven era, this industry has mainly been characterised by technological innovations that have acted as a tool that transformed the apparel industry. Advancements in technology have been seen in numerous ways, for instance, the use of digital printing, 3D knitting, and even the use of automation in designing and production, hence cutting costs and time (Rudrajeet and Jayarathne 2021). Digitisation and automation of several processes associated with generating different types of apparel and related activities, from designing to manufacturing and distribution processes. For instance, the application of robotics in manufacturing is observed in some firms, such as Adidas, with their "Speedfactory" (Adidas 2021). Furthermore, globalisation and especially the development of e-commerce can be marked as one of the primary trends in the apparel industry. E-commerce Internet sales have risen significantly more in the post-COVID-19 period as consumers moved to online purchasing. In the case of apparel, as claimed by eMarketer (2021), apparel e-commerce retail sales crossed USD 759. Hence, the total sales reached USD 5 billion in 2020 and are expected to rise more than USD 1 trillion by 2024.

Within the global apparel industry, the South Asian apparel industry has emerged as one of the largest industries worldwide, consisting of countries such as Bangladesh, India, Pakistan, and Sri Lanka. It enjoys an export reputation and is considered a significant global supplier of textiles and apparel due to huge-scale production, low-cost labour, and export inclination. Doubling the population in seven decades has doubled the result of industrial growth, and carelessness has left large amounts of pollutants in the environment, exploitation of resources, and poor waste management. As a result, numerous players within the context of the industry have endeavoured to pursue improved environmental standards. Thus, the current paper focuses on the South Asian apparel industry.

2.2 *Innovation*

Innovation is the creation and implementation of new ideas, goods or services, technologies, and methods that are new or considerably improved from the existing ones (Edwards-Schachter 2018). It covers both refinements or developments that are slight but sufficient to lead to the creation of new industries or markets. Innovations' importance is apparent as they can be associated with matters such as the economy's growth, competitiveness, and the handling of various issues through the introduction of creativity (Anderson et al. 2014). Regarding environmentalism, innovation is essential since it creates the technologies and practices that reduce environmental impacts and encourage sustainable development (Shabir et al. 2023).

Innovation within the bounds of the apparel industry is the adaptation of new concepts involving products, techniques, or plans, which will recreate the existing pattern, decrease costs, and increase the value added to apparel products (Muthu and Gardetti 2016). (Abbate et al. 2024) notes that customers' tastes have refined; hence, there is a need for innovation to meet their needs and, at the same time, ensure that the firm can sustain a competitive edge within the market, which is constantly dynamic. Technological innovation can be described as the development of new technologies, while process innovation refers to the enhancement of the existing procedures in an organisation (Edwards-Schachter 2018; Muthu and Gardetti 2016). Technological advancements also mean using new tools or programs to enhance production (Edwards-Schachter 2018; Muthu and Gardetti 2016). The other aspect of innovation in the apparel industry is process innovation, which is defined as the improvement of activities taking place throughout the entire process of the supply chain, ranging from sourcing, manufacturing, and delivering to retailing, to deliver better outputs (Edwards-Schachter 2018; Muthu and Gardetti 2016). Product innovation, another aspect of innovation, can be defined as creating new goods and services or enhancing existing ones to satisfy market wants (Edwards-Schachter 2018; Muthu and Gardetti 2016).

2.3 Environmental Sustainability

Environmental sustainability is, therefore, the proper use of the environment without exploiting it to depletion or rendering it substandard for use in the future (Bocken and Short 2021). It entails processes that do not allow the depletion of ecological assets to preserve their capacity to provide services in the present and future. According to the well-known model for sustainability, the triple bottom line, 3BL, environmental sustainability is one aspect of the three aspects of sustainability: people, planet, and profit (MacCarthy and Jayarathne 2012). Environmental sustainability is important because it contributes to preserving ecosystems' capability to support life in its different forms, healthy human life, and perpetual provision of resources needed for wealth and health (Bocken and Short 2021).

There is broad debate on environmental issues, such as growing environmental pollution, over the usage of natural resources that are servilely problematic for both the natural environment and human life, which have been identified in the last decades. All over the world, environmental concerns are a significant challenge faced by every society. The essence of the earth contains the land, air, and water. Because of human activities, the whole balance of the earth's ecosystem has been troubled. Accordingly, 'Greenness' is becoming an essential strategy in a business firm that is a buzzword in today's business world.

Moreover, sustainability has developed as a leading research scheme based on sustainable guidelines introduced by the European Union (European Parliament 2019). Its exhibitions are done through the efforts of firms to develop green strategies for low-carbon landscapes worldwide (Luo et al. 2021). Even in present environments, sustainability has become a prominent factor for the competitive edge of the business world (Luo et al. 2021). Therefore, sustainability has gained prominence and become a rigorous part of public policies (European Parliament 2019). Most businesses also pay more attention to revising their company strategies by including a sense of suitable development (Chladek 2019).

2.4 Relationship Between Innovation and Environmental Sustainability

Innovation is diametrically linked to the environmental sustainability agenda (Ashford and Hall 2011). Innovation creates sustainability through new technologies and approaches that lessen the adverse effects on the environment and use of resources (Ahmad et al. 2023). On the other hand, the call to sustainability can be a source of innovation since companies and societies attempt to find solutions to reduce the harm done to the environment and meet regulatory and market standards for sustainable practices (Hermundsdottir and Aspelund 2021). The kind of relationship created by this sets up a virtuous cycle whereby sustainability issues drive response, from which further progress in sustainability occurs.

The innovation-environment connection within the apparel industry can be defined as the process within which new technologies and procedures are introduced to decrease the level of harm to the environment (Bhandari et al. 2022). This relationship is critical since the applications of sustainable innovations can reduce resource use, waste, and emissions by large margins. Several academic studies have established that firms that invested in sustainable improvements of existing products, services or business processes posted higher environmental and economic outcomes (Govindan et al. 2020).

For instance, technological advancements such as waterless dyeing fabrics in textile processes have drastically reduced water use and the emission of pollutants. Such innovations make the environment more sustainable while having cost reduction and production efficiency advantages for the manufacturers (Arumugam et al. 2024). Also, dramatic innovations in design and production have been observed concerning the principles of circular economy, such as recycling and reusing products (Brydges 2021). H&M is one of the biggest fast-fashion companies worldwide, which demonstrates activities towards circular fashion. This is clearly illustrated by their garment collecting program, whereby after the customers are done with them, they are encouraged to return their garments to be recycled. H&M has also developed a 'Conscious Collection' derived from sustainable resources such as organic cotton and recycled polyester (H&M Group 2021). Adidas has collaborated with Parley for the Oceans to develop shoes and clothes from plastics found in the ocean. Through this campaign, it is possible to minimise the amount of plastic that finds its way to the sea and contribute to the popularisation of recycling and using recycled materials in everyday clothing items, including sportswear (Adidas 2021). Levi's has devised a Water <Less™ technique and claims to have bequeathed a minimum of 96% less water while finishing the jeans (Levi Strauss & Co. 2021).

In South Asia, MAS Holdings, the largest apparel manufacturer in Sri Lanka, has developed the "Eco Go Beyond" programme to create awareness and enhance sustainable development. Sustainable manufacturing is in place through recycled water systems and solar panels (MAS Holdings 2021). Arvind Limited, a textile manufacturing company from India, has implemented the ZLD system in its production line. This initiative helps manage all wastewater, its proper treatment and recycling, reducing water pollution and usage (Arvind Limited 2021). DBL Group in Bangladesh has set up a green factory accredited by the U.S. Green Building Council's Leadership in Energy and Environmental Design program. Energy-efficient systems and equipment, rainwater management, and adequate natural lighting define industrial standards of environmental protection and performance in the factory (Mishra et al. 2024).

However, these initiatives need continuation and further improvement and should be facilitated to achieve environmental sustainability through innovation. The main scope of the current study is what drivers facilitate this context; accordingly, the research purpose is derived and addressed through the following methodological approach.

3 Methodology

According to the nature of the research question, this study is qualitative (Dewasiri et al. 2018). To obtain a broader perspective, semi-structured interviews are conducted with several respondents, including managers of large companies, officials, environmental specialists, and representatives of governmental organisations. Participants in all 37 presenting Sri Lanka, Bangladesh, and India are nominated purposively for a better chance of having adequate knowledge and experience. Bangladesh is the leading apparel exporter in the South Asian context. The second largest country for this industry is India. Due to high industry competition and the local economic crisis, Sri Lanka is a highly vulnerable supplier. This qualitative data is then transcribed and analysed using a method known as thematic analysis; indeed, coding software such as NVivo helps identify patterns or themes. Rand's cross-verification and member-checking techniques improve credibility, and participants can assess the analysis made from their inputs. Further, to enhance the study's credibility, coding and interpreting the data collected by different analysts is adopted to minimise the possibility of a researcher's bias. Another method used to reduce the effects of the researcher's bias is peer debriefing, which makes the exact processes involved in the analysis quite transparent.

4 Analysis and Findings

This section underlines the research findings from the qualitative interview with the key informants, which involved industry experts, top-level management of apparel manufacturing companies, and government officials in South Asia. The data was analysed using thematic analysis, identifying several key themes related to drivers of innovation and environmental sustainability. The section first briefs on some success cases identified concerning innovative initiatives for environmental sustainability and then discusses the specific drivers disclosed through the study.

4.1 Success Cases in the South Asian Apparel Industry

South Asian apparel market is a vital apparel manufacturing region in the world clothing market, including key clothing production countries Bangladesh, India, Pakistan, and Sri Lanka. These countries participate in the global supply chain as essential players and, at the same time, are experiencing environmental issues that are specific to them. This section focuses on the detailed company analysis of organisations in the South Asian apparel industry about the drivers of innovation and environmental sustainability. Through these cases, the study identified how these firms are incorporating sustainability and the challenges and consequences of the processes.

4.1.1 ABC Company in Sri Lanka

ABC company is one of the leading apparel manufacturers in Sri Lanka. This company pays extra attention to the environment and sustainability. The company owns over 50 factories, operates in over 15 countries, and has around 100,000 employees. The company has raised industry standards for achieving sustainability through innovative strategies and technologies. Key such initiatives are;

Eco Go Beyond: The main focus is on the Eco Go Beyond project as it is a diversified multi-disciplinary educational programme on sustainability. This initiative aims primarily at the employees, local communities, and schools in the internal and external sustainability policy. It has enhanced the consciousness of organisations and the populace on equitable use of resources in their daily activities. Thus, many educational programs have encouraged the stakeholders to act more responsibly to preserve the environment so that the company's key initiatives in the sphere of sustainability go far beyond its direct activity.

Green Manufacturing Facilities: The company has invested heavily in establishing and developing environmentally friendly factory plants. These facilities are accredited by the U.S. Green Building Council's LEED Leadership in Energy and Environmental Design. They include efficient lighting, rainwater, and impressive waste management systems, leading to considerable energy and water savings and minimising waste. The LEED certification indicates the company's environmental responsibility as a manufacturer and a means by which other manufacturers can emulate it.

Sustainable Product Development: This has led the company to develop a brand portfolio focusing on the environmental aspects. For example, their "Ocean Plastic" program involves upcycling plastic in the ocean to develop fabric. This aids in conserving plastic waste in the seas while promoting the amount of recycled material used to make apparel. This initiative has dramatically helped minimise cases of pollution within the sea and has served as a benchmark for the utilisation of recycled products, especially within the apparel trade. It also builds the company brand position of being innovative and more sensitive to the environment.

Despite its successes, the company faces challenges, mainly in maintaining the economic viability of sustainable practices and navigating regulatory landscapes across different countries. As the initial investment in green technologies and processes is high, there is often a need to balance these costs with competitive pricing. Consequently, it affects the demand as, for some apparel brands, such prices are expensive; in turn, they place orders with other apparel suppliers in the South Asian region or elsewhere.

4.1.2 XYZ Company in India

This is a large textile manufacturer in India. Its sustainability section revealed that the company has been stewarding sustainability for decades, particularly on water, energy, and raw materials.

Zero Liquid Discharge (ZLD) System: It has installed a ZLD system on its production. This system enables all industrial effluent to be fully treated and recycled back to the production processes, and there is no direct discharge of damaging effluent. It has significantly lowered water pollution levels and become the benchmark for wastewater disposal in the textile industry. This goes a long way in reducing water usage; it also contributes to the reduction of contaminations to the natural environment in the manufacturing processes of textiles.

Sustainable cotton sourcing: The company is attempting to offer farmers solutions to grow organic cotton. This initiative ensures ecological farming and the constant availability of products made of environmentally friendly materials. Local farmers also benefit from its production since it provides them with a steady income and cuts their costs of using chemicals in farming.

Energy-efficient programmes: The company has invested considerably in energy-efficient technologies and alternative energy. This includes using solar panels and biomass to produce energy. Such energy efficiency measures have helped minimise the company's carbon emissions and other related expenses. Using renewable energy sources makes it possible to increase the sustainability of production and strengthen the struggle against climate change. However, the main challenge is that good supply chain management and continuous support from the farmers are required to sustain this practice.

4.1.3 PQR Company in Bangladesh

This organisation is one of the giants in Bangladesh's apparel business and is famous for practising sustainability. It has many operating manufacturing sites and is known for integrating sustainable environmental solutions through novel techniques, processes, and tools.

Green Factory Certification: The company intends to implement a green factory accredited by the U.S. Green Building Council LEED. The factory converts the building features into an environmentally friendly design, layout, and attributes, including efficient water management, renewable energy, and natural materials. The green factory has significantly improved energy efficiency and water and CO_2 usage. Thus, it is an example of effective and environmentally friendly production in Bangladesh while representing the effectiveness of green building systems.

Sustainable Textile Dyeing: The current dyeing technologies adopted ensure that the company uses minimal water and chemicals. This process ranges from the dyes used, such as low-impact dyes, to the process of dying, which usually retains water. In this case, more sustainable technologies have been implemented to minimise environmental contamination in dyeing textiles. They have also brought efficiency to the dyeing process and saved costs.

However, this company needs some help. They are the high cost of efficient technologies and the need for innovation to sustain already adopted policies on sustainable business practices. Besides, the company must operate in the context of various legal requirements and adhere to the norms of the global economy.

4.2 Drivers for Innovation for Environmental Sustainability

4.2.1 Regulatory Policies

Regulative policies were crucial, especially in enhancing innovation and sustainability (Rathnasiri et al. 2024). As interviewees observed, there has also been a quest to embrace sustainability practices prompted by some strict laws on sustainable practices. The following quotation provides an example.

> Our government has recently moved to enforce more stringent environmental standards …. discovered the rules on compliance……Thus, …. to wastewater treatment plants have been imposed to develop cleaner systems. (A Bangladesh government officer)

> …yes, we have made some initiatives to streamline the operations, especially to reduce the waste as instructed by the government authorities …… mainly for water disposal….” (A manager of an Indian Apparel manufacturing company)

> … If regulations make it compulsory, we and everyone should follow it….it is so straight forward as such…. (A team leader in a Sri Lankan Apparel manufacturer)

> It is the rules and regulations. If buyers say something, the industry should follow it. Even the local government should listen as we are a small country. (A government officer, Sri Lanka)

4.2.2 Market Demand and Consumer Awareness

Enhancing consumer awareness of environmentally friendly apparel items has been identified as another driver. Consequently, the market demand for such products increases while creating a promising future for apparel manufacturers and, ultimately, all the apparel supply chain members. The quotations below illustrate this.

> It is needed to have a paradigm shift in customer demand for green products; otherwise, no one will invest in massive innovation projects. (A Sri Lankan apparel manufacturing plant, CEO)

> We give what customers ask, if customers demand such nature of environmentally less harmful products of cause we are ready to change all our practices and invest in different technologies……new machines…. whatever. (General Manager, a leading apparel manufacturer, Bangladesh)

> It can be observed that customers are demanding eco-friendlier products, and this shift in demand is influencing organisations to adapt to these changes. (Apparel Company Manager, India)

> With my experience, I would say this is the industry run by the retailers. The consumers also leverage retailers. If consumers demand such nature of environmentally friendly products, then no option for retailers and accordingly for the rest of the players in the supply chain. (A government officer, Bangladesh)

4.2.3 Technological Advances

As for technological factors, the enablers of innovation and sustainability were mentioned to offer new instruments and approaches to avoid negative environmental influences. It was explicitly noted that technological advancement is essential to reducing resource consumption and achieving environmental sustainability.

> In terms of philosophy, one has to stress technology: better technologies for recycling fabrics, more efficient water-using technologies for dyeing, and many others enable us to minimise our environmental impact. (Sustainability Officer, Sri Lanka)

> This area, I must admit, has registered impressive improvements, especially in technologies of fabric recycling and water economical dyeing processes cutting down environmental impacts considerably. (A team leader of a leading apparel brand in an Indian Apparel Manufacturing firm)

> Technology plays a major role here. We know and have heard of many initiatives; I think everyone would be interested in applying these novelties. However, the main concern is how many of us have the required technologies or technological competency to cope with such novelty.... (A leader of the Capacity Building up team, Sri Lankan apparel company)

> Nothing much to discuss; it is straightway the technology. Because people can tell big stories, but the sustainability can be brought to the reality only if required technology supports, otherwise just a dream or matter of a discussion topic. (A government officer, India)

4.2.4 Economic Incentives

Subsidies and tax holidays for initiatives that support sustainable solutions were listed as one of the appeals that make it even more compelling to carry out innovations. As innovation demands massive investment in novel techniques, processes, materials, etc., providing economic incentives to promote investments that target less environmental damage is necessary. The following selected quotations are evidence of this.

> Promotional measures by the government, such as tax credit on the use of energy-efficient machinery and equipment, among others, have fostered the implementation of sustainable strategies since little capital is contributed towards purchasing new tools. (Government Official, Bangladesh)

> The non-financial technique used by industries that the government compelled to maintain sustainability is: 'Financial incentive from the government includes the tax credit on investments required on energy efficient machinery'. (A Government official in Sri Lanka)

> We are doing a business. Every spending should have a return. So, if there is any financial gain, we can consider having such initiatives targeting sustainability; otherwise, not. (Owner, Apparel Company India)

> We do what our buyers say;... because it is our income source. Some buyers give some extra financial benefits directly or indirectly. Then, we adhere to such processes or technologies, but of course, with such an incentive. (General Manager, Bangladesh Apparel manufacturer)

4.2.5 Corporate Social Responsibility (CSR)

CSR initiatives of companies or the immediate members of the supply chain, mainly the customers, are also explored as drivers to ensure environmental sustainability. An assessment of CSR activities revealed that most of the mentioned CSR projects were inclined to be used as tools for sustainability through creative, novel approaches.

> There are many apparel firms mean while implementing CSR policies to transform sustainability into key strategic business vectors, this transition is not motivated simply by compliance, but notably by the will to associate sustainability with the business image and international purchasing power. (An official member of an apparel association in Sri Lanka)

> Most apparel Giants from developing nations are implementing sustainability into their strategic corporate plans as part of CSR. Such a trend is not exclusive to responding to standard industry norms. However, it is a tactical move in global value merchandising to cultivate a favourable market image that complies with the expectations of global buyers. (Industry Expert, India)

> If financial conditions are supported, it is how the company looks at its corporate citizenship and balances money, people, and the planet; it is like a triangle. It is sustainable if we are responsible for the people and the environment. (Managing Director, Apparel Manufacturer, Bangladesh)

5 Conclusion

The findings from this study reveal several critical drivers of innovation and environmental sustainability in the South Asian apparel industry. These include regulatory policies, market demand, technological advances, economic incentives, and corporate social responsibility (CSR). Each of these drivers plays a crucial role in shaping industry practices and pushing the sector towards more sustainable and innovative solutions.

5.1 Contribution of the Study

This research provides several critical insights that contribute to understanding the innovation and sustainability drivers in the South Asian apparel industry.

1. *Regional Context*: The phenomenon identified in the study underscores how drivers for innovation and sustainability are context-specific, providing propositions that are especially suitable for South Asia.
2. *Comprehensive Framework*: Incorporating pluralistic drivers to superimpose the existing body of knowledge further shows how they complement regulatory policies, market forces, technology adoption, incentives, and CSR policies constituting the industry.

3. *Empirical Evidence*: The study applies interviews among industry players and government authorities and, thus, provides actual cases together with real-life experiences that enrich the investigation and reveal how these drivers work.
4. *Policy Implications*: The research provides valuable advice for enacting supportive conditions in policy-making procedures by representatives of policy-making and innovative industries. This is done to coordinate with market demand and technological advances and follow up with a demarche by lowering the financial barriers through incentives.

5.2 Recommendations for Policymakers in South Asia

Based on the current study's findings, the following suggestions are made for policymakers in the South Asian context.

1. *Strengthen Regulatory Frameworks*: Iterate on enforcing environmental standards to ensure they are applied precisely as the rules require. This means the institution should outline proper compliance measures and periodically check compliance.
2. *Expand Economic Incentives*: Promote the use of grants and subsidies to reduce costs when adopting and fully implementing sustainable technologies.
3. *Promote Consumer Awareness*: Promote public awareness of the effects of sustainable consumption and its advantages, which will, in turn, create a market for environmentally friendly products.
4. *Encourage Collaboration*: Build collaboration between government, industries, and university research institutions to improve innovation and disseminate sustainability.

5.3 Approaches for the Promotion of Innovation in Environmental Conservation

The following managerial implications are derived from the findings of the current study.

1. *Investment in Technology*: Promote and encourage the use of new generation raw materials that, for instance, minimise water consumption and pollution during dyeing and other production processes; advanced, energy-efficient equipment, among other aspects.
2. *Public–Private Partnerships*: Establish partnerships between public and private organisations to address common sustainability issues and develop new solutions.
3. *Research and Development (R&D)*: Promote the funding for research and development to discover new sustainable material forms and execution techniques and back projects targeted at forging regional innovations for local needs.

4. *Sustainability Metrics*: To achieve this, the following interventions should be adopted: Creating awareness of using standard sustainable indexes to enable organisations to determine accurate environmental performance.

While this study provides valuable insights, it can be extended to a longitudinal study to capture how these drivers change over time. Further, this can be repeated in similar contexts, for instance, in other developing nations operating in the apparel industry. Further, this study is worth investigating from the customers' perspectives to explore the motivation of apparel customers for environmental sustainability.

References

Abbate S, Centobelli P, Cerchione R, Nadeem SP, Riccio E (2024) REVIEW Sustainability trends and gaps in the textile, apparel and fashion industries. Environ Dev Sustain 26:2837–2864. https://doi.org/10.1007/s10668-022-02887-213

Adidas (2021) The three loop strategy: Adidas' plan to help end plastic waste. Last Accessed 20 June 2024

Ahmad N, Youjin L, Žiković S, Belyaeva Z (2023) The effects of technological innovation on sustainable development and environmental degradation: evidence from China. Technol Soc 72:102184. ISSN 0160-791X, https://doi.org/10.1016/j.techsoc.2022.102184

Anderson N, Potočnik K, Zhou J (2014) Innovation and creativity in organisations: a state-of-the-science review, prospective commentary, and guiding framework. J Manag 40(5):1297–1333. https://doi.org/10.1177/0149206314527128

Arumugam A, Ramesh Babu V, Sundaresan TD (2024) Sustainable dyeing techniques: advancements and innovations in the textile industry. Melliand Int 30(1/2024):40–42

Arvind Envisol (2021) (Zero Liquid Discharge (ZLD). Arvind Envisol. Last Accessed 20 June 2024

Ashford NA, Hall RP (2011) The importance of regulation-induced innovation for sustainable development. Sustainability 3(1):270–292

Bailey K, Basu A, Sharma S (2022) The environmental impacts of fast fashion on water quality: a systematic review. Water 14:1073. https://doi.org/10.3390/w14071073

Bhandari R, Hassan R, D'Amore G, Tiscini R (2022) Responsible innovation and ethical corporate behavior in the Asian fashion industry: a systematic literature review and avenues ahead. Asia Pac J Manag 41:1129–1173

Bocken NMP, Short SW (2021) Unsustainable business models–recognising and resolving institutionalised social and environmental harm. J Clean Prod 312:127828. ISSN 0959-6526, https://doi.org/10.1016/j.jclepro.2021.127828. (New Business Models)

Brydges T (2021) Closing the loop on the take, make waste: investigating circular economy practices in the Swedish fashion industry. J Clean Prod 293:126245. ISSN 0959-6526, https://doi.org/10.1016/j.jclepro.2021.126245

Chladek N (2019) Why you need sustainability in your business strategy. The Importance of Sustainability in Business. HBS Online

Clean Clothes Campaign (2013) Rana plaza: a preventable tragedy. https://cleanclothes.org. Last Accessed 28 July 2024

Dewasiri NJB, Weerakoon YK, Azeez AA (2018) Mixed methods in finance research. Int J Qual Methods 17(1):1–13. https://doi.org/10.1177/1609406918801730

Earth.org (2023) 10 Concerning fast fashion waste statistics. Earth.org. Last Accessed 27 July 2024

Edwards-Schachter M (2018) The nature and variety of innovation. Int J Innov Stud 2(2):65–79. ISSN 2096-2487, https://doi.org/10.1016/j.ijis.2018.08.004

eMarketer (2021) Global e-commerce update 2021. https://www.emarketer.com. Last Accessed 28 July 2024
European Parliament (2019) Environmental impact of the textile and clothing industry. EPRS Eur Parliam Res Serv PE 633(143)
Fashion Revolution (2021) Who made my clothes? https://www.fashionrevolution.org. Last Accessed 27 July 2024
Geneva Environment Network (2024) Environmental sustainability in the fashion industry–Geneva environment network. Last Accessed 20 June 2024
Global Fashion Industry Statistics-International Apparel (2021) Fashion United
Govindan K, Rajeev A, Padhi SS, Pati RK (2020) Supply chain sustainability and performance of firms: a meta-analysis of the literature. Transp Res Part E Logist Transp Rev 137:101923. ISSN 1366-5545, https://doi.org/10.1016/j.tre.2020.101923
Harsanto B, Primiana I, Sarasi V, Satyakti Y (2023) Sustainability innovation in the textile industry: a systematic review. Sustainability 15:1549. https://doi.org/10.3390/su15021549
Hassan MM, Carr CM (2018) A critical review of recent advancements in removing reactive dyes from dyehouse effluent by ion-exchange adsorbents. Chemosphere 209:201–219
Hermundsdottir F, Aspelund A (2021) Sustainability innovations and firm competitiveness: a review. J Clean Prod 280(1):124715. ISSN 0959-6526, https://doi.org/10.1016/j.jclepro.2020.124715
H&M Group (2021) Circular business models—H&M Group (hmgroup.com). Last Accessed 23 June 2024
Hossain MT, Shahid MA, Limon MG, Hossain I, Mahmud N (2024) Techniques, applications, and challenges in textiles for a sustainable future. J Open Innov Technol Mark Complex 10(1):100230. ISSN 2199-8531, https://doi.org/10.1016/j.joitmc.2024.100230
Innovation for sustainable fashion industry: a design-focused approach toward the development of new business models. Green fashion, environmental footprints and eco-design of products and processes. https://doi.org/10.1007/978-981-10-0245-8_5
International Finance Corporation, World Bank (2024) Sustainable fashion: transforming South Asia's textile industry (ifc.org). Last Accessed 15 Mar 2024
International Labour Organization (ILO) (2021) The future of work in textiles, clothing, leather and footwear. https://www.ilo.org. Last Accessed 27 July 2024
Jayarathne PGSA, Sachithra V, Dewasiri NJ (2021) Sustainable practices through competitive capabilities in the apparel industry: Sri Lankan experience. In: FIIB business review, pp. 1–16. Sage Publication. https://doi.org/10.1177/23197145211060398. (SCOPUS indexed)
Jia F, Yin S, Chen L, Chen X (2020) The circular economy in the textile and apparel industry: a systematic literature review. J Clean Prod 259:120728. ISSN 0959-6526, https://doi.org/10.1016/j.jclepro.2020.120728
Koszewska M (2018) Circular economy—challenges for the textile and clothing industry. Autex Res J 18(4):337–347
Levi's (2021) How Levi's® is Saving Water-Levi Strauss & Co.: Levi Strauss & Co. Last Accessed June 20 2024
Luo Y, Song K, Ding X, Wu X (2021) Environmental sustainability of textiles and apparel: A review of evaluation methods. Environ Impact Assess Rev 86:106497. ISSN 0195-9255, https://doi.org/10.1016/j.eiar.2020.106497
MacCarthy BL, Jayarathne PGSA (2010) Fast fashion: achieving global quick response (GQR) in the internationally dispersed clothing industry. In: Cheng E, Choi J (eds) Springer handbook on innovative quick response programs in logistics and supply chain management, pp 37–60. Springer
MacCarthy BL, Jayarathne PGSA (2012) Sustainable collaborative supply networks in the international clothing industry: a comparative analysis of two retailers. Prod Plan Control Manag Oper 23(4)
MacCarthy BL, Jayarathne PGSA (2013) Supply network structures in the international clothing industry: differences cross different retailer types. Int J Oper Prod Manag 33(7)

MAS Holdings (2021) Sustainable development: eco go beyond by MAS Holdings. Last Accessed 20 June 2024

McKinsey (2024) What is next for Bangladesh's garment industry after a decade of growth? McKinsey. Last Accessed 28 Apr 2024

Mishra U, Rathnasiri MSH, Dewasiri NJ (2024) Net-zero economy and energy sustainability: a nepalese perspective. In: Singh R, Khan S, Kumar A, Luthra S, Chokshi H (eds) Net zero economy, corporate social responsibility and sustainable value creation, pp 145–161. CSR, sustainability, ethics & governance. Springer, Cham. https://doi.org/10.1007/978-3-031-55779-8_9

Muthu SS, Gardetti MA (eds) (2016) Innovation for a sustainable fashion industry: a design focused approach toward the development of new business models

Rathnasiri MSH, Dewasiri NJ, Kumar AA (2024) Policy framework and implementation strategies for Sri Lanka's transition to a net-zero economy. In: Singh R, Khan S, Kumar A, Luthra S, Chokshi H (eds) Net zero economy, corporate social responsibility and sustainable value creation, pp 43–60. CSR, sustainability, ethics & governance. Springer, Cham. https://doi.org/10.1007/978-3-031-55779-8_3

Rudrajeet P, Jayarathne PGSA (2022) Digitalisation in the textile and clothing industry. In: The book the digital supply chain. Paperback ISBN: 978032391614, eBook ISBN: 9780323916158. Elsevier

Shabir M, Hussain I, Işık Ö, Razzaq K, Mehroush I (2023) The role of innovation in environmental-related technologies and institutional quality to drive environmental sustainability. Front Environ Sci 11:1174827. https://doi.org/10.3389/fenvs.2023.1174827

Shaw D, Hogg G, Wilson E, Shiu E, Hassan L (2006) Fashion victim: the impact of fair trade concerns on clothing choice. J Strateg Mark 14(4):427–440

Statista (2021) Apparel market size worldwide from 2011 to 2025. https://www.statista.com/statistics/279757/apparel-market-size-projections-by-region. Last Accessed 27 July 2024

Susitha E, Jayarathna A, Herath HMRP (2024) Supply chain competitiveness through agility and digital technology: a bibliometric analysis. Supply Chain Anal 7:100073. ISSN 2949-8635, https://doi.org/10.1016/j.sca.2024.100073. (SCOPUS indexed)

Sustainable Apparel Coalition (SAC) (2021) About us. https://apparelcoalition.org. Last Accessed 27 July 2024

World Trade Organization (WTO) (2023) World trade statistical review 2021. https://www.wto.org. Last Accessed 28 July 2024

The Cobra Effect: Exploring the Dark Side of Innovation and Environmental Sustainability in Developing Nations

Oyenuga Michael Oyedele

Abstract Nations facing an ongoing pursuit of progression and sustainability often utilize innovation as a mechanism toward economic expansion while ensuring that the local environment is protected. Innovative practices, when implemented sometimes result in grave consequences; the term "Cobra Effect" is assigned to such an event. Like a historical anecdote where the attempt to rid Delhi of cobras resulted in an uptick in killing them, this illustrates how innovation can be detrimental if unaccompanied by adequate institutional capability or incentives. It is essential to explore the dark side of innovation and environmental sustainability in emerging economies, across a spectrum of literature by delving into how well-intended efforts to push things forward can and do backfire, making situations worse while introducing innovations. It offers insights through literature analysis and cases about challenges organizations face in deploying innovative solutions by focusing on resource-scarce settings, socioeconomic divides, and institutional obstacles. This chapter underscores the importance of a nuanced appreciation of the Cobra Effect in the developing world. It brought to the fore the fact that innovation can lead to social harm, health hazards, and environmental and ecological degradation, among others which can be detrimental to the people whose lives are supposed to be taken care of by those in power. When such complexities around innovation in developing countries are addressed, policymakers will be conscious of the illustrative indications that can support truly sustainable and conscientious ways of advancing life in developing economies.

Keywords Cobra effect · Dark side · Developing nations · Innovation · Sustainability

---

O. M. Oyedele (✉)
School of Business, Woxsen University, Hyderabad 502345, Telangana, India
e-mail: oyenuga.michael@woxsen.edu.in

R. Singh et al. (eds.), *Zero Carbon Industry, Eco-Innovation and Environmental Sustainability*, World Sustainability Series,
https://doi.org/10.1007/978-3-031-80220-1_15

1 Introduction

In the modern world, innovation is well-valued for its contribution, to expansion triggering technology and supporting evolution. It is playing an essential role in the solving of problems like utilization rate intensity, pollution control, and waste management (Schot and Steinmueller 2018). The concept of the economy (CE) is intended to help address some of these social challenges by transitioning from a linear 'take make disposal' process, into an alternative circular mindset 'make use return' (Geissdoerfer et al. 2017). In the Circular Economy Action Plan 2020 for the European Union (EU), targets have been implemented for which material and product will be reborn from end-of-life to reduce or preclude an increase in CO_2 emissions throughout its lifecycle events–emission-savings values retained as a league toward policy interventions: production-consumption techniques optimized through increased use of more efficient by-products with greater durability-to-use times.

However, it is important to point out that innovation may result in undesired outcomes (Bu et al. 2021; Manning et al. 2018; Tang et al. 2022). In the developing world, innovation is rarely a straightforward path, and fears of unscrupulous activities and clandestine processes could impede progress and hinder sustainable development efforts, as explored in studies by Damanpour et al. (2009) and Manning et al. (2018).

2 Cobra Effect

The origin of the concept is traced back to British rule in India (1858–1947) when the colonial government paid rewards for killing cobras—only causing a spike in the cobra population (Theerthaana and Arun 2021). This term represents the negative impact that meaningful actions can have especially as it relates to innovation and sustainability. Various fields benefit from this reflection by approximating a glimpse into the hardships of the case of The Cobra Effect. In the context of education, that is exactly what happened; the Cobra Effect becoming an important example for handling and solving problems with great caution in order not to make things worse. This emphasizes the importance of targeted intervention to limit adverse events. In the realm of disaster risk reduction, the Cobra Effect is exemplified by the potential backfiring of containment strategies, emphasizing the importance of testing protocols to mitigate unintended consequences effectively. Within medicine, we see the analogy of this in looking at measures for effectiveness and risk abatement when evaluating therapeutic strategies for things like rheumatoid arthritis. The Cobra strategies of disease management upend the smooth functioning of healthcare interventions: benefits must be balanced with harms.

Also, this concept works well for vaccine development by utilizing computational optimization to increase the immune response and efficacy of such protection against diseases. This approach explores the intricate interplay of innovation with unintended consequences and underscores how critical it is to assess and quantify risks (benefits, harms) when creating vaccines.

However, we should always take innovation with a grain of salt–the Cobra effect gives us an angle that illustrates it: reminding the importance of thinking about potential ramifications when trying to push some progress and change. It is a story about the silver lining of innovation in new financial products and services. Although financial innovations may foster risk sharing and efficiency gains, it is important to be cautious of side effects such as market distortions or increased exposure to risks (Beck et al. 2012). Indeed, the Cobra Effect in public sector innovation serves as an example reflecting that we must address the side effects of processes even more carefully—particularly concerning where outcomes shape public value and society (Meijer and Thaens 2020). Further, regarding practices the risk of demurral or unethical behavior in fostering innovation becomes more obvious after all cultures have a fail tale like the Cobra Effect.

In such a context, a more open, honest and ethical innovation process as well as acting to avoid material amounts for results is necessary in order that trust can be appropriately sustained (Kapoor and Henderson 2023). Appreciating the harmful side of innovation is essential to both risk mitigation and implementation success for all types of innovation (Thajil and Al-Abrrow 2023). An analysis of the Cobra Effect in complex ecosystems even hints that beneficial relations between inputs on business-to-business innovation can produce negative consequences. The dynamics within innovation ecosystems also play a fundamental role in the success of a technological change endeavour as they shape how interconnected relationships and perspectives on what is possible in terms of e.g. typesetter (and related) knowledge affect potential outcomes positively, but similarly negatively—hence expressing why any systemic understanding about pathways beyond isolated components need to be nurtured with due complexity (Noordhoff et al. 2011) The Cobra Effect serves a harsh reminder of the balance between progress and unintended consequences, calling upon stakeholders to be vigilant in their pursuit towards achieving excellence.

3 What is Innovation?

Throughout history innovation has been conceived, defined, interpreted and understood in different ways but what is it? Innovation is a multifaceted concept that plays a pivotal role in driving progress, fostering growth, and spurring transformation across various sectors and industries. Defined as the generation, acceptance, and implementation of new ideas, processes, products, or services, innovation serves as a catalyst for change and a driver of competitiveness in today's dynamic and evolving landscape (Baregheh et al. 2009). The concept of innovation transcends traditional boundaries, encompassing a spectrum of activities that range from incremental improvements to radical breakthroughs.

Numerous definitions of innovation are widely utilized in various academic, industrial, governmental, and service sectors. Academic literature that is accessible has a wide range of topics, which means it can be multi-disciplinary and draws on other fields for analysis (Fagerberg et al. 2005; Malerba and Brusoni 2007). Several definitions of innovation (Table 1) have been explored in the literature to fully understand what makes up an idea about this concept. These definitions ensure that innovation has many faces and is subject to wide interpretation, driven partly by the rather loose style of literature available on the topic (Fagerberg et al. 2005; Linton 2009). The literature provides us with various models, theories and frameworks that we can use to understand what innovation is (in addition to analyzing the components of innovation). Efforts have been made at various stages to come up with a common definition of innovation (OECD/Eurostat 2005), and this could be an advantage by producing clarity and providing focus. Nevertheless, this view can be criticized (Wolfe 1994). Innovation in this study is a concept that has been derived from an extensive review of literature and consideration of the multi-component aspects that make up innovation. This aims to encourage academic research and make a significant addition to knowledge since innovation is recognized as a fundamental element in achieving sustainable development goals, with its role in driving progress towards environmental sustainability being particularly emphasized (Yiğit 2021). Businesses involved in innovation are now emphasizing the effects of their advancements indicating a transition, towards making decisions with a greater focus, on environmental awareness (Shkola et al. 2021).

4 Types of Innovation

According to Taylor (2017), different types of innovation play a role, in helping organizations evolve and adapt to the evolving market needs. The categories include total innovation, which brings about changes for the organization and caters to a new user group, expansionary innovation, which involves offering an existing service to a new user group, evolutionary innovation, which provides a new service, to the current user group of an organization and developmental innovation where the organizations services are adjusted or enhanced for its existing user group (Walker et al. 2002).

5 The Importance of Innovation and Environmental Sustainability in Developing Countries

Innovation is widely acknowledged as a driver of sustainability, across economic and environmental aspects highlighting its crucial role in promoting sustainable development (Astor et al. 2021; Zemlickienė and Turskis 2020). In developed nations the convergence of innovation and environmental sustainability plays a vital role in nurturing long-term growth and prosperity (Vitunskienė and Dapkuvienė 2022) by enhancing both physical and human capital (Nazeer et al. 2022) especially given the

Table 1 Various definitions of innovation

S/N	Definition	Source
1	Creation of new combinations of existing resources	Godin (2008)
2	Implementation of a new or significantly improved product (good/service) or process (method/practice/relationship)	OECD/Eurostat (2005)
3	Innovation is the creation and implementation of new processes, products, services and methods of delivery which result in significant improvements in outcomes, efficiency, effectiveness or quality	Mulgary and Albury (2003)
4	The successful exploitation of new ideas or ones that are adopted from other sectors or organizations	National Audit Office (NAO) (2009)
5	Creation and application of good ideas	OECD/Eurostat (2005)
6	A continuous and dynamic process in which ideas are transformed into value	Confederation of British Industry (CBI)/QUINETIQ (2008)
7	The successful introduction of new services, products, processes, business models and ways of working	Economic and Social Research Council (ESRC) (2008)
8	The development (generation) and/or use (adaption) of new ideas or behaviors	Damanpour and Schneider (2009)
9	The introduction of new elements into a service—new knowledge, new organization, new management/skills	De Vires et al. (2014)
10	Innovations are in a significant way new and disruptive towards the routines and structures prevailing	Evers et al. (2014)
11	Innovation is the process by whichnew ideas turn into practical value in the world	NESTA (2012)
12	Innovation has been thought of as falling on a continuum from evolutionary to revolutionary	Oyenuga et al. (2019)

interconnected nature of economic, social and environmental issues in these areas (Youssef et al. 2018). International bank funding is recognized as a catalyst for sustainable development in these countries (Муршудли 2023) while green innovation and finance are underscored as essential elements of sustainable development that emphasize the importance of eco-friendly practices in economic endeavours (Tolliver et al. 2020).

The goals of development stress the significance of integrating growth, social inclusivity and environmental sustainability as guiding principles for both advanced and developing nations (Garcia Mogollón et al. 2021). However, achieving sustainability, in developing countries relies heavily on innovations and effective institutions (Gao and Fan 2023). Sustainability indices provide a tool to evaluate a country's advancement, in environmental and social aspects highlighting the importance of taking a comprehensive approach to development. As a method of progress implementing environmental conservation policies embracing advancements and adopting industrial practices are crucial for effectively reducing carbon emissions and fostering economic growth in regions such as East Africa. Meanwhile, economies such as China has given emphasis on cutting of carbon emission for making sustainable economic and social progress that highlighted how environmental sustainability had been put at the core of national development agendas.

The 2030 Agenda for Development insists on the use of natural resources and the introduction of green technologies to achieve economic, social and environmental development. Certain developing nations have shown improvement in enhancing quality in areas, like water and sanitation sustainability following the implementation of Sustainable Development Goals (SDGs). Furthermore, it is essential, for medium enterprises (SMEs) to integrate sustainability into their business strategies to adopt more innovative and eco-friendly approaches. This shift can give companies an edge in emerging and developing countries where sustainability practices are typically less advanced. In Latin America, companies emphasize innovation as a driver of Environmental, Social and Governance (ESG) initiatives leading to positive outcomes for both businesses and the communities they serve. To make meaningful contributions to environmental and social sustainability, organizations must embrace innovation focused on sustainability goals. This approach is gaining momentum as it has the potential to drive change and address sustainability issues effectively. Emphasizing the development of solutions that align with sustainability objectives is vital for companies aiming to have an impact on sustainable development efforts. Additionally, organizations can utilize advancements, in environmental technology innovation to achieve sustainability goals and advance development initiatives.

Developing technology and fostering innovation plays a vital role in advancing sustainable development especially in regions such, as India facing significant environmental challenges (Pandey and Pandey 2023). The concept of eco-innovation is widely acknowledged as a driver for supporting sustainability initiatives sparking transformations and shaping policies aimed at promoting sustainability (Urbaniec and Tomala 2021). Also, by introducing energy innovations and improving performance, countries like Nigeria can bolster their sustainability outcomes (Noja et al. 2022; Nazeer et al. 2022).

The impact of innovation on an environment is multifaceted, underscoring the importance of advancements for sustainable development. These impacts vary across country groupings such as the BRICS and G7 nations (Santana et al. 2015; Hossin 2023). Notably, green patents are identified as drivers of economic growth. They serve as tools, for policymakers and innovators in advancing conscious technologies and practices (Rakić et al. 2022).

In Asia, the implementation of sustainable energy innovations in countries like China and Bangladesh is crucial for advancing sustainability goals and addressing energy challenges in a rapidly evolving global context (Kopishynska 2024; Jang et al. 2015; Goncharenko and Shapoval 2021). Policy instruments for eco-innovation in Asian countries are highlighted as critical tools for implementing sustainable development practices and fostering environmentally conscious innovation (Jang et al. 2015).

6 The Dark Side of Innovation

Innovation can provide both beneficial and detrimental consequences, and both positive and negative outcomes are often unequally distributed. Recently, policymakers have overlooked the fact that innovation has a certain orientation and have assumed that "innovation is good" and therefore, "more innovation is better" (Coad et al. 2021). When countries are compared based on metrics like R&D investment or patenting, it is generally considered that higher values indicate greater performance. Gaining a comprehensive understanding of the negative aspects of innovation and how these drawbacks are dispersed might provide valuable insights for developing more effective policies (Grigorescu et al. 2019). Innovation has largely been championed as a force for good: fostering economic growth, scientific discovery and social progress. But innovation has another side, as well—the dark one that requires focus; specifically in the context of developing countries. The following are the dark sides of innovation.

6.1 End-of-Product-Life Considerations

In developing countries, the end-of-life considerations of products represent a significant challenge associated with innovation. While progress is commonly associated with innovation dealing with the disposal and handling of items toward the end of their lifespan poses difficulties in areas with limited resources and infrastructure. Improper discarding of products and technologies such as electronic devices may result in environmental contamination, health risks and the depletion of resources. The challenges related to end-of-life considerations are exacerbated by the lack of efficient treatment procedures and the limited capacity of waste management facilities in developing countries creating a cycle of negative impacts that disproportionately affect vulnerable communities (El-Ramady et al. 2021). In developing nations, there is a pressing issue of dealing with growing amounts of waste without having the infrastructure and regulations in place to handle waste disposal effectively. The introduction of new solar panel technology has raised concerns regarding the disposal of heavy metals, namely lead, which is a very toxic substance affecting the nervous system. Several Sub Saharan African countries have limited capacities for waste disposal, and there are doubts over the long-term environmental efficacy of solar panels. Impoverished individuals, who are devoid of necessary equipment, protective measures, and political advocacy, frequently undertake the recycling of hazardous items. While innovations targeting specific types of pollution can lead to significant overall improvements, it is important to acknowledge that they may also give rise to new problems affecting various individuals. It is crucial not to overlook these costs, despite the overall positive impact.

6.2 *Environmental and Energy Rebounds*

Energy rebound describes scenarios where anticipated environmental advantages stemming from the deployment of low-carbon innovation are counterbalanced by heightened consumption rates, thereby risking an escalation in total emissions (Herring and Roy 2007). Energy rebounds refer to the surge in energy usage often triggered by efficiency improvements that can reduce energy prices, subsequently driving up energy or fuel consumption. Understanding the nuances between these two concepts is crucial for grasping the complex, unintended consequences of low-carbon innovations: While both may stem from heightened efficiency, an environmental rebound relates to situations where the decrease in environmental impact per unit, such as lower emissions per activity, leads to amplified consumption levels overall, which might eclipse the initial efficiency benefits. Conversely, energy rebound is a narrower term focusing on increased energy consumption as a direct consequence of efficiency improvements leading to reduced energy costs. Examples of environmental rebounds include Akimoto et al. (2022), who points out that the realization of the emission-reduction potential of ride- and car-sharing with fully autonomous vehicles is directly linked to a reduction in the number of cars on the road. However, the authors also point out that the large-scale adoption of this low-carbon innovation might generate increased demand for this form of travel; hence, estimated emission-reduction effects may be in jeopardy. Similarly, Richter et al. (2022) also saw increased road traffic and related emissions through the uptake of autonomous vehicles with low occupancy rates. Marletto (2019) warns that an unregulated diffusion of autonomous vehicles could increase transport demand, resulting in uncertain emission reductions. Also, the broad deployment of artificial intelligence in different industrial sectors to solve environmental sustainability problems must be viewed critically due to the technology's current high ecological footprint (Kopka and Grashof 2022).

In the context of the global South, the diffusion and adoption of energy-efficient, low-carbon innovation may also lead to energy rebounds. An example of this is Hills et al. (2018), who discuss how the implementation of the renewable energy system based on solar PV in an indigenous and vulnerable island community in Fiji resulted in an energy rush where each household tried to maximize its energy consumption due to the low pricing of electricity. They argue that placing such low-carbon innovation in a traditional village without properly discussing how energy use should be organized or educating villagers about energy-consumption behaviours heightens the risk of a negative energy rebound. In this sub-category of reviewed articles, the energy rebound, thus, is not merely an increase in energy use but also includes socioeconomic factors that may lead to broader sustainability issues.

6.3 Moral Licensing Effect

The moral licensing effect is an important aspect of innovation in developing countries. Moral licensing is a cognitive operation which focuses on the license because when people perform positive behaviours they get licensed to do subsequent negative and unethical deeds (Blanken et al. 2015). This consequence leads to more meaningful insights for the introduction and enactment of innovation in emerging economies which beckons attention towards ethical decision making with due responsibility. Moral licensing can also lead to the creation of irresponsibility and related misconduct, within corporate social responsibility (CSR), or organizational behaviour toward withholding CSR in favour of maximizing profits (Bouzzine and Lueg 2022). Where innovation is concerned, this could mean that the positive thing done might create a moral 'bank account' with a sense of entitlement to future moves. It can skew ethical implications for new actions forward or even in the present time to accommodate previous ones as well. Even more concerning is the absence of regulatory frameworks in developing countries which may mean that moral licensing can be rampant and unchecked. Prior research has explored moderating factors on moral licensing effects, including social sharing, temporal distance and unrelated personal accomplishments or good deeds as well as individual differences in general (Wen and Hu 2023). Moral licensing may give rise to unethical practices or create shortcuts in the innovation processes, especially for resources-starved success-striving developing countries. If progress demands innovation, then there can be no successful pursuit of the future without a bold ethic that abstains from foreclosure fluctuation problems. In addition, moral licensing can also interact with misconduct in many other areas such as scientific research and work behaviour (Zhang et al. 2022). Engagement by stakeholders responsible for innovation in these territories should take into account the presence of a moral licensing effect and promote strategies to help avoid its negative effects on new ethical standards as well as responsibility or sustainability units. In the quest for accelerating innovation, fueled by a desire to promote economic growth and societal advancement in developing countries (Xiao and Su 2022), it is equally important—if not more so now than before—that we beguile these dark sides of innovation towards moral licensing. The creations of a culture that raises ethical awareness, embraces accountability and responsible innovation practices can be achieved by developing countries in negotiating the minefield on innovations while adhering to moral integrity, security and societal wellness.

6.4 Marginalization and Infringement of Procedural Justice and Recognition Justice

Developing countries often find that the dark side of innovation shows in its most destructive way: marginalizing procedural justice and recognition justice. Procedural justice is about the fairness and transparency of steps that are taken in terms of decision-making, while recognition justice focuses on recognizing as well as respecting individuals for what they have done and who they are within an organization. But that poses the risk of negative externalities, particularly given developing countries' innovation infrastructures. Procedural injustice in delivering fairness may leave some agents aside from taking part in decision-making procedures or have an equal chance to contribute at any process that enables innovation. The worst situation regarding procedural justice is the enhanced effects of long-term inequalities that might have led to larger power dynamics and hierarchies in developing countries which can hamper full participation including the probability of a better decision-making process. This exclusion has the potential to curtail a broad range of diverse perspectives and fresh ideas that would otherwise propel sustainable development and growth. In addition, the violation of recognition justice can decrease others' value in their contributions to work and reduce willingness to carry out innovative behaviour ultimately. Yet in some poorer countries where recognition within training is influenced by social norms and the culture of practice, not acknowledging contributions can act as an obstacle to building innovation resilience (Noerchoidah and Harjanti 2021). As a solution, there is a need for one more step to be taken towards eliminating the obstacles as mentioned above and fostering innovative solutions addressing local challenges effectively—such recognition deprivation results in demotivated individuals who eventually end up being disengaged over time. It is the interaction between procedural justice and recognition justice that can encourage favourable conditions for innovation in developing countries. This may encourage people to be more innovative when they feel their voices are being heard and that their contributions add value—or conversely to become disengaged in decision-making processes if the perception of fair treatment is not their. On the contrary, failure in these aspects can debilitate effort and bring about distrust for participation in innovation initiatives thereby hampering progression towards entrenching inside-out development (Wyse et al. 2021).

6.5 Health Hazards

Innovation, however promising it may be in terms of opportunities for change, inherently has dark sides that can have devastating public health consequences particularly when introduced at scale to low and middle-income countries. Technological development and innovation are moving fast which means sometimes faster than public health systems can manage the associated risks. Which can have unintended consequences that in turn affect the health and welfare of people living there. Evolving threats and changes in public health, even under usual circumstances like novel risk factors to populations of developing countries (including emerging issues) are among the major worries about inventions. This, for example, happens when new technologies are adopted without proper regulation or oversight (such as environmental pollution and occupational hazards). This is exacerbated in many developing countries where a paucity of public health infrastructure and legal frameworks to regulate innovation further complicates the task. Corruption can also contribute to the public health risks associated with the implementation of health innovations if exercised in implementing them unethically. It can also erode confidence in health systems when resources are used poorly, fraud is pervasive and procurement and distribution of health technologies and services lack transparency. The end result is poor care, compromised patient safety and ultimately harm done to individuals and communities. Additionally, growth and new technologies found in developing countries will sometimes works against the priorities of public health. For example, innovation-fueled industrial growth and urbanization can result in environmental degradation, poor air quality and sanitation imperatives for public health. When economic interests are given precedence over health considerations (a view that overlooks the fact that being able to afford healthcare is a key). Changes in healthcare practice and technology offer valuable improvements but are also always accompanied (and perhaps complicated) by ethical or equity challenges. This could further widen health disparities within and across countries, as unequal access to emerging healthcare solutions such as advanced medical treatments or digital health technologies is compounded. Such inequitable access to health innovations may only serve to exacerbate the marginalization of vulnerable populations and undermine progress towards universal health coverage and sustainable development goals.

6.6 Environmental Degradation

Technology in developing countries is often the issue that lies with the environment. Studies focusing around sustainability have addressed the symbiotic relationship between technology, economic growth and environmental issues (Gao and Fan 2023). The challenges for developing countries lie in how best to properly balance development and environmental protection. Things like increasing industrial energy consumption, carbon emissions from burning fossil fuel and continuous exploitation are all creating environmental damage (Gao and Fan 2023). Efforts towards industrialization and modernization can result in habitat loss, pollution, deforestation and loss of biodiversity worsening problems (Nilsson et al. 2018). Despite addressing needs, technological innovations may inadvertently harm the environment (Stukalo et al. 2020). Reliance on fuels without regulations can increase emissions and air pollution impacting ecosystems and climate patterns (Gao and Fan 2023). Expansion of industries and cities due to innovation can lead to land degradation, water contamination and increased waste production that threaten sustainability (Nilsson et al. 2018). Additionally, rapid technological progress along with automation adoption, in industries may further strain the environment. Rising energy usage, from automated operations, coupled with waste disposal practices may result in soil pollution, water shortages and air quality degradation causing harm, to the environment (Ibekwe 2024; Oladele et al. 2023). Uncontrolled utilization of resources driven by advancements can exhaust reserves, disturb the ecological equilibrium and undermine ecosystems, especially in less developed nations (Suki et al. 2022).

6.7 Societal Harm

Although innovation brings advancements and improvement, it may drag its negative impacts resulting in the disadvantages superseding or outweighing the advantages causing harm to society (Oyenuga et al. 2023). One of the primary means innovation becomes harmful for society in developing countries is if it instead exacerbates social inequalities. Creativity, or any other area that looks like technology and the financial industry are examples of sectors where it is believed that innovation can be very harmful for the rich/poor divide. As new technologies are used, people who have access to them disproportionately benefit with marginalised communities falling even further behind (Aghion and Jaravel 2015). In turn, this can widen the gap between those who are already most privileged in society and restrict social mobility more generally. In addition, the break-neck speed that innovation means some jobs can vanish pushing people towards unemployment, especially in areas and sectors where they are replaced by automation or technological advances. In developing countries, with very limited job opportunities to begin with, the introduction of disruptive technologies that eliminate the demand for manual labour can lead to significant levels of unemployment and subsequent social unrest (Leitner et al. 2016). This also adds pressure on social welfare systems and widens the gap with poverty and inequalities within society. It is also important to remember that innovation can lead to negative impacts on the environment, especially in developing countries where regulations are minerals compared with developed nations (e.g., the US), so firms may not be as ethical or sustainable socio- and environmentally conscious as their counterparts from more environmentally proactive areas of the world. Research, innovation and development unhindered by the environment are responsible for environmental liability such as air pollution, resource shortages or loss of biodiversity affecting both the affected area's cradle health as well as its component communities depending on them (Eichler and Schwarz 2019). These effects can have a lasting major impact on society including long-term health consequences, displacing communities and loss of biodiversity. In addition, profit-based innovation in developing countries may neglect social utility and ethical concerns. While chasing growth and competitiveness, companies might engage in labour exploitation, violating human rights or ignoring their social responsibilities (Saka-Helmhout et al. 2021). This can result in sustainable social unrest, business ethics dilemmas and erosion of trust between businesses with society to the detriment of community values.

6.8 Economic Hazard

One concern is that innovation may widen the gap between groups (Souza 2024). In countries where income inequality is high, the benefits of innovation may not reach everyone equally. The introduction of technologies and methods can displace industries and jobs, especially in sectors that struggle to adapt quickly. This could lead to job cuts, income opportunities and increased disparities within society (Sweet and Maggio 2015). Additionally focusing much on innovation might take resources away from sectors like infrastructure, healthcare and education in developing countries where these areas are already lacking. This shift could slow down progress and worsen social issues hindering sustainable growth (Maradana et al. 2017). Another drawback of embracing innovation in developing nations is the reliance on technologies and expertise. While embracing ideas can lead to transformations relying too heavily on foreign innovations or technology transfers might hinder the growth of local industries' independent capabilities. This dependency could weaken sovereignty making domestic industries reliant, on sources for innovation (Guo 2024) and hindering their progress (Donate et al. 2015). Additionally, the rapid pace of innovation sometimes surpasses the frameworks and institutional capabilities of developing nations. Insufficient protection of intellectual property rights lax enforcement measures and limited oversight can create openings for individuals to exploit. This can result in issues like theft of property, unfair competition and distortions, in the market impeding advancement and diminishing the potential benefits of innovation (Tabrizian 2019).

6.9 Problem-Shifting Effects

A key limiting factor for innovation is; the concept of "problem-shifting effects" This concept describes what happens accidentally as an unexpected harm when we do something innovative in practices, technologies or policies. Problem-shifting effects in developing countries can take the form of disasters that weaken rather than strengthen society, environmental and economic straits. A significant problem is that developing countries might be tempted to settle for the short-term (most convenient/innovative) solution in contrast with what developed nations believe to be long-term actions and ethics. It can result in the problems simply being displaced or shifted, rather than fully resolved. An example is in the pursuit of economic growth by guiding industrialization paths, and underestimating environmental problems that later result with pollution, natural resource depletion etc. Additionally, once we consider the social domain then problem-shifting effects can also occur. For example, the introduction of new technologies or innovative practices could cause traditional ways to be disrupted; this would in turn potentially create social disruptions as well. Innovation can increase social divides, in some circumstances or create new exclusions (with costs that are higher than average) especially if whole groups of the population cannot benefit and change with it. The issues of problem-shifting can also be magnified in the context of developing countries by poor indicatory frameworks, insufficient resources for monitoring and enforcement or lack of awareness or access to some information that would have permitted actors—including governments—to anticipate possible negative consequences. This can lead to innovation benefits being unequally shared in a worst-case scenario, leaving marginalized communities with most of the negative outcomes.

7 Conclusion

This study on the dark side of innovation suggests that innovation per se includes a host of hazards and negative externalities capable of overshadowing beneficial aspects. Specific dark sides covered in this chapter—end-of-product life challenges, environmental and energy rebound effects, moral licensing impacts, negative marginalization consequences, health intoxication problems causing societal harm, economic risk problems, all underscore that innovation for sustainable development is multifaceted and complex. Paying attention to end-of-product life is about the implications of innovation in terms of environmental waste more than anything else: it results into higher amount for resource depletion and pollution. This shows how, in some cases, environmental- and energy-centric rebounds may create incentives or increases that ultimately divert innovations intended to improve resource use (or efficiency) toward greater resource consumption or increased energy delivery. The moral licensing effect challenges the ethic of innovation, asking whether do-good actions in one area may excuse bad outcomes from another. Marginalization and breaches in procedural justice, as well as recognition of social injustice by less formalized mechanisms through which innovation is achieved, were shown to have important implications for the societal acceptance of innovations to equitable benefits distribution from innovation processes and also that even when such systems exist they might be overused or misused. These examples of health risks, environmental damages and social harm illustrate one type of negative outcome from innovation that focuses on the impacts on human wellbeing, ecosystems or society—namely those adverse consequences on public health implicating economic systems. Concerns regarding economic hazards and problem-shifting effects vary depending on the type of innovation being considered but together demonstrate some risks related to innovations (economic instability or market distortion) which can lead to unintended consequences as an anticipation from the well-intentioned initiative. The dark sides of innovation are just part and parcel in the perennial struggle for sustainable development, which might as well serve as a timely reminder: this requires developing nations to take an all-encompassing approach that appreciates its potential repercussions along with putting measures into place. Hence, greater care is needed by policymakers, businesses and others to make sure that the innovation does not lead to potentially worse systems than those that already exist but are going in a sustainable direction. This might include the integration of social responsibility, environmental stewardship, and principles of good governance into their innovation processes and further ensuring meaningful engagement with communities who are most affected or have a stake in what they are developing.

References

Aghion P, Jaravel X (2015) Knowledge spillovers, innovation and growth. Econ J 125(583):533–573. https://doi.org/10.1111/ecoj.12199

Astor E, Rius-Sorolla G, Seoane F (2021) Is innovation a driver of sustainability? An analysis from a Spanish region. Sustainability 13(16):9286. https://doi.org/10.3390/su13169286
Baregheh A, Rowley J, Sambrook S (2009) Towards a multidisciplinary definition of innovation. Manag Decis 47(8):1323–1339. https://doi.org/10.1108/00251740910984578
Beck T, Chen T, Lin C, Song F (2012) Financial innovation: the bright and the dark sides. SSRN Electron J. https://doi.org/10.2139/ssrn.1911677
Blanken I, Ven N, Zeelenberg M (2015) A meta-analytic review of moral licensing. Pers Soc Psychol Bull 41(4):540–558. https://doi.org/10.1177/0146167215572134
Bouzzine Y, Lueg R (2022) Csr, moral licensing and organizational misconduct: a conceptual review. Organ Manag J 20(2):63–74. https://doi.org/10.1108/omj-09-2021-1355
Bu J, Luo Y, Zhang H (2021) The dark side of informal institutions: how crime, corruption, and informality influence foreign firms' commitment. Glob Strateg J 12(2):209–244. https://doi.org/10.1002/gsj.1417
Coad A, Nightingale P, Stilgoe J, Vezzani A (2020) Editorial: the dark side of innovation. Ind Innov 28(1):102–112. https://doi.org/10.1080/13662716.2020.1818555
Confederation of British Industry (CBI)/QUINETIQ (2008) Excellence in service innovation. CBI
Damanpour F, Schneider M (2009) Characteristics of innovation and innovation adoption in public organisations: assessing the role of managers. J Public Resour Theory 19:495–522. https://doi.org/10.1093/jopart/mun021
Damanpour F, Walker R, Avellaneda C (2009) Combinative effects of innovation types and organizational performance: a longitudinal study of service organizations. J Manage Stud 46(4):650–675. https://doi.org/10.1111/j.1467-6486.2008.00814.x
De Vires HA, Bekkers VJ, Tummers LG (2014) Innovations in the public sector: a systematic review. In: Ottawa IRSPM conference, pp 9–11
Donate M, García-Pardo I, Pablo J (2015) HRM practices for human and social capital development: effects on innovation capabilities. Int J Hum Resour Manag 27(9):928–953. https://doi.org/10.1080/09585192.2015.1047393
Economic and Social Research Council (ESRC) (2008) Innovation research initiative distributed projects cost specification. ESRC, Swindon
Eichler G, Schwarz E (2019) What sustainable development goals do social innovations address? A systematic review and content analysis of social innovation literature. Sustainability 11(2):522. https://doi.org/10.3390/su11020522
El-Ramady H, Brevik EC, Elbasiouny H, Elbehiry F, Amer M, Elsakhawy T, Omara AE, Mosa AA, El-Ghamry AM, Abdalla N, Rezes S, Eid, Y (2021) Planning for disposal of COVID-19 pandemic wastes in developing countries: a review of current challenges. Environ Monit Assess 193(9). https://doi.org/10.1007/s10661-021-09350-1
Evers A, Ewert B, Brandsen T (eds) (2014) Social innovations for social cohesion: 77 cases from 20 European cities 2010–14. European Commission, Nijmegen
Fagerberg J, Mowery DC, Nelson RR (eds) (2005) The Oxford handbook of innovation. Oxford University Press, Oxford
Gao X, Fan M (2023) The role of quality institutions and technological innovations in environmental sustainability: panel data analysis of BRI countries. PLoS ONE 18(6):e0287543. https://doi.org/10.1371/journal.pone.0287543
Garcia-Mogollón A, Malagón-Sáenz E, Garcia-Pacheco N, García-Mogollón J (2021) Effects of climate change and its impact on some variables that influence the quality of life of populations in Latin America. Aibi Revista De Investigación Administración E Ingeniería 8(S1):340–358. https://doi.org/10.15649/2346030x.2483
Geissdoerfer M, Savaget P, Bocken NM, Hultink EJ (2017) The circular economy–a new sustainability paradigm? J Clean Prod 143:757–768
Godin B (2008) Innovation: the history of a category. montreal: project of the intellectual history of innovation. Working paper no. 1, p 26

Goncharenko N, Shapoval V (2021) Eco-innovation financing as an element of a "green" economy formation in the globalization conditions of sustainable development. Green Blue Digit Econ J 2(2):15–23. https://doi.org/10.30525/2661-5169/2021-2-3

Grigorescu A, Maer-Matei M, Mocanu C, Zamfir A (2019) Key drivers and skills needed for innovative companies focused on sustainability. Sustainability 12(1):102. https://doi.org/10.3390/su12010102

Guo L (2024) How does digital transformation improve corporate carbon emission reduction performance? An empirical study on data from listed companies in China. Sustainability 16(8):3499. https://doi.org/10.3390/su16083499

Herring H, Roy R (2007) Technological innovation, energy efficient design and the rebound effect. Technovation 27(4):194–203. https://doi.org/10.1016/j.technova-tion.2006.11.004

Hills J, Michalena E, Chalvatzis K (2018) Innovative technology in the pacific: building resilience for vulnerable communities. Technol Forecast Soc Chang 129:16–26. https://doi.org/10.1016/j.techfore.2018

Hossin M (2023) Examining sustainable development goals: are developing countries advancing in sustainable energy and environmental sustainability? Environ Sci Pollut Res 31(3):3545–3559. https://doi.org/10.1007/s11356-023-31331-9

Ibekwe K (2024) Energy efficiency in industrial sectors: a review of technologies and policy measures. Eng Sci & Technol J 5(1):169–184. https://doi.org/10.51594/estj.v5i1.742

Jang E, Park M, Roh T, Han K (2015) Policy instruments for eco-innovation in Asian countries. Sustainability 7(9):12586–12614. https://doi.org/10.3390/su70912586

Kapoor H, Henderson S (2023) Innovative deception across cultures, pp 143–173. https://doi.org/10.1007/978-3-031-28206-5_7

Kopishynska K (2024) Prerequisites and key measures for the implementation of sustainable energy innovations in china. Probl Syst Approach Econ 1(94). https://doi.org/10.32782/2520-2200/2024-1-2

Kopka A, Grashof N (2022) Artificial intelligence: catalyst or barrier on the path to sustainability? Technol Forecast & Soc Change 175:121318. https://doi.org/10.1016/j.techfore.2021.121318

Leitner K, Warnke P, Rhomberg W (2016) New forms of innovation: critical issues for future pathways. Foresight 18(3):224–237. https://doi.org/10.1108/fs-07-2014-0050

Linton JD (2009) De-babelizing the language of innovation. Technovation 29:729–737. https://doi.org/10.1016/j.technovation.2009.04.006

Manning P, Stokes P, Visser M, Rowland C, Tarba S (2018) Dark open innovation in a criminal organizational context: the case of madoff's Ponzi fraud. Manag Decis 56(6):1445–1462. https://doi.org/10.1108/md-05-2017-0535

Malerba, F. & Brusoni, S., Eds. (2007) Perspectives on Innovation. Cambridge University Press, Cambridge. https://doi.org/10.1017/CBO9780511618390

Maradana R, Pradhan R, Dash S, Gaurav K, Jayakumar M, Chatterjee D (2017) Does innovation promote economic growth? Evidence from European countries. J Innov Entrep 6(1). https://doi.org/10.1186/s13731-016-0061-9

Marletto G (2019) Who will drive the transition to self-driving? A socio-technical analysis of the future impact of automated vehicles. Technol Forecast Soc Change 139:221–234. https://doi.org/10.1016/j.tech-fore.2018.10.023

Meijer A, Thaens M (2020) The dark side of public innovation. Public Perform Manag Rev 44(1):136–154. https://doi.org/10.1080/15309576.2020.1782954

Муршудли Ф (2023) Green banking for sustainable development. Foresight STI Gov 17(2):82–94. https://doi.org/10.17323/2500-2597.2023.2.82.94

Mulgary G, Albury D (2003) Innovation in the public sector. Strategy Unit, Cabinet Office, London

National Audit Office (NAO) (2009) Innovation across Central Government. The stationery Office, London

Nazeer M, Tabassum U, Alam S (2022) Environmental pollution and sustainable development in developing countries. Pak Dev Rev 589–604. https://doi.org/10.30541/v55i4i-iipp.589-604

NESTA (2012) Plan 1: the case for innovation led growth. NESTA, London, p 17

Nilsson M, Chisholm E, Griggs D, Howden-Chapman P, McCollum D, Messerli P, Neumann B, Stevance AS, Visbeck M, Stafford-Smith M (2018) Mapping interactions between the sustainable development goals: lessons learned and ways forward. Sustain Sci 13(6):1489–1503. https://doi.org/10.1007/s11625-018-0604-z
Noerchoidah N, Harjanti D (2019) Exploring the relationship between procedural justice and innovative work behavior in hospitality industry. Jurnal Manajemen Dan Kewirausahaan 21(1):21–31. https://doi.org/10.9744/jmk.21.1.21-31
Noja G, Cristea M, Panait M, Trif S, Ponea C (2022) The impact of energy innovations and environmental performance on the sustainable development of the EU countries in a globalized digital economy. Front Environ Sci 10. https://doi.org/10.3389/fenvs.2022.934404
Noordhoff C, Kyriakopoulos K, Moorman C, Pauwels P, Dellaert B (2011) The bright side and dark side of embedded ties in business-to-business innovation. J Mark 75(5):34–52. https://doi.org/10.1509/jmkg.75.5.34
OECD/Eurostat (2005) The Oslo manual. OECD, Paris
Oladele TO, Oyenuga MO, Adoga GJ (2023) Green management practices and organisational efficiency in higher institution. Bp Int Res CritS Linguist Educ (BirLE) J 6(3), 276–290 https://doi.org/10.33258/birle.v6i3.7716
Oyenuga MO, Marcus GO, Ahungwa AI (2023) Do consumers care about green marketing practices? Insight from a developing nation. Bp Int Res CritS Inst J (BIRCI-J) 6(3):1424–1436. https://doi.org/10.33258/birci.v6i3.7665
Oyenuga MO, Andah RA, Marcus GO, Agabi AU (2019) Effects of customer relationship management on product innovation in Nigeria. A case study of Xerox HS Nigeria Limited. Am J Theor Appl Bus 5(4):113–126. https://doi.org/10.11648/j.ajtab.20190504.15
Pandey M, Pandey M (2023) Green technology & innovation: its implications on the sustainable development in Indian context. Vestn Samara Univ Econ Manag 14(1):87–98. https://doi.org/10.18287/2542-0461-2023-14-1-87-98
Rakić M, Krstić B, Berezjev L (2022) Green patents as a determinant of sustainable economic growth. Econ Sustain Dev 6(2):1–15. https://doi.org/10.5937/esd2202001j
Richter M, Hagenmaier M, Bandte O, Parida V, Wincent J (2022) Smart cities, urban mobility and autonomous vehicles: how different cities needs different sustainable investment strategies. Technol Forecast Soc Chang 184:121857. https://doi.org/10.1016/j.techfore.2022.121857
Saka-Helmhout A, Chappin M, Rodrigues S (2021) Corporate social innovation in developing countries. J Bus Ethics 181(3):589–605. https://doi.org/10.1007/s10551-021-04933-x
Santana N, Rebelatto D, Périco A, Moralles H, Filho W (2015) Technological innovation for sustainable development: an analysis of different types of impacts for countries in the BRICS and G7 groups. Int J Sustain Dev & World Ecol 1–12. https://doi.org/10.1080/13504509.2015.1069766
Schot J, Steinmueller WE (2018) Three frames for innovation policy: R&D, systems of innovation and transformative change. Res Policy 47(9):1554–1567
Shkola V, Prokopenko O, Stoyka A, Nersesov V, Sapiński A (2021) Green project assessment within the advanced innovative development concept. Stud Appl Econ 39(5). https://doi.org/10.25115/eea.v39i5.5135
Souza P (2024) The impact of environmental innovation and national culture on ESG practices: a study of Latin American companies. Academia Revista Latinoamericana De Administración 37(2):246–264. https://doi.org/10.1108/arla-11-2023-0187
Stukalo N, Lytvyn M, Petrushenko Y, Omelchenko Y (2020) The achievement of the country's sustainable development in the conditions of global threats. E3s Web Conf 211:01029. https://doi.org/10.1051/e3sconf/202021101029
Suki N, Sharif A, Afshan S, Jermsittiparsert K (2022) The role of technology innovation and renewable energy in reducing environmental degradation in Malaysia: a step towards a sustainable environment. Renew Energy 182:245–253. https://doi.org/10.1016/j.renene.2021.10.007
Sweet C, Maggio D (2015) Do stronger intellectual property rights increase innovation? World Dev 66:665–677. https://doi.org/10.1016/j.worlddev.2014.08.025

Tabrizian S (2019) Technological innovation to achieve sustainable development—renewable energy technologies diffusion in developing countries. Sustain Dev 27(3):537–544. https://doi.org/10.1002/sd.191801.008

Tang T, Cuesta L, Tillaguango B, Alvarado R, Rehman A, Bravo-Benavides D, Zárate N (2022) Causal link between technological innovation and inequality moderated by public spending, manufacturing, agricultural employment, and export diversification. Sustainability 14(14):8474. https://doi.org/10.3390/su14148474

Taylor S (2017) What is innovation? A study of the definitions, academic models and applicability of innovation to an example of social housing in England. Open J Soc Sci 05(11):128–146. https://doi.org/10.4236/jss.2017.511010

Thajil K, Al-Abrrow H (2023) Effect of personality styles on positive and negative innovation: the mediating role of emotional intelligence and abusive supervision using the structured equation model. Int J Organ Anal 32(4):624–644. https://doi.org/10.1108/ijoa-12-2022-3530

Theerthaana P, Arun C (2021) Did the double lockdown strategy backfire? Cobra effect on containment strategy of COVID-19. Int J Disaster Risk Reduct 65:102523. https://doi.org/10.1016/j.ijdrr.2021.102523

Tolliver C, Fujii H, Keeley A, Managi S (2020) Green innovation and finance in Asia. Asian Econ Policy Rev 16(1):67–87. https://doi.org/10.1111/aepr.12320

Urbaniec M, Tomala J (2021) Eco-innovation development in selected European and Asian countries: a comparative analysis. https://doi.org/10.34659/2021/4/28

Vitunskienė V, Dapkuvienė A (2022) Research and innovation as drivers of sustainability and competitiveness of the bioeconomy: a comparative analysis in BIOEAST countries. Rural Dev 2019 2021(1):447–452. https://doi.org/10.15544/rd.2021.077

Walker R, Jeanas E, Rowlands R (2002) Measuring innovation—applying the literature-based innovation output indicator to public services. Public Adm 80:201–214. https://doi.org/10.1111/1467-9299.00300

Wen N, Hu T (2023) License is "suspended": the impact of social sharing on curbing moral licensing. J Consum Mark 40(6):712–720. https://doi.org/10.1108/jcm-06-2021-4711

Wolfe R (1994) Organizational innovation: review, critique and suggested research directions. J Manage Stud 31:405–431

Wyse S, Das R, Hoicka C, Zhao Y, McMaster M (2021) Investigating energy justice in demand-side low-carbon innovations in Ontario. Front Sustain Cities 3. https://doi.org/10.3389/frsc.2021.633122

Xiao D, Su J (2022) Role of technological innovation in achieving social and environmental sustainability: mediating roles of organizational innovation and digital entrepreneurship. Front Public Health 10. https://doi.org/10.3389/fpubh.2022.850172

Yiğit S (2021) An empirical perspective on the relationship between innovation performance and sustainable development. Ege Akademik Bakis (Ege Acad Rev) 47–57. https://doi.org/10.21121/eab.874020

Youssef A, Boubaker S, Omri A (2018) Entrepreneurship and sustainability: the need for innovative and institutional solutions. Technol Forecast Soc Change 129:232–241. https://doi.org/10.1016/j.techfore.2017.11.003

Zemlickienė V, Turskis Z (2020) Evaluation of the expediency of technology commercialization: a case of information technology and biotechnology. Technol Econ Dev Econ 26(1):271–289. https://doi.org/10.3846/tede.2020.11918

Zhang N, Guo M, Jin C, Xu Z (2022) Effect of medical researchers' creative performance on scientific misconduct: a moral psychology perspective. BMC Med Ethics 23(1). https://doi.org/10.1186/s12910-022-00876-8

Zero Carbon Industry Challenges and Opportunities in Enhancing Urban Water Sector for Environmental Sustainability and Innovation

Wasswa Shafik

Abstract As the entire globe is focusing on sustainable development goals ensuring zero carbon operations industries by 2050, the chapter explores the imperious aspects between zero carbon principles and the urban water sector, examining the challenges, opportunities, and prospects. It commences with the fundamental concepts of zero carbon industry (ZCI) and its triggering acquaintances with environmental sustainability, emphasising their significance in tackling water-related concerns in metropolitan areas. It then describes the sampled impediments that hinder the effective implementation of zero carbon in the urban water sector, containing infrastructural limitations, technical gaps, public perception barriers and regulatory challenges. Moreover, the chapter highlights several abundant potentials for embracing zero carbon industrialisation, such as resource recovery, water recycling and reuse, deployment of sustainable infrastructure, collaborative partnerships, and the achievement of industry advantages. Furthermore, it scrutinises disruptive, contemporary technologies that are poised to transform the urban water environment by accenting the mobilisation of zero Carbonite and the optimisation of water sources. This chapter envisions potential developments for zero-carbon initiatives in the urban water sector, accentuating the need for more research, policy development, and strategic interventions to support sustainable water management practices. This chapter stresses the substance of collaborative efforts, progress, and continuous advances in ZCI principles. To conclude, it draws significant critical insights obtained from the study and best practices from successful ZCI and urban water campaigns to address challenges, maximise opportunities, and pave the way for a more resilient and sustainable urban water future.

Keywords Zero carbon industry · Urban water sector · Environmental sustainability · Sustainable development · Innovation · Climate change

---

W. Shafik (✉)
School of Digital Science, Universiti Brunei Darussalam, Brunei Darussalam BE1410, Brunei Darussalam
e-mail: wasswashafik@ieee.org

Dig Connectivity Research Laboratory (DCRLab), Kampala 600040, Uganda

R. Singh et al. (eds.), *Zero Carbon Industry, Eco-Innovation and Environmental Sustainability*, World Sustainability Series,
https://doi.org/10.1007/978-3-031-80220-1_16

1 Introduction

Water is a fundamental need for living things and is cardinal when it comes to cities (or urban) locations, as it does an essential function in maintaining human life, sustaining financial activities, and maintaining environmental wellness (Rissman 2024b). Urban water application includes the diverse selection of processes and methods associated with managing water sources within cities, including water systems, circulation, reuse, and disposal. As city populations continue to grow rapidly worldwide, the need for water in cities is rising, placing pressure on existing water facilities and environments. Effective management of city water sources is necessary for making certain accessibility to secure and reputable water materials, decreasing contamination, and guarding public health and environmental top quality (Rissman 2024a). However, the intricacy of city water systems, coupled with the difficulties posed by climate change, population growth, and source shortage, highlights the need for ingenious methods to optimise water application and improve sustainability in urban areas (Guo et al. 2024).

To appreciate the importance of zero carbon industry (ZCI) operations, influenced by increased carbon emissions leading to global warming and its content, it has been reported that fifty gigatons of different greenhouse gases are emitted into the air annually. To stop the massive effects of climate changes led by global warming, ZCI advocates reducing these emissions to zero (Wang et al. 2024a). However, for every nation to shut down industries that produce these emission-producing operations is immensely challenging due to the national and international industry and industrial governing laws. The ideal approach to operating amidst all these laws and restrictions is zero carbon could be the only option, which translates to each emission gas molecule that is emitted in the air also removing an equal portion from the air, making our zero carbon or zero net emissions (Wu et al. 2024a).

Zero carbon largely requires us to make significant changes at present, through using fewer effective approaches and even implementing the current and future technological approaches, for instance, swapping each greenhouse gas emission (carbon-intensive fuels) for clean ones since over 80% of the power we use is attained in the same setting (Tsai and Lin 2024). More still, to capture the already emitted gases in the air and have them permanently and properly. Zero carbon is the perfect approach to environmentally sustainable development and innovation. Another approach could be using electric-based approaches instead of the current approaches since they are mainly used in different developing countries alongside other clean energy technologies like sustainable bioenergy, clean hydrogen, and carbon capture utilisation and storage (Lee et al. 2024).

Urban locations are increasingly facing obstacles connected to water utilisation due to swift population growth, automation, and urbanisation. Efficient monitoring of water sources in cities is vital for meeting the requirements of homeowners, sectors, and environments while maintaining long-sustainable sustainability. Urban water utilisation includes different aspects, including water, wastewater treatment, stormwater monitoring, and water conservation procedures (Wang et al. 2024b).

Cities typically count on centralised water systems that draw out water from remote resources, such as rivers or tanks, and disperse it to city customers via comprehensive networks of pipelines. Nonetheless, this traditional method of the water system is commonly ineffective. It can lead to substantial losses through leaks and evaporation, highlighting the requirement for innovative options to maximise water usage in city locations (Kamel et al. 2024).

Achieving sustainability in metropolitan water management requires stabilising the requirements of present and future generations while lessening adverse ecological effects. Sustainable water monitoring techniques in cities prioritise the efficient use of water resources, the security of high-quality water, and the preservation of natural communities (Abudu et al. 2024). One instance of sustainable city water monitoring is the execution of decentralised wastewater treatment systems, such as built wetlands or decentralised wastewater treatment plants, which can properly deal with wastewater close to the source and minimise the burden on centralised treatment facilities. Sustainable urban-style approaches, such as low-impact growth and eco-friendly infrastructure, purpose to resemble natural hydrological procedures, enhance stormwater seepage, and reduce metropolitan overflow contamination (Sama 2024).

Several cities worldwide have applied ingenious ZCI situation campaigns to deal with water obstacles and advertise sustainability. As an example, the city of Amsterdam in the Netherlands has established a comprehensive approach to shutting the water loop called the "Amsterdam Method." This strategy includes campaigns such as aquifer storage and healing, where dealt with wastewater is penetrated under zero carbon aquifers for later retrieval during periods of water scarcity (Zakkour et al. 2024). Consequently, Amsterdam has significantly decreased its dependence on freshwater resources and minimised the effects of environmental change-induced water scarcity. Correspondingly, the city of San Francisco in the USA has implemented decentralised rainwater gathering systems and greywater recycling activities to boost neighbourhood water supplies and minimise dependence on imported water (Wu et al. 2024b).

1.1 Understanding Zero Carbon Industry and Environmental Sustainability

There are several terms that come into play when referring to sustainable development, including "carbon," "climate," "neural," "positive," "negative", and "zero" industrialisation in technological innovation and environmental sustainability. Carbon neutral refers to the amount of gas emissions equal to the amount of emission off settings reduction through credits such as renewable energy generation (Ejaz et al. 2024). However, this carbon–neutral does not cut appropriately how much carbon is in the air, opening circles to zero carbon or net zero carbon since it has been seen as inter-used. The international agenda to zero carbon emissions by 2050 was discussed in the Paris Climate Summit; meeting this target will help to reduce and limit global warming to 1.5 °C this century (Paliampelou 2024). Zero carbon also focuses on neutralising emissions but largely starts with how much carbon ends up in the air. This can be attained by paying attention to activities such as industrialisation, manufacturing, transportation and other related activities and their gas emissions in relation to activities. Having a close relationship with national and international laws that focus on sustainable development goals[1] (Yueyang and Yuan 2024).

Zero carbon climatic and industry practices use a myriad of ecological advantages, consisting of the preservation of natural resources, the decrease of greenhouse gas discharges, and the prevention of contamination. The fostering of closed-loop recycling systems in the garment industry can dramatically minimise the ecological influence of textile manufacturing by drawing away textile waste from landfills and incineration (Moalosi et al. 2024). Brand names such as Eileen Fisher and Patagonia have implemented take-back programs that enable consumers to return utilised apparel and reuse it into brand-new garments, thus shutting the loop on material circulations and decreasing source removal. Through minimising reliance on material materials and reducing waste generation, ZCI situation methods aid in mitigating the environmental impact of industrial activities and influence the conservation of communities and biodiversity (Gupta et al. 2024).

ZCI efforts can additionally yield significant social benefits, such as activity creation and area empowerment, and boost accessibility to essential products and services. The establishment of community-based reusing cooperatives in developing countries gives livelihood chances for marginalised populations while addressing waste management obstacles (Singh et al. 2024). These cooperatives engage in activities such as sorting, handling, and offering recyclable materials, therefore generating revenue for local neighbourhoods and minimising dependence on informal waste picking (Zhang et al. 2024). ZCI efforts that promote the sharing industry climates, such as car-sharing and bike-sharing programs, enhance accessibility to transportation solutions, decrease traffic jams, and boost air top quality, thereby enhancing the total quality of life that can be enhanced by six illustrated factors in Fig. 16.1.

Nevertheless, zero Carbon industry condition principles offer many financial benefits, including expense savings, resource effectiveness, and task development. Transitioning from a linear to a zero Carbon industry version, organisations lower production costs, enhance resource usage, and take advantage of new revenue streams (Rissman 2024b). The adoption of zero carbon supply chain techniques by firms like Philips and Interface has caused significant financial savings through material reuse and decreased waste. In addition, the ZCI can boost advancement and entrepreneurship, driving financial development and developing new opportunities for financial investment and work. ZCI climate concepts, factories can enhance their competitiveness, resilience, and long-term stability in a significantly resource-constrained globe (Tsai and Lin 2024).

Regardless of its prospective advantages, the shift to a ZCI climate deals with several difficulties, including technical barriers, regulatory restraints, and behavioural

[1] https://sdgs.un.org/goals.

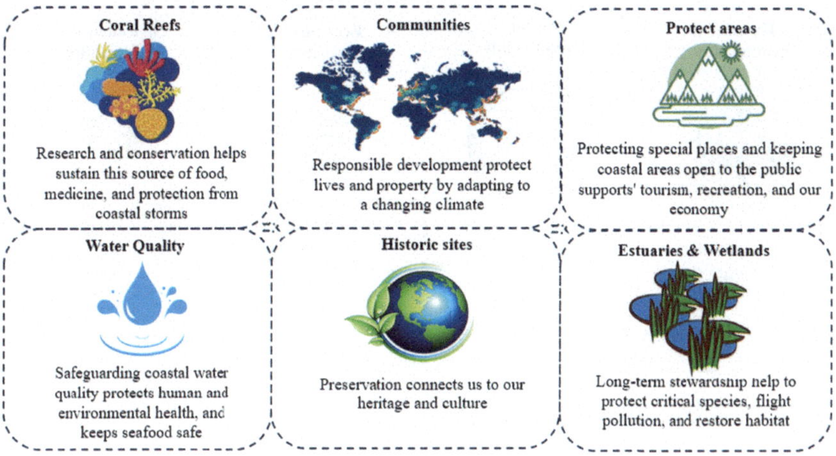

Fig. 16.1 Sustainable enabling points to support the water sector

obstacles. Numerous industries lack the essential infrastructure and modern technologies for efficient source recovery and recycling, and existing policies might hinder development and financial investment in zero Carbon industry situation initiatives (Wu et al. 2024a). In addition, changing consumer habits and cultural standards towards more sustainable usage patterns can be tough, calling for education and learning, awareness-raising, and incentives to encourage behaviour adjustment (Rissman 2024a). Getting over these difficulties calls for collaboration and innovation across sectors, including the federal government, business, the academic community, and civil culture, to develop and deploy services that promote zero carbonify and sustainability (Wang et al. 2024c). The shift to a zero Carbon industry climate concerted efforts and collective activity from federal governments, businesses, the academic community, and civil society to overcome challenges, seize chances, and recognise the complete possibility of zero carbon to drive sustainable advancement and enhance the quality of life for all, four different sustainability are presented in Fig. 16.2.

2 Challenges in Implementing Zero Carbon Industry in the Urban Water Sector

Applying ZCI principles in the metropolitan or city water sector is crucial for dealing with the pressing challenges associated with water shortage, contamination, and resource depletion while promoting sustainable advancement, as depicted.

Fig. 16.2 Top water sector sustainability

2.1 *Infrastructure Constraints*

Updating and retrofitting existing water facilities to accommodate zero carbon water systems poses a momentous obstacle in the urban water markets. Lots of city locations possess maturing water facilities originally designed for straight water flow, which makes complex integration with zero carbon methods such as water recycling and resource recovery (Ejaz et al. 2024). Retrofitting centralised wastewater treatment plants to involve decentralised treatment devices or executing water-reusing facilities might demand comprehensive pumping terminals, adjustments to pipes, and treatment centres. These upgrades can be pricey, technically complicated, and disruptive to the active city framework (Paliampelou 2024). Conversely, attending to facility restrictions is critical for enhancing the strength and performance of the city water supply.

2.2 *Regulatory Barriers*

Different nations have different controlling frameworks, which presents one more substantial difficulty in applying ZCI concepts in the city water market. The laws and plans presented might not totally support or incentivise the fostering of urban water practices, presenting barriers to innovation and financial investment (Yueyang and Yuan 2024). The guidelines controlling water top quality criteria and discharge licenses may not appropriately make up for the reuse of treated wastewater or the recovery of beneficial resources from wastewater streams. Conquering regulatory hurdles requires joint initiatives between policymakers, regulatory agencies, industry stakeholders, and civil society to create supportive structures that encourage ZCI and water options while making sure of compliance with safety, health, and ecological requirements (Barbosa 2024).

2.3 Contemporary Technological Limitations

The schedule of appropriate modern and water disruptive technologies is important for carrying out industry climate concepts in urban water. On the other hand, several existing technologies may be pricey, energy-intensive, or not appropriate for urban atmospheres, depending on the capacity and consumption demands (Moalosi et al. 2024). Advanced modern water technologies such as membrane filtering or reverse osmosis may require extensive power inputs and functional costs, limiting their usefulness for widespread fostering. Developing industrial, scalable, and energy-efficient innovations for water recycling, nutrient recovery, and decentralised treatment is vital for recognising the capacity of zero carbon and water supply in city locations (Gupta et al. 2024). Innovation and ecological research play an essential role in technological advancement and getting over technological restrictions.

2.4 Financial Constraints

Financing zero carbon industralisation and facility upgrades poses a considerable challenge for towns and energies, especially in resource-constrained settings. Water system and ZCI frequently require significant advanced financial investments in framework, technology, and functional costs (Wu et al. 2024b). Protecting financing from public sources, private investors, and innovative funding mechanisms such as eco-friendly bonds or public–private partnerships is critical for overcoming financial constraints and enabling the change to zero Carbon water systems (Zhang et al. 2024). Governments can play a crucial duty in giving monetary incentives, gives, and low-interest lending to support zero Carbon water initiatives. Furthermore, leveraging industry sector investment and discovering different funding mechanisms can aid in bridging funding spaces and speed up the fostering of zero Carbon industry situation methods in the city water market, the possible development of the sector is illustrated in Fig. 16.3 (Zakkour et al. 2024).

2.5 Public Perception and Acceptance

Public perception and acceptance of recycled water and other zero carbon and water solutions position considerable obstacles to execution. False impressions and issues about water's high quality, safety, and health and wellness dangers might deter public assistance for water recycling and reuse efforts. Building depends on self-confidence among stakeholders, and robust public education and outreach efforts are needed to raise understanding about the safety and benefits of recycled water (Rissman 2024a). Transparent communication concerning water treatment procedures, quality control steps, and regulative oversight is essential for getting over obstacles to public

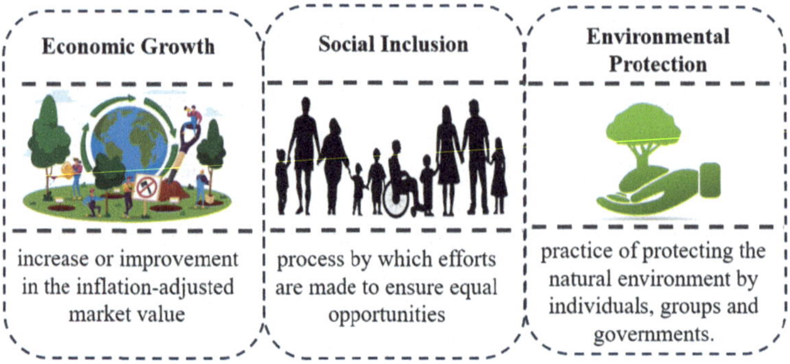

Fig. 16.3 Water sector development strategies

approval. Demonstrating effective case studies and showcasing the environmental and industry advantages of zero carbon water methods can assist in changing attitudes and foster greater public support for sustainable water management efforts in metropolitan areas (Rissman 2024b).

2.6 *Capacity and Expertise*

Constructing the needed capacity and knowledge to strategy, apply, and handle zero carbon water systems is necessary for effective implementation. Nevertheless, numerous municipalities and water utilities might not have the technological proficiency and institutional capacity to create and run zero carbon water systems effectively (Ejaz et al. 2024). Purchasing labour force development, training programs, and knowledge-sharing platforms can aid in constructing the ability and proficiency needed to support the change to zero Carbon water monitoring methods in city locations. Collaboration with academic institutions, research study organisations, and sector partners can promote modern technology transfer, skill advancement, and knowledge exchange (Yueyang and Yuan 2024). Advertising interdisciplinary partnerships and fostering a society of development within water energies can help cultivate the expertise required to address sophisticated obstacles and drive sustainable water management options ahead.

2.7 *Integration and Coordination*

Accomplishing smooth assimilation and control among multiple stakeholders is vital for the effective application of ZCI and climate concepts in the urban water market. Zero carbon water systems include complex communications between water, wastewater treatment, stormwater administration, and other framework systems, needing coordination amongst government firms, energies, organisations, and communities (Barbosa 2024). Conquering silos and promoting cooperation throughout fields and territories is vital for enhancing regulatory procedures, sharing ideal techniques, and making the most of the efficiency and efficiency of zero carbon water initiatives in metropolitan areas. Collaboration and synchronisation are necessary for recognising the complete capacity of zero ZCI concepts in the urban water sector and promoting sustainable advancement (Barnett et al. 2024).

2.8 Industry Viability

Making sure the financial stability of urban water projects presents one more considerable challenge in the metropolitan water industry. While zero Carbon industry climate methods have the potential to produce sustainable expense savings and create brand-new revenue streams, they often require substantial advance financial investments. They may have longer repayment periods than standard water administration strategies (Aryai and Goldsworthy 2024). The complete industry benefits of zero carbon emission in water supply may not always be captured or quantified, making it difficult to warrant financial investments from a simple financial point of view. Overwhelming industry obstacles require a thorough assessment of the costs and benefits of zero-carbon water activities, including the worth of resource healing, ecological externalities, and long-sustainable durability (Regadera-Macías et al. 2024).

2.9 Behavioral Change and Stakeholder Engagement

Advertising behavioural adjustment and cultivating stakeholder involvement are crucial for the successful execution of zero carbon reduction concepts in the metropolitan water field. Stakeholder resistance, lack of recognition, and social norms might impede the adoption of zero carbon industry developments and techniques, even in the presence of encouraging plans and incentives (Wu et al. 2024a). Water users may be reluctant to accept recycled water for drinkable purposes because of concerns regarding preference and odour or viewed health and wellness threats. Connecting stakeholders through participatory decision-making processes, public appointments, and multi-stakeholder collaborations can aid in developing consensus, addressing issues, and fostering ownership of ZCI and urban water initiatives (Zakkour et al. 2024).

3 Opportunities for Zero Carbon Industry in the Urban Water Sector

Realising these opportunities that the zero-carbon industry comes within the urban water sector exposes opportunities for enhancing resource effectiveness, resilience, and sustainability, as presented.

3.1 *Water Recycling and Reuse*

Water recycling and reuse represent essential opportunities for climate methods and ZCI within metropolitan water management. Applying sophisticated treatment technologies, communities can treat wastewater to fulfil various quality standards ideal for non-potable and even safe and clean usages. Apportioning with wastewater can be used for watering parks, fairways, and agricultural fields, minimising the need for freshwater sources and minimising pressure on city water system systems (Rissman 2024b). Furthermore, direct potable reuse (DPR) modern technologies permit the secure and reputable recycling of dealt with wastewater back right into the drinking water. Execution of water recycling and reuse systems not only saves water resources but likewise improves drought resilience, minimises contamination, and cultivates a much more sustainable water cycle within metropolitan environments (Singh et al. 2024).

3.2 *Resource Recovery*

Resource healing from wastewater streams provides substantial industry and environmental possibilities. Nutrients such as phosphorus and nitrogen which are abundant in wastewater, can be recuperated and repurposed for use in agriculture, aquaculture, and fertiliser manufacturing. Additionally, organic matter in wastewater can be changed into beneficial resources such as biogas with anaerobic food digestion or bioenergy manufacturing procedures (Zhang et al. 2024). Capturing and valorising these resources, cities can minimise contamination, generate renewable resources, and develop brand-new income streams from waste streams. Source healing not only diverts useful products from landfills but likewise adds to the ZCI and environmental sustainability of metropolitan water administration systems (Regadera-Macías et al. 2024). Seven practical designs for proper water sector design, maintenance, and sustainability are presented in Fig. 16.4.

Fig. 16.4 The seven-water sector practical design

3.3 Green Infrastructure Deployment

The release of the green framework supplies multifaceted advantages for city water administration and sustainable advancement. Environment-friendly infrastructure remedies such as rain yards, bioswales, and porous sidewalks help to capture, treat, and infiltrate stormwater overflow, reducing the risk of flooding, erosion, and air pollution in urban areas (Kamel et al. 2024). Furthermore, green facilities improve city biodiversity, enhance air quality, and supply entertainment amenities for communities. With the assimilation of eco-friendly infrastructure right into urban planning and development, smart cities can boost water resilience, mitigate environmental change effects, and promote sustainable city settings that benefit both individuals and nature (Tsai and Lin 2024).

3.4 Collaborative Partnerships

Collaborations between federal government companies, water utilities, services, academia, and civil culture companies are crucial for advancing ZCI campaigns in the urban water field. These partnerships assist in knowledge sharing, modern technology transfer, and resource mobilisation, enabling cities to utilise cumulative competence and sources to address complicated water obstacles (Lee et al. 2024). Public–private partnerships can facilitate financial investment in water-reusing projects, while academic-industry collaborations can drive study and technology in water treatment modern technologies. Promoting collaboration and collective action, cities can maximise the impact of ZCI initiatives and increase the change to sustainable water administration techniques (Guo et al. 2024).

3.5 Industry Benefits

Embracing ZCI climate principles in the urban water sector can generate substantial industry advantages, including cost financial savings, activities development, and organisation possibilities. ZCI and sustainable water systems supply opportunities for lowering operational costs, enhancing source usage, and producing revenue from recycled water and recuperated resources (Wang et al. 2024a). For example, water recycling tasks can minimise the need for expensive investments in new water system infrastructure and energy-intensive treatment procedures. Furthermore, zero-carbon practices boost technology and entrepreneurship, driving industry growth and generating brand-new markets for sustainable water technologies and solutions (Guo et al. 2024).

3.6 Policy Support and Regulatory Frameworks

Policy assistance and governing structures are crucial for creating an enabling environment for zero-carbon industry climate initiatives in the city water market. Governments can play a crucial function in developing helpful plans, setting targets, and providing monetary motivations to advertise water recycling, source recovery, and green facilities implementation (Paliampelou 2024). Laws mandating the use of recycled water for irrigation or industrial procedures can promote the need for recycled water and drive financial investment in water recycling facilities. Policy frameworks that promote integrated water monitoring, decentralised services, and multi-stakeholder collaboration can promote the shift to zero-carbon water systems and advertise sustainable urban water administration techniques (Yueyang and Yuan 2024).

3.7 Zero Carbon Water Systems Integration

Integrating zero-carbon water principles across the whole water cycle, from supply to treatment to distribution, cities can produce closed-loop systems that minimise source loss and make the best use of resource healing. Decentralised treatment facilities can be strategically placed to treat wastewater close to the source, minimising the demand for comprehensive transportation networks and centralised treatment plants (Moalosi et al. 2024). Moreover, incorporating water recycling facilities with existing water system systems allows the direct reuse of dealing with wastewater, decreasing reliance on freshwater resources and improving water safety. Through comprehensive planning, design, and implementation of urban water supply, cities can maximise water monitoring methods, conserve resources, and reduce the effects of urbanisation on water communities (Ejaz et al. 2024).

4 Disruptive Technologies for Zero Carbon Industry in the Urban Water Sector

Within this section, innovative, disruptive technologies that are revolutionising the urban water field, supplying transformative options to resolve pressing obstacles and advancing ZCI situation concepts are demonstrated.

4.1 Advanced Water Treatment Technologies

These technologies stand for a paradigm change in urban water administration by using precise and efficient methods for detoxifying wastewater to fulfil strict, high-quality standards. One-of-a-kind in their capacity to remove impurities and contaminants at molecular degrees, technologies such as membrane layer purification, reverse osmosis, and advanced oxidation processes (AOPs) guarantee the manufacturing of premium recycled water ideal for numerous applications, consisting of potable reuse (Moalosi et al. 2024). In urban setups, these innovations play a critical role in closing the water loop by treating wastewater to a degree that it can be safely reintroduced into the water system. The benefits of innovative water treatment modern technologies prolong past water reuse; they contribute to sustainable development by lowering pressure on freshwater resources, lessening contamination, and enhancing water resilience when faced with growing urbanisation and climate adjustment, sustaining the accomplishment of sustainable growth objectives related to environmental protection, public health and water security (Gupta et al. 2024).

4.2 Decentralized Water Treatment Systems

Unique in their capacity to deal with wastewater at or near the point of generation, these systems get rid of the demand for substantial conveyance networks and central treatment plants, lowering energy consumption, water loss, and infrastructure costs. Technologies such as modular treatment units, compact filtration systems, and onsite reusing centres can be customised to water quality requirements and neighbourhood problems, making them suitable for diverse city settings (Zhang et al. 2024). Decentralizing water treatment, cities can improve water strength, enhance solution distribution, and reduce the dangers connected with centralised facilities failures or disturbances. Decentralised systems promote neighbourhood involvement and empowerment by involving homeowners in regional water administration decisions and campaigns (Regadera-Macías et al. 2024). This contributes to sustainable development by enhancing water security, durability, and inclusivity in urban areas.

4.3 Smart Water Management Technologies

Smart water monitoring modern technologies utilise information analytics, sensor networks, and internet-of-things to optimise water use, detect leaks, and improve operational effectiveness in urban water supply. Special in their capability to supply real-time insights into water intake patterns, these technologies make it possible for utilities to recognise ineffectiveness, expect infrastructure failures, and maximise resource allowance (Wang et al. 2024b). Applications such as advanced metering infrastructure (AMI) and anticipating analytics aid utilities in lowering water losses, enhancing solution distribution, and enhancing the general durability of the water framework. Smart water monitoring technologies right into metropolitan water systems bearing ZCI integration can help cities achieve several benefits, including decreased water consumption, reduced functional costs, and enhanced customer fulfilment (Zakkour et al. 2024). In addition, clever water monitoring contributes to sustainable growth by advertising reliable resource usage, mitigating water deficiency, and lowering environmental impact.

4.4 Nutrient Recovery Technologies

Ingenious solutions for removing beneficial nutrients such as nitrogen and phosphorus from wastewater streams and repurposing them for advantageous reuse. Technologies such as struvite precipitation, ion exchange, and biological nutrient elimination promote the recovery of nutrients from wastewater treatment plants, transforming them right into important resources for agriculture, aquaculture, and plant food production. Valorising nutrients from wastewater, cities can minimise ecological effects, create revenue from resource recovery, and promote zero-carbon nutrient cycles within urban water systems (Paliampelou 2024). Additionally, nutrient recovery modern technologies add to justifiable development by enhancing dirt fertility, promoting agricultural efficiency, and decreasing dependence on synthetic fertilisers.

4.5 Water-Energy Nexus Solutions

Distinct in their capacity to incorporate water and power monitoring, these remedies promote harmony and effectiveness in the city water supply. Technologies such as energy-efficient pumps, renewable energy integration, and power recovery systems help energies lower power consumption, reduce operational expenses, and minimise greenhouse gas emissions (Barbosa 2024). Energy healing from wastewater treatment processes such as anaerobic digestion or micro-hydro turbines reduces power costs and contributes to renewable energy generation. Optimizing the water-energy nexus, smart cities attain several advantages, including improved source efficiency, expense savings, and environmental stewardship. Water-energy nexus solutions play an influential role in advertising sustainable advancement by cultivating combination, resilience, and improvement in metropolitan water monitoring (Gupta et al. 2024).

4.6 Innovative Water Reuse Applications

Technologies such as aquifer storage and recovery and direct potable reuse, as well as toilet-to-tap systems, facilitate the safe and reputable reuse of treated wastewater for alcohol consumption and water augmentation. With executing ingenious water reuse applications, cities can improve water strength, boost drought strength, and advertise sustainable water management practices (Aryai and Goldsworthy 2024). In addition, ingenious water reuse technologies contribute to sustainable development by reducing water scarcity, decreasing environmental effects, and enhancing water security in urban areas. Innovative water reuse applications play a crucial duty in achieving ZCI and sustainability in the city water market while attending to essential water obstacles and promoting public health and wellness (Regadera-Macías et al. 2024).

4.7 Blockchain and Distributed Ledger Technologies (DLTs)

Distinct in their ability to offer safe and unalterable record-keeping, these modern technologies enable transparent monitoring of water purchases, payment, and conformity with regulations. Applications such as blockchain-based water trading systems and smart contracts promote peer-to-peer water trading for legal human rights, advertising efficient allocation of water sources and incentivising water preservation. Leveraging blockchain and distributed ledger technologies, cities boost their count on integrity and performance in water monitoring, promoting a much more sustainable and resilient metropolitan water industry (Wu et al. 2024b). Furthermore, blockchain-enabled remedies lower administrative overhead, enhance water administration processes and encourage stakeholders to join water management decisions.

5 Future Directions of Zero Carbon Industry in the Urban Water Sector

The future of the ZCI situation in the urban water market holds enormous prospects for transformative modification driven by technical developments, ingenious methods, and advancing social requirements.

5.1 Integration of Nature-Based Solutions

In the future, city water monitoring will significantly include nature-based solutions (NBS) to address water obstacles sustainably. NBS, such as green roof coverings, rainfall yards, and marshes, use eco-friendly alternatives to typical infrastructure, offering benefits like stormwater administration and biodiversity conservation (Sama 2024; Shafik 2024a). Cities will decrease pollution, improve water user durability, and advertise community services, cultivating zero carbon in the city water industry. These options mimic natural processes, enhancing water top quality, decreasing runoff, and alleviating the city warmth island result. NBS can boost area well-being by giving recreational rooms, improving air quality, and sustaining metropolitan biodiversity (Abudu et al. 2024). Embracing NBS in urban water administration stands for a shift towards an even more sustainable and durable water supply, lining up with ZCI situation concepts and adding to the general health wellness and sustainability of metropolitan settings.

5.2 Zero Carbon Water Systems Design

These systems will integrate decentralised treatment, water recycling, and source healing to make the best use of efficiency and sustainability. By accepting zero carbon, cities can minimise reliance on freshwater resources, reduce power intake, and alleviate pollution. Cutting-edge technologies and modelling tools will certainly enable cities to replicate and optimise zero-carbon water systems for ideal performance and environmental influence. In addition, zero-carbon water systems will advertise strength by diversifying water sources, decreasing susceptibility to environmental modification, and improving adaptive capability (Wang et al. 2024c). Deliberately water systems with zero carbon in mind, cities can foster sustainable development, improve water protection, and produce much healthier and more durable urban environments.

5.3 Digitization and Smart Water Networks

The future of metropolitan water management will certainly be defined by digitisation and the adoption of clever water networks. Advanced sensor innovations, data analytics, and expert systems such as artificial intelligence will certainly transform exactly how cities handle their water. Smart water networks will allow real-time monitoring, predictive modelling, and adaptive administration of water systems, enhancing resource use and boosting functional effectiveness (Gupta et al. 2024; Shafik 2024b). These technologies will certainly discover leaks, recognise anomalies, and enhance water distribution networks, reducing water losses and enhancing system dependability. In addition, smart water networks will certainly advertise sustainability by decreasing energy intake, lowering carbon emissions, and boosting water quality. There will be more accomplishments and several advantages, including improved water administration, lowered functional costs, and boosted strength to climate modification and urbanisation stress (Singh et al. 2024).

5.4 Community Engagement and Social Equity

Future instructions for the zero Carbon industry situation in the metropolitan water sector will certainly prioritise area involvement, social equity, and inclusivity. Participatory techniques, co-design processes, and community-based water administration efforts will certainly encourage locals to add to decision-making and form the future of metropolitan water supply. Furthermore, equitable access to water services, cost, and ecological justice will certainly be the main considerations in the design and implementation of zero-carbon water projects (Aryai and Goldsworthy 2024). By involving neighbourhoods in water administration decisions and efforts, cities can construct, count on, foster partnerships, and guarantee that the advantages of zero carbonity are equitably dispersed among all stakeholders.

5.5 Resilience and Climate Adaptation

The future of zero Carbon industry climate in the city water market will certainly focus on building strength and environmental adaptation right into water systems to resolve the expanding difficulties posed by climate change, population growth, and urbanisation. Climate-resilient facilities, such as an environment-friendly framework, water-sensitive urban design, and decentralised water management services, will certainly minimise the effects of extreme climate occasions, sea-level surges, and water shortages (Wang et al. 2024a). By incorporating durability and adjustment steps right into zero Carbon water preparation and management, cities can improve their capacity to hold up against shocks and tensions while promoting sustainable advancement and zero carbon. Moreover, durable water systems can enhance social, financial, and environmental durability, providing numerous benefits such as boosted public health, industry stability, and community services (Wu et al. 2024a; Shafik 2024c).

5.6 Innovative Water Governance and Policy Frameworks

This includes the establishment of governing incentives, criteria, and accreditation systems to advertise water recycling, source recovery, and sustainable water techniques. Collaborative governance versions, such as multi-stakeholder systems and watershed-based administration techniques, will help with coordinated decision-making and stakeholder interaction throughout different levels of federal government, markets, and communities (Abudu et al. 2024). With adaptive and comprehensive administration procedures, cities can conquer obstacles to zero carbon-free water implementation projects, address competing interests, and build consensus on sustainable water management strategies. Furthermore, reliable policy structures can provide essential legal and institutional assistance for zero carbon water initiatives, making certain conformity with laws, safeguarding public health, and safeguarding environmental high quality (Kamel et al. 2024).

6 Lessons Learned and Conclusion

As we came to the end of the chapter, we sampled lessons learned from the chapter and the conclusion.

6.1 Lessons Learned for the Chapter

By embracing ZCI, urban water management climate principles, cities transform these obstacles right into possibilities for enhancing water strength, promoting environmental stewardship, and driving financial development.

- The urban water industry is faced with various interconnected difficulties, including water shortage, pollution, maturing framework, and environmental change influences. However, within these difficulties exist substantial opportunities for development, resource healing, and sustainable advancement.
- Resolving the challenges and using the changes in the city water field calls for incorporated and alternative approaches that consider the entire water cycle. ZCI

principles stress the interrelation of water resources, waste streams, and ecosystems, highlighting the relevance of embracing integrated water management strategies.

- Advanced water treatment, modern technologies, decentralised water supply, clever water administration solutions, and digital platforms allow cities to improve resource usage, operational efficiency, and resilience to water-related challenges.
- Public–private collaborations, multi-stakeholder partnerships, and community engagement efforts promote knowledge sharing, innovation transfer, and resource mobilisation, making it possible for cities to utilise cumulative experience and sources to address complicated water difficulties.
- Policy assistance and regulatory frameworks are essential for enabling the environment to zero-carbon industry situation techniques in the city water industry. Nations and agencies play a crucial function in developing helpful policies, setting targets, and providing monetary incentives to promote water recycling, resource healing, and green framework implementation.
- The excursion toward the ZCI, and other climatic change management prospects in the urban water sector is a recurring process that requires continuous discovery, adjustment, and improvement. Cities must remain available for testing, innovation, and learning from both successes and failures. Surveillance, examination, and understanding-sharing systems are vital for recording lessons discovered, identifying ideal techniques, and scaling up effective treatments.
- Educating citizens regarding water conservation and sustainable techniques is vital for fostering a society of water stewardship in city areas. Public understanding campaigns and neighbourhood engagement initiatives can equip people to adopt water-saving behaviours and assistance and industry climate initiatives in the city water industry.
- Ensuring the industry practicality of zero-carbon water projects is crucial for their successful application. Cutting-edge funding devices and cost–benefit analyses can activate funding and attract private financial investment, opening the industry capacity for zero-carbon water management for sustainable urban development.
- ZCI state approaches to regional problems and priorities must be tailored for effectiveness. Cities must analyse neighbourhood water resources, frameworks, and institutional capacities to identify context-specific solutions and promote durability in the metropolitan water supply.

6.2 Conclusion

The adoption of round industry situation principles in the metropolitan water industry represents an encouraging path in the direction of attending to pushing water challenges while advertising sustainability and resilience in cities. Despite the complex obstacles and challenges, the chapter highlights many opportunities for development, collaboration, and transformative adjustment. With nature-based options, advancing technological innovations, promoting collaborations, and promoting policy assistance and public awareness, cities can unlock the capacity of urban water management to enhance water safety and security, reduce air pollution, and promote industry prosperity. Additionally, embracing flexibility, constant knowing, and inclusivity are crucial for browsing uncertainties and optimising the benefits of round industry climate initiatives in varied metropolitan contexts. As cities continue to grapple with expanding urbanisation, environmental modification effects, and resource constraints, the essential change in the direction of ZCI and urban water supply is becoming increasingly evident. Through concerted efforts and collective activity, cities can lead the way in the direction of an extra sustainable and resilient future, where water is valued, saved, and managed, attuned to nature and the needs of present and future generations.

References

Abudu H, Wesseh PK, Lin B (2024) Climate bonds toward achieving net zero emissions and carbon neutrality: evidence from machine learning technique. J Manag Sci Eng 9(1). https://doi.org/10.1016/j.jmse.2023.10.001

Aryai V, Goldsworthy M (2024) Real-time high-resolution modelling of grid carbon emissions intensity. Sustain Cities Soc 104. https://doi.org/10.1016/j.scs.2024.105316

Barbosa FC (2024) Zero carbon emission aviation fuel technology review-the hydrogen pathway. SAE Technical Papers. https://doi.org/10.4271/2023-36-0029

Barnett PR, Hmeidat NS, Zheng B, Penumadu D (2024) Toward a circular economy: zero-waste manufacturing of carbon fiber-reinforced thermoplastic composites. NPJ Mater Sustain 2(1). https://doi.org/10.1038/s44296-024-00006-y

Ejaz U, Taj A, Sohail M, Alanazi AK, Abo-Dief HM (2024) Toward a zero waste approach: utilization of sugarcane bagasse for dye removal and multienzymes production. J Basic Microbiol 64(2). https://doi.org/10.1002/jobm.202300529

Guo D, Li B, Yu W, Baroutian S, Young BR (2024) A system engineering perspective for net zero carbon emission in wastewater and sludge treatment industry: a review. Sustain Prod Consum 46. https://doi.org/10.1016/j.spc.2024.02.033

Gupta NC, Tanwar R, Dipesh Kaushik A, Singh R, Patra AK, Sar P, Khakharia P (2024) Perspectives on CCUS deployment on large scale in India: insights for low carbon pathways. Carbon Capture Sci Technol 12. https://doi.org/10.1016/j.ccst.2024.100195

Kamel E, Pittau F, Dal Verme LM, Scatigna P, Iannaccone G (2024) A parametric integrated design approach for life cycle zero-carbon buildings. Sustain (Switz) 16(5). https://doi.org/10.3390/su16052001

Lee J, Eom J, Park J, Jo J, Kim S (2024) The development of a machine learning-based carbon emission prediction method for a multi-fuel-propelled smart ship by using onboard measurement data. Sustainability 16(6). https://doi.org/10.3390/su16062381

Moalosi R, Rapitsenyane Y, Sealetsa OJ (2024) Africa's transition towards net zero in the transport industry. In: Research journeys to net zero: current and future leaders. https://doi.org/10.4324/9781003380566-17

Paliampelou I (2024) Net zero targets and governance: a literature review (2009–2021). In: Handbook on corporate governance and corporate social responsibility. https://doi.org/10.4337/9781802208771.00014

Regadera-Macías AM, Morales-Torres S, Pastrana-Martínez LM, Maldonado-Hódar FJ (2024) Optimizing filters of activated carbons obtained from biomass residues for ethylene removal in agro-food industry devices. Environ Res 248. https://doi.org/10.1016/j.envres.2024.118247

Rissman J (2024a) Introduction: what is zero-carbon industry? In: Zero-carbon industry. https://doi.org/10.7312/riss20420-001

Rissman J (2024b) Zero-carbon industry. In: Zero-carbon industry. https://doi.org/10.7312/riss20420

Sama A (2024) Introducing graphene-enhanced technologies to the regulated market to mitigate carbon emissions. Environ Prog Sustain Energy 43(1). https://doi.org/10.1002/ep.14338

Shafik W (2024a) Industry 4.0 technologies' opportunities and challenges for realising net-zero economy. In: Net zero economy, corporate social responsibility and sustainable value creation: exploring strategies, drivers, and challenges, pp 19–41. Springer, Cham. https://doi.org/10.1007/978-3-031-55779-8_2

Shafik W (2024b) Industry revolution 4.0 and beyond: abilities and perils for sustainable zero net economy development. In: Powering industry 5.0 and sustainable development through innovation, pp 292–316. IGI Global. https://doi.org/10.4018/979-8-3693-3550-5.ch020

Shafik W (2024c) Shaping the next generation smart city ecosystem: an investigation on the requirements, applications, architecture, security and privacy, and open research questions. In: Smart cities: innovations, challenges and future perspectives, pp 3–52. Springer Nature Switzerland, Cham. https://doi.org/10.1007/978-3-031-59846-3_1

Singh VK, Sharma V, Jha NK, Jamwal A, Agarwal R (2024) Multimodal transportation and net zero emission world: an emerging research agenda. In: IFIP advances in information and communication technology, vol 698. AICT. https://doi.org/10.1007/978-3-031-50192-0_20

Tsai WH, Lin WH (2024) Production decision model for the cement industry in pursuit of carbon neutrality: analysis of the impact of carbon tax and carbon credit costs. Sustainability 16(6). https://doi.org/10.3390/su16062251

Wang Q, Zhu K, Guo P, Zhang J, Xiong Z (2024a) Key issues and solutions in the study of quantitative mechanisms for tropical islands zero carbon buildings. Appl Sci 14(4). https://doi.org/10.3390/app14041659

Wang Q, Zhu K, Guo P, Zhang J, Xiong Z, Wu H (2024b) Analysis of the development trends and key challenges of zero carbon buildings in tropical islands. Adv Econ Manag Res 9(1). https://doi.org/10.56028/aemr.9.1.132.2024

Wang Z, Dong B, Yin J, Li M, Ji Y, Han F (2024c) Towards a marine green power system architecture: integrating hydrogen and ammonia as zero-carbon fuels for sustainable shipping. Int J Hydrog Energy 50. https://doi.org/10.1016/j.ijhydene.2023.10.207

Wu M, Sun W, Meng X, Kang J, Yang Y (2024a) Natural marmatite photocatalyst for treatment of mineral processing wastewater to help zero wastewater discharge. J Environ Sci 142. https://doi.org/10.1016/j.jes.2023.07.006

Wu Y, Li K, Fu X (2024b) An integrated zero-sum game and data envelopment analysis model for efficiency analysis and regional carbon emission allocation. Decis Anal J 10. https://doi.org/10.1016/j.dajour.2023.100387

Yueyang W, Yuan PF (2024) A parametric approach towards carbon net zero in agricultural planning. In: Computational design and robotic fabrication, vol Part F2072. https://doi.org/10.1007/978-981-99-8405-3_26

Zakkour P, Kuijper M, Dixon P, Haszeldine RS, Towns M, Allen M (2024) Carbon storage units and carbon storage obligations: a review of policy approaches. Int J Greenh Gas Control 133. https://doi.org/10.1016/j.ijggc.2024.104087

Zhang Y, Qian K, Wang Q, Wang J, Feng Y, Zhang J (2024) Comparative study on international zero-carbon building certification system under the vision of carbon neutrality. Lect Notes Electr Eng 1118. https://doi.org/10.1007/978-981-99-8878-5_29

The manufacturer's authorised representative in the EU is Springer Nature Customer Service Centre GmbH, Europaplatz 3, 69115 Heidelberg, Germany. If you have any concerns regarding our products, please contact ProductSafety@springernature.com

Printed and bound by CPI Group (UK) Ltd, Croydon, CR0 4YY
01/10/2026
02225914-0009